Caribbean Tsunamis

Advances in Natural and Technological Hazards Research

VOLUME 20

The titles published in this series are listed at the end of this volume.

Caribbean Tsunamis

A 500-Year History from 1498-1998

by

KAREN FAY O'LOUGHLIN

and

JAMES F. LANDER

Cooperative Institute for Research in Environmental Sciences,
University of Colorado, Boulder, Colorado, U.S.A.

In cooperation with

National Oceanic and Atmospheric Administration,
National Geophysical Data Center, Boulder, Colorado 80305, U.S.A.

KLUWER ACADEMIC PUBLISHERS
DORDRECHT / BOSTON / LONDON

A C.I.P. Catalogue record for this book is available from the Library of Congress.

ISBN 978-90-481-6467-7

Published by Kluwer Academic Publishers,
P.O. Box 17, 3300 AA Dordrecht, The Netherlands.

Sold and distributed in North, Central and South America
by Kluwer Academic Publishers,
101 Philip Drive, Norwell, MA 02061, U.S.A.

In all other countries, sold and distributed
by Kluwer Academic Publishers,
P.O. Box 322, 3300 AH Dordrecht, The Netherlands.

Printed on acid-free paper

Printed in the Netherlands.

TABLE OF CONTENTS

PREFACE

In the past 500 years, the Caribbean has had devastating tsunamis that have caused incalculable damage. Since tsunamis are quite rare, they are often overlooked as a hazard until a disastrous event occurs. The Caribbean region is an area of relatively high seismicity, and although tsunamis are not the chief natural hazard, they have the potential to produce catastrophic regional disasters. A history of tsunami occurrences within this broad region is essential for helping to define the diverse characteristics of the local hazard. The purpose of *Caribbean Tsunamis—A 500-Year History from 1498–1998* is to provide information about tsunamis, great and small, and to create greater awareness of the infrequent but destructive effects of major tsunamigenic events.

This 500-year overview broadly characterizes the nature of tsunamis in the Caribbean Sea, while bearing in mind both scientific aspects as well as potential interest by the many governments and populations likely to be affected by the hazard. The book endeavors to assist in the complex task of establishing an understanding of the need for public awareness, from school children to government officials, about the tsunami hazard. This compendium contributes data, essential information, and anecdotal material to inform diverse groups of people about this natural hazard—many of whom do not even realize the need for such awareness. Comprehension of the nature of tsunamis and past effects is crucial for the awareness and education of populations at risk.

The aim of this book is to aid both the scientific research community and the people of the many nations across the Caribbean. An effort has been made to meet the differing levels of familiarity with the many facets of this comprehensive topic. Concepts and terms are defined in language suitable for all. Despite waves of information, the idea has been neither to inundate those unfamiliar with the field nor to be unscholarly for scientists and other experts. It is hoped that an intelligible combination has been achieved between scholarly work and a readable book.

The book comprises five chapters discussing tsunamis and their effects plus the 500-year history of Caribbean tsunamis, which furnishes a detailed chronicle of 127 tsunamis reported since 1498 in the Caribbean region. For each event, the likelihood of its occurrence has been assessed and a validity rating ascribed. Eliminating the 3 tsunami reports judged as invalid occurrences brings to 124 the tally of tsunamis evaluated in the history. Of this remaining 124 reports, 53 have been evaluated as authentic occurrences; another 8 were probable tsunamis; 19 may or may not have occurred; and 44 were unlikely or the information is so meager that to ascertain if tsunamis actually occurred needs further research.

This compilation is rendered in a layout intended to ease access to such a breadth of detail, over such a vast region, to such a diverse audience. The format enables browsing for features of interest, locations, dates, validity ratings, and source references. Data on runup heights, intensity and magnitude ratings, fatalities, inundation, and recession from the seashore are given along with descriptive reports of tsunamis, tsunamigenic earthquakes, volcanic eruptions, tectonic triggers, and tsunami and seismic effects.

Caribbean Tsunamis—A 500-Year History from 1498–1998 contains updated information on an astonishing 2,556 tsunami-related deaths during the past 150 years. This death toll amounts to almost five times greater than during the same period for the combined tsunami-prone regions of Hawai'i, Alaska, and the U.S. West Coast. Even using the most conservative assessments, Caribbean fatalities are still of great consequence. Excluding these updates, the lowest estimate of tsunami fatalities for the 150-year period totals 719 victims—one-and-a-half times greater than the combined total of the other regions—a significant loss of life. Over the centuries, countless numbers of people are unaccounted for and untold hundreds of thousands have been left homeless.

The potential for a major catastrophe is revealed by the extensive devastation caused by a significant number of disastrous tsunamis reported in the history of the Caribbean. To gain insight for designing effective early warning and mitigation plans, an understanding of the nature of the hazard is needed, based on research into tectonics and volcanism, source mechanisms, prior events, and damaging effects. Data and information from the historical record can be of use for scientists, researchers, engineers, architects, and land use planners to help:

- Discern locations with similar tectonic regimes and vulnerability.
- Identify where to effect emergency response options and evacuation plans that could significantly impact loss of life.

Although initial implementation has begun, as yet no region-wide warning system is fully in place. This catalog of events can aid in ascertaining the needs of future development plans by stimulating:

- Research into historical literature, archives, and marigraph records.
- Development of computer simulations to replicate hazardous geophysical conditions.

Tectonic forces continually build stress—until the inevitable release of strain that may trigger a tsunamigenic earthquake. The lack of a major tsunami in the past 57 years is due to a relative lack of relief of the built-up energy. The potential extent of the stress release grows as time elapses. The long period without relief of seismic stress buildup only increases the ominous threat of a devastating tsunami that could result from a sudden seafloor cataclysm.

It is the risks, the local hazard, and any special circumstances of past tsunamis that can be garnered from this compendium. When data are lacking or specifics vague, particularly in the pre-instrumental period, it is the descriptions that can render the best understanding of the diverse tsunami and earthquake effects. Learning from past events is vital to understanding the tsunami hazard in this seismically active region.

An analogy for the value of using the 500-year compendium of Caribbean tsunamis may be found in comparing it to a light bulb. What is important about a light bulb is not the filament or glass, but its power to radiate light. What is important about the Caribbean Tsunami History is not the data or specifics rendered, but what may be learned about tsunamis.

For the general populace, the need for awareness is paramount. The necessity for awareness cannot be overemphasized. A major tsunami will strike the Caribbean, which could result in incalculable devastation. To save themselves from disastrous consequences, people need know the three most helpful "natural warning" signs of a tsunami—if a person:

- Feels an earthquake tremor while near the coast.
- Hears a roaring sound.
- Notices unusual movement of the sea or a recession from the shore.

Moreover, understanding other critical aspects of the hazard is a must:

- Tsunamis cannot be outrun.
- People need to evacuate coastal areas immediately.
- Waters rushing back to the sea may be as dangerous as the initial surge.
- The second wave to strike may be even larger than the first.
- Intervals up to an hour may occur before a subsequent wave arrives.
- Returning too soon to check on damages can cost a person's life.
- Backwash may contain debris of all kinds and can be lethal.
- Tsunamis can create havoc with port infrastructure, buildings, shipping, and, not least, can be deadly to people.

A unique opportunity exists to ascertain potential risk areas and establish hazard warning and mitigation systems before the next, now overdue, major tsunami. The impending threat from a devastating tsunami in this seismically active region is greater now than in the past.

Given the substantial growth in population, coastal development, and tourism-related industries, along with rising sea levels worldwide, unimaginable disaster looms ominously. With the prospect of perilous tectonic activity, the potential danger from a catastrophic tsunami cannot be underestimated.

津波

ACKNOWLEDGMENTS AND NOTES

We wish to recognize the late Professor Mohammed Ibrahim El-Sabh for his focus on establishing an understanding of the need for public education and awareness about the tsunami hazard, is a great loss since his untimely death in February 1999. His invitation to submit this extended history of Caribbean tsunamis was instrumental to our effort. Dr. El-Sabh of the University of Rimousky, Quebec, Canada, was the founding President of the Natural Hazards Society and Founding Editor of *Natural Hazards, Journal of the International Society for the Prevention and Mitigation of Natural Hazards.* Professor El-Sabh's dedication to aid the need for awareness and education about natural hazard management was incalculable in its importance and will carry forward into the future as his legacy.

We would gratefully like to acknowledge Dr. Tad Murty, Baird & Associates, whose kind support and encouragement was invaluable. His outstanding commitment to tsunami research—both for the scientific research community and for the people of the Caribbean—has been a tremendous help throughout this undertaking.

The work was made possible through the auspices of the University of Colorado, Cooperative Institute for Research in Environmental Sciences (CIRES) in cooperation with the National Oceanic and Atmospheric Administration (NOAA), National Geophysical Data Center (NGDC). We wish to thank Dr. Lowell S. Whiteside for his efforts on many phases of this project.

Translation of certain texts from the original French, Spanish, or German were derived primarily from texts available in José Grases' excellent 1971 catalog, *La Sismicidad Histórica del Caribe.* It is humbly hoped that clarification of uncertain details was achieved and not further obfuscation.

The bibliography includes many works that were not examined. To have had these assembled at the commencement of this project would have facilitated the research greatly. It is hoped that their inclusion here may assist those seeking to do further research. We express regrets for any works inadvertently omitted.

Caribbean Tsunamis—A 500-Year History from 1498–1998 updates all preliminary Caribbean tsunami catalogs compiled by the authors. This book also updates any books, catalogs, web sites, or other locations where information or data may have been derived from work by the authors leading up to the publication of this book. New or updated data, additional reports, corrections, and comments will be gratefully accepted. Inquiries and updates should be directed to: karen.oloughlin@noaa.gov.

This compendium can only be thought of as a starting place. In reality, it will be outdated the minute the presses cease. Nevertheless, as it stands, this book can serve to assist the people of the Caribbean in raising public awareness about tsunamis and the catastrophic disasters that occurred in historic times.

In Memoriam

Christopher John O'Loughlin

October 13, 1976 – July 2, 2002

I would like to dedicate this book to my son, Chris, whose valiant four-year battle with Lymphoma was lost this past summer. Christopher's irrepressible spirit nearly conquered this virulent Lymphoma/Leukemia during three long years of chemotherapy that included illness-of-the-day, life-threatening crises, and vigilance against germs. His relapse necessitated a bone marrow transplant. Rare as it is for a parent to be a match as a donor, we were blessed with this miracle. It was my honor to donate my own bone marrow to my 25-year-old son, a humbling and amazing privilege. Chris triumphed each day over the profusion of life-threatening difficulties that ensued. He seemed to have beaten the terrible odds. The ominous cancer returned. Alas, my emergent stem cells could not kill off every one of his tenacious leukemia cells. The unrelenting cancer overwhelmed all efforts to defeat its ruthless advance, not unlike the menacing approach of a looming tsunami.

Karen Fay O'Loughlin

CHAPTER 1

THE CARIBBEAN

1.1. Islands and countries affected by Caribbean tsunamis

The locations affected by tsunamis encompass the entire Caribbean region. The Caribbean Sea is a part of the Atlantic Ocean, bounded by the coasts of Central America, South America, and the West Indies. The Greater Antilles and Lesser Antilles island chains constitute the West Indies. The Greater Antilles lie along the northern boundary of the Caribbean Sea and include Puerto Rico, Hispaniola (Dominican Republic and Haiti), Jamaica, Cuba, and the Cayman Islands. The Lesser Antilles comprise both the Leeward and Windward Islands. Together they form the long arc of the West Indian archipelago.

Encircling the Caribbean are the continental countries of Venezuela, Colombia, Panamá, Costa Rica, Nicaragua, Honduras, Guatemala, Belize, and Mexico. Numbers of small islands belonging to these countries dot the Caribbean, mostly along the coasts. However, no major islands lie in the broad expanse of the Venezuelan, Colombian, and Yucatán Basins of the vast Caribbean Sea.

Lying north of the Windward Passage, between Cuba and Haiti, is Great Inagua, the only Bahamian island that may be considered part of the great island chain that forms the boundary of the Caribbean Sea. Great Inagua, positioned like a finial atop the Windward Passage, lies near Grand Turk and the Caicos island group. The Turks and Caicos Islands, situated adjacent to the Atlantic Ocean, are sometimes considered to be on the rim of the Caribbean Sea. The Bahamas, which are otherwise looked upon as part of the West Indies, do not belong to the Caribbean island group.

The Leeward Islands, sometimes called The Leewards are so named because they are in the lee of, or sheltered from, the northeast trade winds. The Windward Islands, also called The Windwards, which lie to the south of The Leewards, are exposed to the strong trade winds. The eastern shores of the entire arc of the West Indian archipelago, typically, are battered by rough surf from the Atlantic. The calmer waters of the lee side of the West Indies, or western shores, are more protected; that is, except for the occasional tsunami. The Leeward Islands are separated from the Windward Islands between Martinique and Dominica, at Canal de Martinique (Martinique Passage) at approximately 15° North Latitude. The Lesser Antilles encompass the British Virgin Islands, the British West Indies, the United States Virgin Islands, Antigua and Barbuda, the

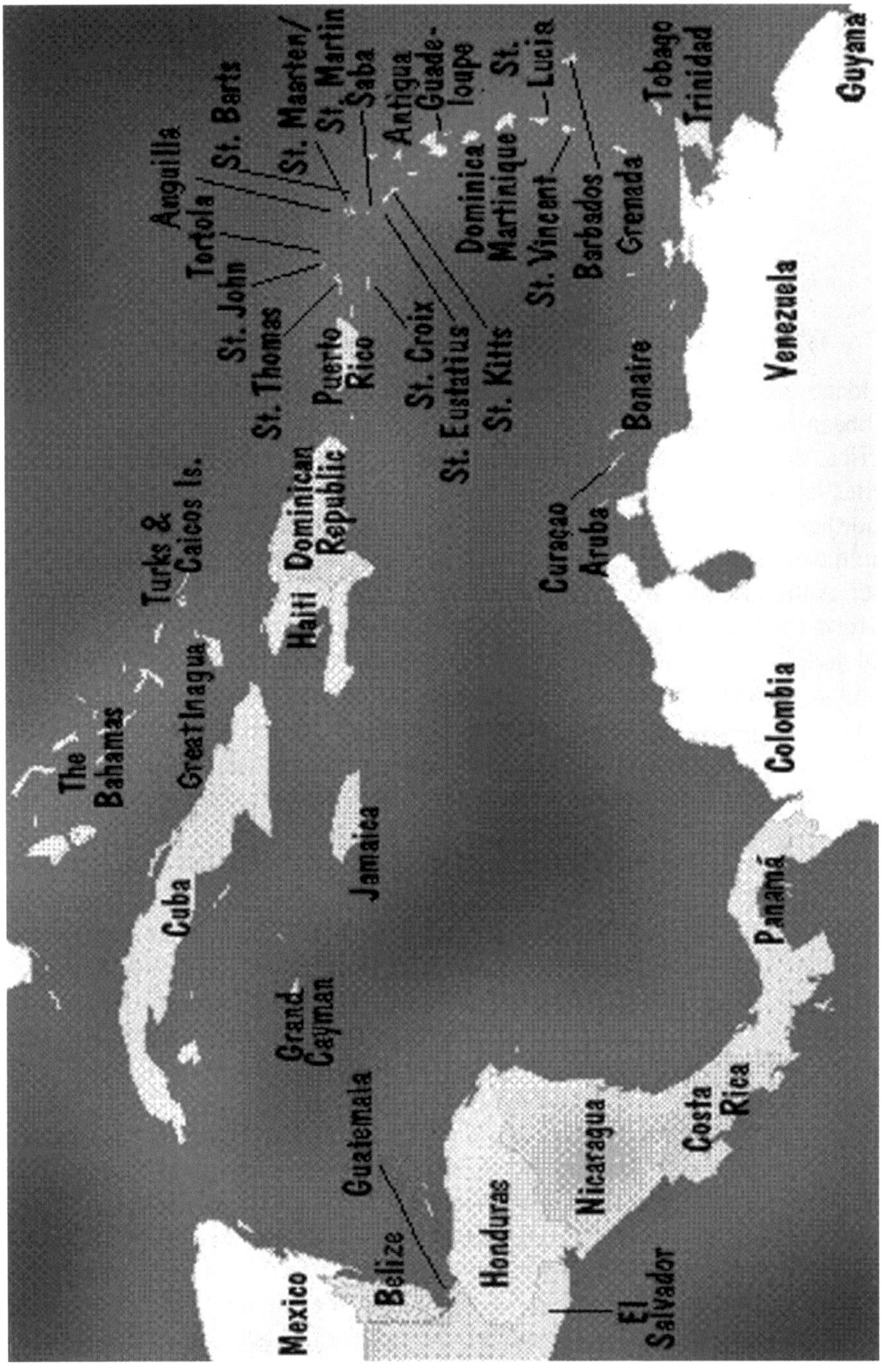

Figure 1. The Caribbean region encompasses some 33 countries or political units made up of eight continental countries and 38 islands in the West Indies and their many small islets plus the island groupings of The Grenadines and the Dependencias Federales (Venezuelan Dependencies).

Nederlandse Antillen (Netherlands Antilles), Saint Christopher–Nevis, the Antilles Françaises (French West Indies), Dominica, Martinique, Saint Lucia, Barbados, Saint Vincent and The Grenadines, and Grenada, which lies at the southern-most tip of this long island archipelago.

The Southern Antilles, consisting of Trinidad and Tobago, the Dependencias Federales, Aruba, Curaçao, and Bonaire, are also part of the Lesser Antilles. See Appendix A for a list of the Islands of the Caribbean. Appendix A also lists the country or political association of an island or group of islands, and some of their smaller islets. Listed also are some of the other islands mentioned in the historical texts. Figure 1 shows a map of the islands of the Caribbean and the continental countries bordering the Caribbean Sea.

1.2. Early names

During the early history of the "Charibee," islands were sometimes called by the name of the principal settlement, as with San Juan for Puerto Rico, Road Town for Tortola, Bassin (Christiansted) for Saint Croix, and St. John's for Antigua. Nueva Esparta was the name for the island grouping of Isla de Margarita, Isla Cubagua, and Isla Coche off the coast of Venezuela near Península de Araya. A degree of uncertainty in the historic record is attributable to differing languages, previous names, and variant spellings. Archaic or alternate names found in the sources are noted in Appendix A. Figure 2 is a very old illustration of the Caribbean region drawn in 1511 by Pietro Martaire d'Anghiera, known as Peter Martyr.

The rendering of this ancient map of the Caribbean provides an interesting look at the state of exploration and mapping at the time. The 1511 date of this map is worthy of note in light of the 1498 tsunami at Boca de la Sierpe, Pedernales, Venezuela. The map is also noteworthy in that it predates the events of 1530, 1541, and 1543 in this vicinity of Venezuela.

Despite the indistinct lettering, an attempt to describe what is portrayed on the map continues here in a more-or-less clockwise direction. *Isla de Cuba, Jamaica, Isla de Española* [Hispaniola], and *San Juan* [Puerto Rico] show considerable detail. They are very large in proportion to the rest of the map. The long island chain comprising both The Leewards and The Windwards appears to be entirely missing—or the Antilles correspond to the sweep of islets drawn above Hispaniola, culminating in The Grenadines, with the flip of the tail cheerfully Barbados.

Relative distances appear to be either stylistically drawn or as yet unknown—the Atlantic Ocean being particularly abbreviated. *El Estrecho* [The Strait], undoubtedly the Strait of Gibraltar, is in close proximity to *La Bermuda* [Bermuda] and *Canarias* [Islas Canarias, the Canary Islands]. The Canaries, located off the northwest coast of Africa, lie entirely across the Atlantic Ocean from Bermuda, which is situated off the Carolinas, part of the East Coast of the United States. On the map, Bermuda and the Canary Islands are, in turn, shown quite near *San Juan* [Puerto Rico].

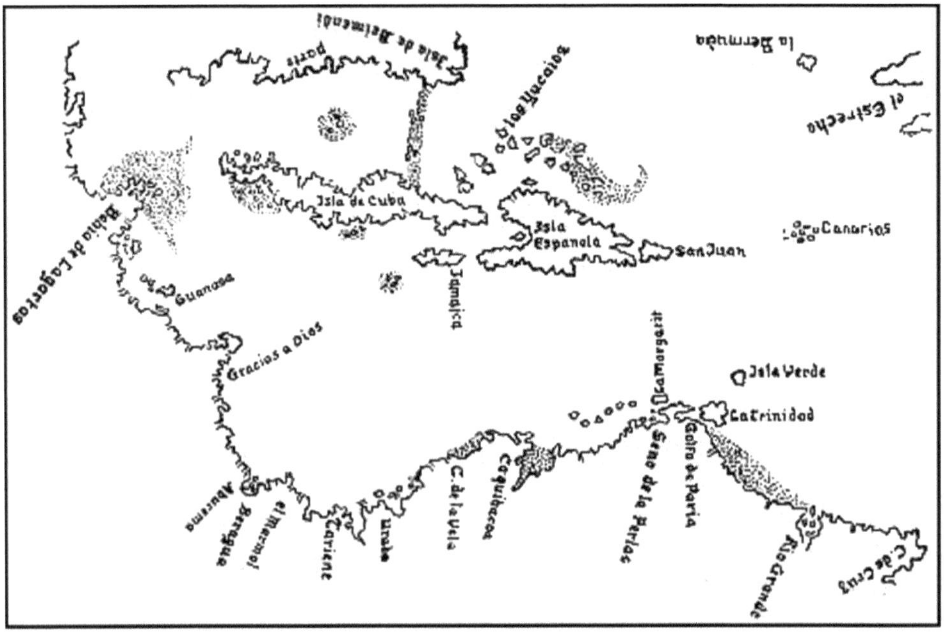

Figure 2. Map of the Caribbean region drawn in 1511 by Peter Martyr. This early map is interesting for both its detail and its lack of detail.

Isla Verde, probably the Cape Verde Islands off the coast of West Africa, is situated nearly on top of *La Trinidad* [Trinidad], while Trinidad is appropriately just off the coast of Venezuela. *La Margarit* [Isla de Margarita] is the primary island of Nueva Esparta. *Sena de la Perlas* is without doubt referring to the opulent pearl beds of Isla Cubagua, discovered by Spanish explorers circa 1499. A formation resembling Golfo de Cariaco appears beneath what appears to illustrate Península de Araya and Península de Paria. The traditional account of the inhabitants in the region, carried by word-of-mouth from generation to generation, recounted that the cataclysm that cleft the mountains was very recent at the time of Christopher Columbus' third voyage, which took place in 1498. The immense breach that remained was deluged by seawater for 68 kilometers, thus forming the Gulf of Cariaco.

From *La Margarit*, a small chain of islands, apparently depicting the Southern Antilles, leads credibly westward toward Golfo de Venezuela and *Cabo de la Vela,* Colombia. *Uraba* is undoubtedly Golfo de Urabá, Colombia, and *Tariene* may well be Darién, Panamá. *Gracias a Dios* anchors the Atlantic seaboard of Central America, while *Guanasa* probably represents Isla de Guanaja in Golfo de Honduras.

The large triangular group of islets enclosing the point of a peninsula near *Bahía de Lagartos* may be those situated off Belize. The coastline continuing upwards then peters out, perhaps somewhere along the Peninsula de Yucatán. Alternate interpretations are clearly possible. Aligned with Cuba, directly across what would be Canal de Yucatán, the triangle of islets may enfold Península de Yucatán itself. Depending on the proportions projected by Peter Martyr, the coastline may pass along Golfo de Campeche, to the west past the Yucatán where it abruptly ends, mistakenly drawn here running more-or-less to the north. Knowing the location of *Bahía de Lagortos* would help to solve some of the proportional difficulties of this fascinating ancient map.

The coastline may alternatively portray the entire sweep of the Gulf of Mexico north as far as the Gulf Coast of the United States, as abbreviated as the Atlantic Ocean is. The small peninsula shown just near where the coastline ceases uncannily resembles the Mississippi River delta. The *"ports"* to the north of Cuba may be along the out-of-proportion southern coastline of the United States. Not quite twenty years since the first explorations by Europeans, the Gulf of Mexico remained largely unexplored, thus the apparent confusion as to where (or if?) the two broken coastlines adjoin.

Isla de Beimendi, feasibly the United States mainland, remains a mystery. Could it be an early name for Bimini? It could represent the southern tip of the Florida peninsula, although way out of proportion; albeit not as hugely disproportionate as if *Beimendi* corresponded to the tiny isle of Bimini. Resembling the drawing fairly well, the southern coastline of Florida may have been abundant in *"ports."*

The round grouping of islets above Cuba may represent some of the many keys in the Straits of Florida. Fortunately, no confusion is apparent concerning the group of islets below Cuba, clearly The Caymans. The nearly vertical swath of islets rising from Cuba may be Archipiélago de Camagüey, shown continuous with the many reefs and cays of The Bahamas or the Florida Keys. The group of somewhat larger islands running northeast from Cuba towards *Los Yucaios* may be Andros and The Bahamas, with the curving sweep of Bahamian isles partially misplaced.

Is *Los Yucaios* an early name for the Caicos Islands? The somewhat larger islands running northeast from Cuba may be Great Inagua, Little Inagua, and Providenciales and other islands of Turks and Caicos. Under this scenario, depending on the relative proportion of this section of Martyr's map, the sweep of islands fanning down toward Hispaniola might represent, not Bahamian cays, but the arc of the Lesser Antilles—notably displaced.

津波

CHAPTER 2

THE TSUNAMI HAZARD

2.1. Characteristics of tsunamis

Tsunamis are comparatively rare. Since many of them are small and do very little damage, they are often overlooked as a natural hazard until a disastrous event occurs. A tsunami is a series of several succeeding ocean waves usually launched by an undersea earthquake that increase in height near the shore. A tsunami can hit the shoreline line as a destructive wall of water. Large tsunamis can be a menacing threat to life and property for anyone living near the ocean. The Caribbean region is an area of relatively high seismicity and, although tsunamis are not the major natural hazard in the Caribbean, they have the potential to produce catastrophic regional disasters.

A tsunami is a gravitational sea wave usually generated by large, rapid disturbances of the ocean floor (Bates and Jackson, 1987). The word tsunami (soo-NAH-mee), from the Japanese words *tsu* (harbor) and *nami* (wave), has replaced such terms as "seawave" and "tidal wave." "Harbor" wave is not precisely correct either, but the term "tsunami" has become the adopted idiom.

Still often called "tidal waves," tsunamis are not caused by the gravitational pull of the moon or sun, although high tides can significantly add to their runup above mean water level. Nor are wind-generated storm waves, originating from the violence of Caribbean hurricanes, tsunamis. However, it has been hypothesized that the atmospheric pressure changes during hurricanes put so much stress on the seafloor that they may serve as submarine earthquake triggering mechanisms. Although reports of earthquakes occurring during hurricanes are often assumed to be spurious, seismically generated tsunamis have been triggered during the passage of hurricanes.

A tsunami can be confused with another form of wave called a seiche (pronounced 'saysh' in both singular and plural). Seiches are short-period wave oscillations that may last for several hours, set up in lakes, bays, harbors, or other enclosed bodies of water by seismic phenomena, atmospheric disturbances, or tsunamis.

A seismic event can produce two basic types of tsunamis that are classified by their distance from the tsunamigenic source. A tsunami generated more than 1,000 kilometers from the affected area is called a teletsunami. A major earthquake can generate a teletsunami that travels great distances, often propagating across entire oceans. "A powerful tsunami has a very long reach: it can transport destructive energy from its source to coastlines thousands of kilometers away," according to González (1999a).

Near-field or local tsunamis are generated at distances of less than 1,000 kilometers from the area affected and strike the shoreline from their point origin more quickly than teletsunamis. Most areas are affected by such local tsunamis. A tectonic source event can provide the energy to trigger landslides, volcanic eruptions, or other seismic disturbance that can give rise to near-field tsunamis.

Classified as shallow waves or long waves, tsunamis are oceanic gravity waves in which the wavelength is much larger than the water depth. As such, propagation is mainly affected by large-scale depth changes (Mercado and McCann, 1998). As the seafloor collapses, for instance along a fault rupture, the drop in water level forms a "hollow" or concave deformation of the sea surface. Water rushes in to fill the depression and then spreads in all directions. The slumping of the seabed thus launches the series of highly energized waves. The sudden displacement of ocean water may be triggered by tectonic movement, submarine landslides, volcanic eruptions, or even meteoric impact.

The crest-to-crest measurement, or tsunami wavelength, can be hundreds of kilometers in length when in deep ocean water. Unlike wind-generated waves that move the surface of the ocean, the entire water depth is moved by tsunamis (González, 1999a). The amplitude of a tsunami propagating across the ocean usually ranges no more than 30 to 60 centimeters. This creates a sea-surface slope so gentle that it makes them virtually undetectable at sea. In the open ocean, waves may reach speeds exceeding 700 kilometers per hour and could easily keep pace with a Boeing 747 (González).

2.2. Tsunami hazard

The awesome destructive power of a tsunami occurs on its approach to shore. As depth decreases, the shallower bottom causes the wavelength to become shorter and thus the amplitude greater, as depicted in Figure 3. A tsunami dramatically increases in height at the shore, its impelling force intensifying and overwhelming anything in its path.

Tsunamis of 3 to 6 meters can be deadly or tremendously destructive, depending where the tsunami strikes. Tsunamis larger than 10 meters in height are uncommon since large-scale depth changes from tectonic processes produce sea surface displacement generally no greater than 10 meters. Although extremely rare, immense tsunamis have been known to rise over 30 meters, caused by landslides or some impact trigger.

The majority of destructive tsunamis are local, traveling less than 30 minutes from their place of origin to the areas where they can inflict loss of life and significant property damage (Lander, 1986; Blackford, 1996). Ninety percent of all fatalities occur from tsunamis that strike within 200 kilometers of the site of origin.

The three phases of a tsunami are the generation and propagation of the series

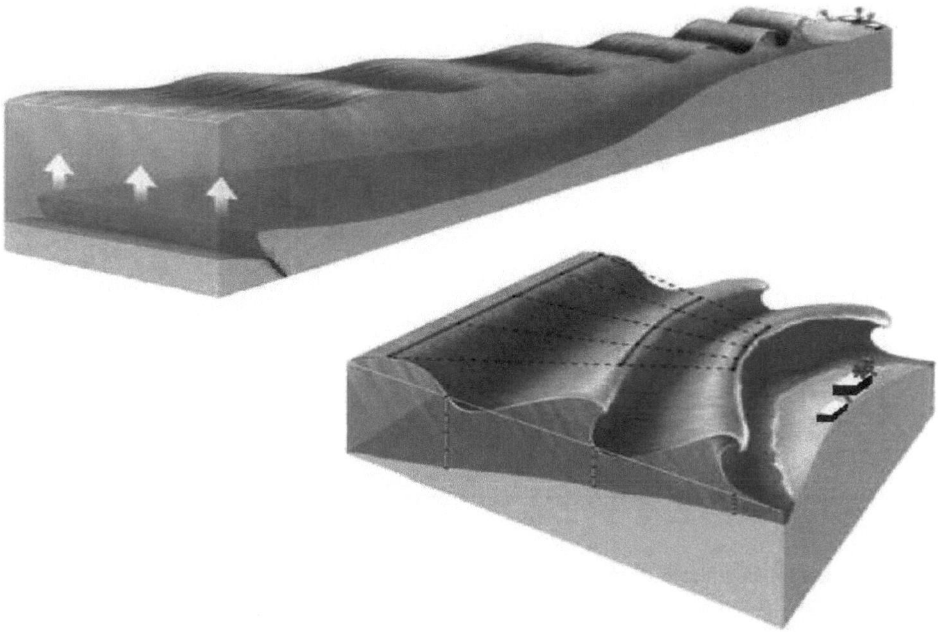

Figure 3. Tsunamis propagate in all directions from the source of the trigger mechanism, generally an underwater disturbance on or near the ocean floor, such as the fault rupture depicted in the top illustration. The long crest-to-crest wavelength decreases as the bottom becomes shallower during its approach to shore. As the wavelength shortens, the amplitude increases. A wave can dramatically mount in height and powerfully surge ashore. (Figure modified from artwork by Jennifer C. Christiansen for the National Oceanic and Atmospheric Administration.)

of waves, and the inundation or flooding of the seacoast. Generation is the process by which a seafloor disturbance, such as movement along a fault, reshapes the sea surface into a tsunami (González). Propagation is the phase during which the tsunami travels as a gravity wave through the full depth of the water from the seafloor to the surface of the ocean. Inundation is the process of seawater movement as it overflows the land along the seashore. The movement of seawater inland is measured both as runup, the maximum level in height above mean sea level reached by the surge, and also as the distance that the inundation reaches inland.

Accurate simulations are crucial for disaster mitigation plans and forecasting potential waveheights and locations most likely to be affected. Since each location has a unique form of the hazard, it is vital to know about the previous effects of the differing nature of tsunamis at many diverse locations (Lander and O'Loughlin, 1999c). To understand the characteristics of tsunamis, their phases, their action along the coastline, and where they may strike, knowledge of geophysical, oceanographic, and hydrodynamic processes is needed. Based on

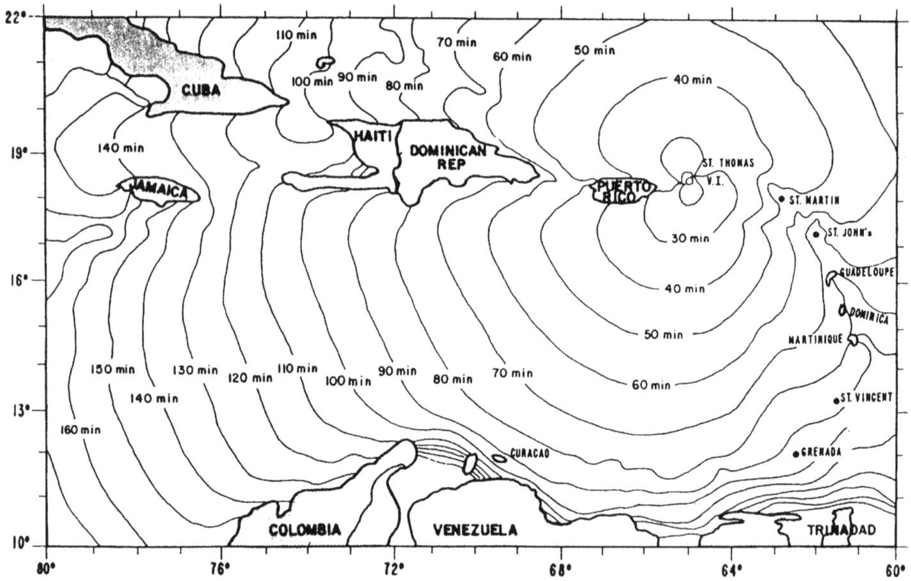

Figure 4. Isochrons indicate 10-minute wavefronts showing travel times in the Caribbean Basin for a hypothetical tsunami triggered near Saint Thomas in the United States Virgin Islands. This plot exemplifies choosing a specific locale as a sink. These computer-simulated wavefronts approximate the travel times for the notable tsunami of November 18, 1867, which propagated in all directions from its point of origin 15 to 20 kilometers southwest of Saint Thomas.

the physics of ocean waves, mathematical algorithms are used to simulate wave action in computerized models for studying tsunamis. It is difficult to predict tsunami heights because the nature of each tsunami differs depending on many hydrodynamic and geophysical components. These components incorporate source mechanisms, seismicity, length of the path, obstructions between the tsunami origin and site affected, sea floor bathymetry, angle of the seabed approaching the shoreline, near-shore topography and geology, dry land elevation, and local features.

Accurate modeling is essential to understanding the potent threat of tsunamis. Knowing the time needed for a wave to propagate from the tsunamigenic source to any location—and thus the amount of time available before the onset of the waves—can greatly facilitate warning, emergency response, and evacuation efforts. Comprehension of coastal behavior and tsunami risks is needed by engineers, city and land-use planners, and public officials responsible for tsunami warning and mitigation systems. Looking into the record of Caribbean tsunamis in historic times can aid these potentially life-saving efforts.

Modeled wavefronts created by computer simulations can give travel times to any port. Plots can be generated choosing a specific locale as a sink to create wavefront maps. Figure 4 depicts travel times in 10-minute isochrons as wavefronts emanating from St. Thomas Harbor.

Waveheight and propagation are proportional to the magnitude of the displacement. During plate movement intense forces build and release. If plate movement becomes locked, the built up strain must eventually give way. Plates may lurch suddenly into new positions, causing an earthquake. Epicenters are located directly above earthquake foci, which are the hypocenters where the energy release occurs. Tsunamigenic earthquakes are those associated with tsunamis. Kanamori (1972) said "tsunami earthquakes" generate much larger tsunamis than expected from seismic waves.

Crustal plates react with each other with several types of tectonic plate interaction; these may be, but are not necessarily tsunamigenic:

- New crust can be formed from material (usually volcanic) issuing from a rift in the ocean floor. Shallow earthquakes are associated with diverging plates.
- Plates can slide by each other. Shallow earthquakes are associated with such transform faults.
- Plates can ram each other, causing the crust to buckle. Earthquakes are typically shallow.
- One plate can sink beneath another plate; the subducting plate gradually breaks apart or melts. Deep-foci earthquakes associated with subduction under continents are not tsunamigenic. Shallow earthquakes associated with subduction can cause tsunamis.

Tsunamis tend to occur when there is significant displacement of the bathymetry. This typically occurs with shallow sub-oceanic earthquakes associated with rapidly subducting plates. Some of the largest earthquakes are due to plate subduction, as with the massive 1755 Lisbon earthquake. The epicenter of this mammoth seismic convulsion was placed off the Iberian Coast at the Gorringe Bank, near the Azores–Gibraltar Ridge. Myles (1985) said the 1755 cataclysm was probably caused by the huge crustal boundary of the Eurasian Plate subducting the African Plate:

> Somewhere along the north edge of the African Plate, the infinitesimally slow process of subduction had quickened; a major segment of the crust was slipping. The pressure built and built, trillions of tons of it, the stored energy doubling and trebling with the passing seconds, reaching incalculable levels.

He determined that this immense continental plate boundary essentially extends from Asia Minor to the Azores–Gibraltar Ridge in the Atlantic.

Tectonic displacement of the ocean floor caused by earthquakes, crustal plate movement, or volcanic eruptions may be tsunami trigger-mechanisms or tsunamis may be generated by submarine landslides or subaerial pyroclastic flows or by volcanic debris, glacial ice, or other landslides issuing into the sea. Vertical fault-block shifts, slumping of submarine canyon walls, or other submarine collapse can cause the initial sea surface deformation that launches the succession of oceanic gravity waves.

In discussing the three physical processes of a tsunami: generation, propagation, and inundation, González (1999a) spoke of propagation in terms of linear wave theory, which predicts that the deeper the water and the longer the wave, the faster the tsunami.

> This dependence of wave speed on water depth means that the refraction by bumps and grooves on the seafloor can shift the wave's direction, especially as it travels into shallow water. In particular, wavefronts tend to align parallel to the shoreline so that they wrap around a protruding headland before smashing into it with greatly focused incident energy.

Tsunami generation and inundation are more difficult to model with computer simulations, since they are less well understood than propagation (González). Inundation, when a tsunami floods over the land, is the most difficult to model: "A tsunami may run ashore as a breaking wave, a wall of water, or a tidelike flood." What's more,

> The wave height is now so large that linear theory fails to describe the complicated interaction between the water and the shoreline. Vertical run-up can reach tens of meters, but it typically takes only 2 or 3 meters to cause damage. Horizontal inundation, if unimpeded by coastal cliffs or other steep topography, can penetrate hundreds of meters inland. Both kinds of flooding are aided and abetted by the typical crustal displacement of a subduction zone earthquake, which lifts the offshore ocean bottom and lowers the land along the coast.
>
> This type of displacement propagates waves seaward with a leading crest and landward with a leading trough (the reason a receding sea sometimes precedes a tsunami). Not only does the near-shore subsidence facilitate tsunami penetration, but, according to Raissa Mazova of the Nizhny Novgorod State Technical University in Russia and Costas Synolakis of the University Southern California, both theoretical predictions and field surveys indicate that coastal run-up and inundation will be greater if the trough of the leading wave precedes the crest.

2.3. Caribbean seismic hazard

Bands of earthquake activity exist at the periphery of the Caribbean Plate where it grinds against the surrounding plates. In the Treatise by naturalist, A.S. Örsted (1867), on the "Natural Features of the Westindia Islands," he commented on the seismicity and volcanism circumscribing the Caribbean Sea. Örsted noted instances of volcanic activity subsiding, followed closely by disastrous earthquakes, "the earthquakes continuing months or years until their cessation, when volcanoes again commenced to erupt." Örsted concluded:

> From the volcanoes on the eastern coast of Central America and the various volcanic features in the larger Antilles, it becomes apparent that the whole Caribbean Sea is surrounded by a vast volcanic fissure, which in the Lesser Antilles from Grenada to Sint Eustatius finds a vent through a range of volcanoes.

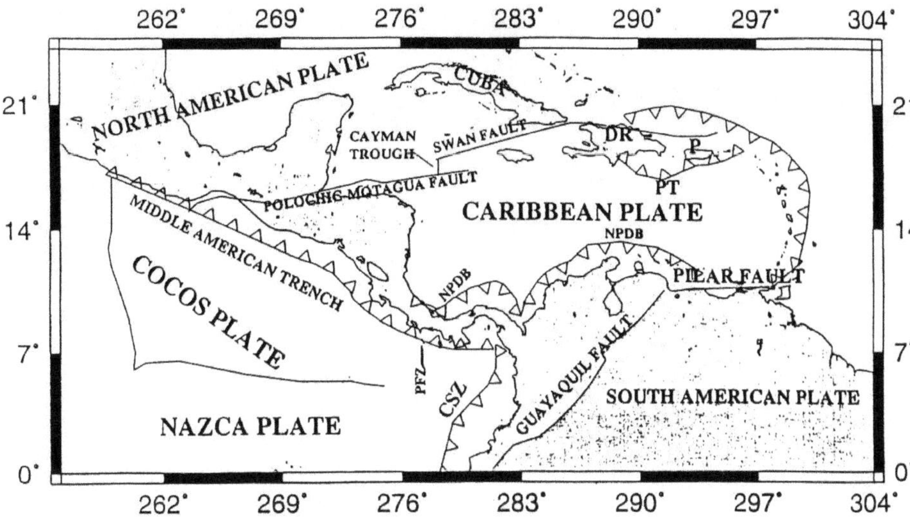

Figure 5. Generalized geology of the Caribbean region showing the tectonic setting. The principal plate zones are the North Panama Deformed Belt (NPDB), the Panama Fracture Zone (PFZ), and the Colombia Subduction Zone (CSZ). *PT* stands for the Puerto Rico Trench, *DR* is the Dominican Republic, and *P* is Puerto Rico. (From Arce *et al.*, 1998, based on Calais *et al.*, 1992.)

Dillon *et al.* (1996) report shallow-focused seismicity marking a distinct boundary. To the west, the Cocos Plate is being driven northeast and subducted beneath the Caribbean Plate. The principal plate zones for the Caribbean Sea are shown in Figure 5. Arce *et al.* (1998) observed, "The tectonic setting of Central America is given by the interaction of the Cocos, Nazca, and Caribbean Plates. The Cocos Plate subducts under the Caribbean Plate along the Middle America Trench." Some places usually considered part of the Caribbean: The Bahamas, Cuba, and the Yucatán, lie outside the band of earthquakes, and therefore off the Caribbean Plate (Dillon *et al.*, 1987).

The roughly rectangular Caribbean Plate is flanked by the North and South American Plates. It slides eastward at about two centimeters per year relative to the North American Plate (Dillon *et al.*, 1999). "Motion along its northern boundary (in the Plate Boundary Zone) is dominantly strike-slip, with perhaps a small component of shortening. In contrast, on its eastern boundary, the Caribbean Plate overrides the North American Plate." Earthquakes tend to be shallow in this subduction zone. The crust of the descending North American Plate begins to melt into molten magma, which rises to form volcanoes that have become the island arc of the Lesser Antilles. According to Dillon *et al.* (1999), the northeastern Caribbean Plate region has high seismicity and tends to have large earthquakes.

Taber (1922b) described the two most persistent fault zones in the northeastern Caribbean as parallel for 2,000 kilometers and 100 to 150

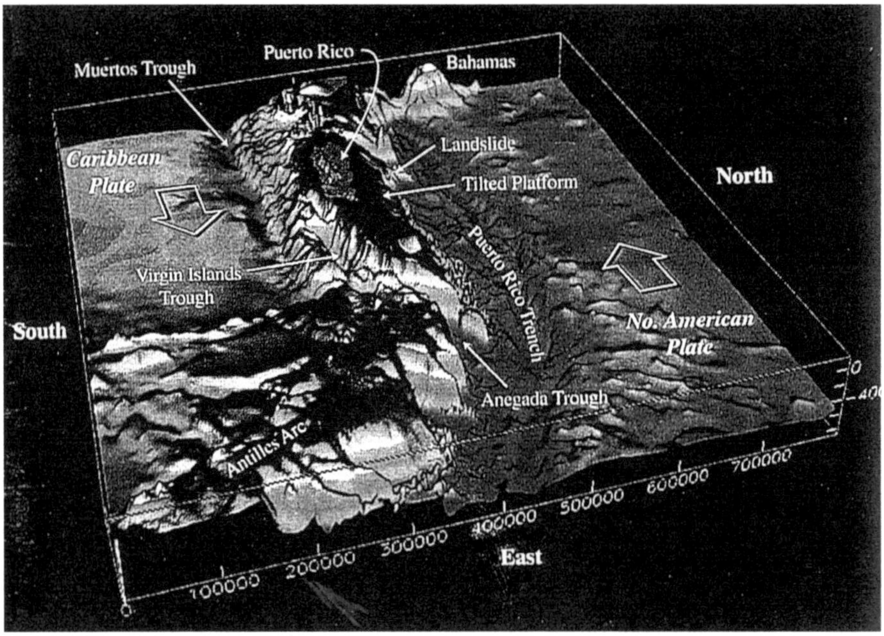

Figure 6. Morphology of the northeast Caribbean region viewed from the east. The Virgin Islands, Puerto Rico, and eastern Hispaniola exist on microplates in an active Plate Boundary Zone, delimited by the Puerto Rico Trench to the north and the Anegada Trough and Muertos Trough to the south. This Plate Boundary Zone lies between the generally westward moving North American Plate and eastward moving Caribbean Plate. (Figure from Dillon *et al.*, 1999; Digital data courtesy of Troy Holcombe and Lisa Taylor, National Geophysical Data Center, Boulder, from the International Bathymetric Chart [1–09] published by Instituto Nacional de Estadística, Geografía e Informática, Mexico, 1992.)

kilometers apart, with a long, relatively narrow strip in between. The narrow ridge rises at its eastern end to form the highest peaks of the Antilles, the Cordillera Central of Hispaniola. Characterized by continuing displacements, these two major fault zones were designated as the Swan Island–Jamaica–South Haiti Fault Zone and the Cayman Islands–Sierra Maestra–North Haiti Fault Zone. Taber depicted a third major fault zone along the northern scarp of the Puerto Rico–Virgin Islands Ridge rising precipitously from the Brownson Trough, 8,526 meters deep.

The controls on earthquakes are poorly understood, since the motions of these tectonic plates are poorly known (Dillon *et al.,* 1996). The tectonic setting "probably includes interaction of the major Caribbean and North American Plates, at great depth, associated with the semi-independent movement of microplates in the broad Plate Boundary Zone." As depicted in the three-dimensional plot in Figure 6 from Dillon *et al.* (1999), Puerto Rico plus the Virgin Islands to its east and eastern Hispaniola to its west exist on microplates in an active Plate Boundary Zone (Dillon *et al.,* 1999). Delimited by the Puerto

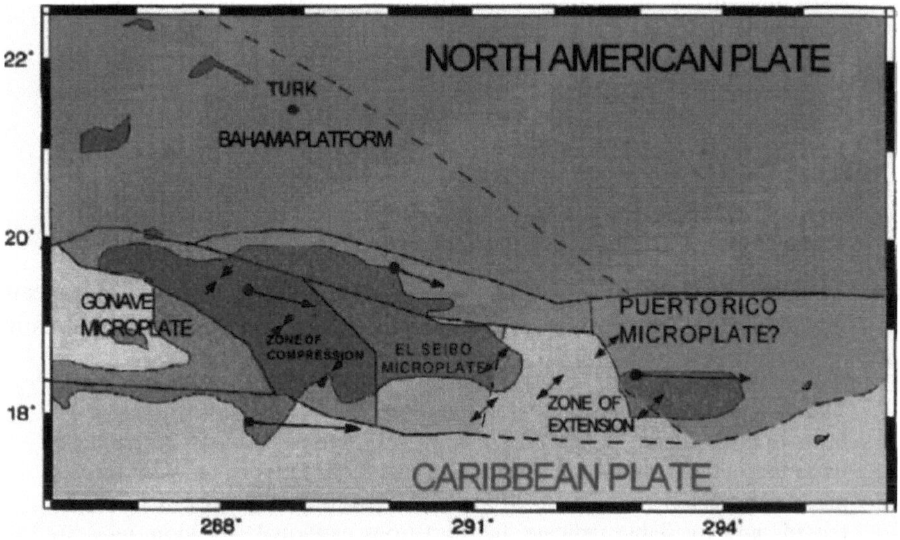

Figure 7. A complex set of interblock movements occurs among the plates, platelets, and tectonic blocks in the narrow Plate Boundary Zone. The arrows show the direction of movement within this zone. Figure courtesy Mercado and McCann (2000).

Rico Trench to the north and the Anegada and Muertos Troughs to the south, the Plate Boundary Zone encompasses most of the Greater Antilles plus the Virgin Islands, except for St. Croix, which is part of the Caribbean Plate tectonic environment. According to Holcombe *et al.* (1977), "It has been proposed that the St. Croix Ridge was once part of the Greater Antilles Platform, the two elements having been subsequently separated by rifting or transtensional movement, which opened up the Virgin Islands Trough."

This complicated zone of shifting microplates lies sandwiched between the generally westward moving North American Plate and eastward moving Caribbean Plate. A complex set of secondary movements occurs among the plates, platelets, and tectonic blocks comprising the Plate Boundary Zone (Mercado and McCann 1998), illustrated in Figure 7. Interblock movements within this zone are characterized by Mercado and McCann as even less precisely understood.

Meyer (1996) summarized the four seismic source zones acting in the Caribbean:

> The Eastern Caribbean Subduction Zone [1], with possible magnitudes of 8.5; the faulting systems [2] that limit the Caribbean Plate to the south (coast of Venezuela) and to the north (Jamaica, southern Cuba, Hispaniola), both of which have historical records of tsunamigenic earthquakes, but probably do not pose regional threat due to their strike-slip nature (records

> show the almost simultaneous occurrence of shaking and tsunami, thus confirming their very local nature; and, as the third type of source, the Panama–Columbia Shallow Underthrusting Zone [3], where only the Panama segment has generated tsunamis in historical times (Mendoza and Nishenko, 1989). The most likely source of a Caribbean-wide tsunami seems to be, then, the Lesser Antilles Subduction Zone [4]. A detection and warning system for this situation, technologically now possible and available, could be operating efficiently in the source region and give timely warnings (about 4 hours across the Sea), even close to sites.

Taber (1922b) remarked, "Since the region became known to the Europeans, seismicity has been high; prior to that time history is a blank." An interesting commentary emerges as, summing up, Taber forewarns:

> The conclusion that destructive earthquakes in the Greater Antilles, have been limited almost entirely to a few well-defined belts is of practical importance, for in the future earthquakes and their accompanying sea-waves may be expected to originate from time to time along these belts. The records now available indicate that disastrous earthquakes seldom recur in exactly the same place except after long periods of time; therefore, those places along the zones of active faulting near which severe earthquakes have not originated during historic time are to be regarded as seismically most dangerous.
>
> Unfortunately, in the mountainous Antilles, most larger towns are built on the least stable ground—the water-logged alluvium of the narrow coastal plains or playas. Formerly, as a precaution against easy attack from the sea, it was customary to locate towns farther back, on the lower slopes of the hills, but in recent years, this has ceased to be a controlling factor. The danger from seismic sea-waves is of course greatest along the flat low shores.

Along the northern and eastern subduction zones, the return-event period for the most significant, destructive, or deadly tsunamis averages from 0 to 72 years. The average interval is approximately 21 years. Appendix B summarizes the 13 major tsunamis used for this tally:

- 1690, Nevis and St. Thomas, U.S. Virgin Islands
- 1692, Jamaica
- 1751, Dominican Republic
- 1755, Saba, Antigua, Martinique, Barbados, and Dominican Republic
- 1770, Haiti
- 1842, Haiti and St. John, U.S. Virgin Islands
- 1867, Puerto Rico, Virgin Islands, Guadeloupe, and the Antilles
- 1902, Martinique and St. Vincent
- 1906, Venezuela
- 1907, Jamaica
- 1918, Puerto Rico
- 1946, Dominican Republic and Puerto Rico
- 1946, Dominican Republic and Puerto Rico

It has now been 57 years since the last major tsunami in this relatively seismically active region of the Caribbean, more than double the "average" interval between major tsunamis [2.7 times as great]. Although no such thing as an "average" interval really exists—tectonic forces being what they are—nevertheless, such a reckoning can help to convey a sense of urgency about the growing lapse of time since the last disastrous tsunami.

Tectonic forces are continually building stress. Eventual relief of built-up strain is certain. The potential extent of the inevitable energy release grows as time elapses. The long period without relief of seismic stress buildup only increases the ominous threat of a major tsunami that could result from a sudden seafloor cataclysm.

With the prospect of perilous tectonic activity, the potential for a catastrophic tsunami is greater now than in the past. The vast area of the Caribbean is at severe risk for a terrible disaster. Future tsunamis in the Caribbean may cause significantly greater destruction than ever before for a number of reasons:

- Worldwide rise in sea levels
- Enormous increases in population
- Growth of tourism-related industries
- Numerous cruise ships stopping at Caribbean ports
- Copious numbers of beach-goers and pleasure boaters
- Extensive development of coastal areas
- Increased harbor and transportation infrastructure
- Greater shipping in the region, with the ominous prospect of petroleum and other toxic spills
- Intensified building of houses, resorts, and other structures near shorelines

津波

CHAPTER 3

ABOUT THE 500-YEAR TSUNAMI HISTORY

3.1. Details about the 1498 to 1998 Caribbean Tsunami History

The Caribbean Tsunami History is a comprehensive 500-year compendium of 127 tsunamis from the earliest available report dated 1498, through 1998. The intent is to portray as accurately as possible the occurrence of tsunamis, both great and small. Many still do not consider a wave a tsunami unless it is gigantic. Most tsunamis are not so. Nor are they all detrimental. Many small tsunamis can go entirely unnoticed, depending on where and when they occur.

Lesser tsunamis are poorly identified in the historic record. However, knowledge about minor tsunami effects, as well as how often they tend to occur, is critical to understanding the hazard. Where it has been inferred that at least small waves may have been produced, the texts were cited as possible instances. Further evaluation of sources as well as the seismicity of a region may reveal multiple occasions when smaller tsunamis occurred. Such smaller tsunamis would almost certainly not have been chronicled in the historic past. In the post-instrumental era, a train of lesser waves recorded at a marigraph station may confirm the occurrence of a small or very distant tsunami.

Earthquake catalogs, primary sources of information about tsunamis, usually mentioned waves or inundations only if they played a significant part in a seismic event or if they were in some way exceptional. Frequently, information simply was not included in the catalogs for the sake of brevity. Oftentimes civil authorities may not have reported tsunamis or earthquakes unless great damage was done or municipal services were disrupted. Scherer (1912) formed the opinion that many notable earthquakes were not reported: "Moreover at this time [1700's], in order for an earthquake to be recorded in history, it was necessary that it upset in a very effective manner the affairs of the colony."

Additional events are likely to be found by searches of local historical data, such as newspapers, libraries, diaries, memoirs, colonial archives, school and university history departments, government and church archives, museums, and historical societies throughout the islands and seaboard of the Caribbean.

Marigraph stations past and presently operating, along with any archived records, need to be located and researched. Given the long and still continuing colonial history, many accounts should be available from European historic archives and ship logs of countries that had colonies or trade in the Caribbean. Piecing together an understanding of a specific feature, locale, topography, or type of hazard can be enhanced by examining details in the Caribbean Tsunami History and applying the knowledge gained to the particular emphasis. To allow browsing for special features of interest, the Caribbean Tsunami History is set up as a table with columns that encompass two broad categories:

1. Origin Data:
   - Date; Time
   - Latitude; Longitude
   - Source Area
2. Tsunami Data:
   - Location of Disturbance
   - Maximum Wave Amplitude in Meters
   - Description of Tsunami and Seismic Effects

The section on tsunami and seismic effects furnishes reported details about tsunamigenic origins, seismic disturbances, source mechanisms, withdrawal from the shore, inundation and maximum heights, intensities, fatalities, damage, and other particulars gleaned from source materials. This study has been compiled from published sources and personal communication with knowledgeable individuals. Sources are referenced by author and year of publication.

The extended bibliography contains the primary references as well as many relevant works and secondary sources. An author's primary and secondary references may be included to assist in accessing still earlier information for those seeking to do further research. The bibliography lists a great number of titles unearthed while compiling the list of references, many discovered too late to be retrieved or examined. As they became known, it was deemed relevant to include them in the growing list.

References were alphabetized and then separated into two lists. Textual references cited are presented first, followed by a section of source materials not directly referenced as well as other relevant works. Newspapers are listed alphabetically within the bibliography with thumbnail descriptions. Multiple works by the same authors for a given year, carried over to either list as appropriate. Thus, an author of five works in 1850 may have two directly referred to in the text, 1850a and 1850c, and the remaining three, 1850b, 1850d, and 1850e, placed in the list of unreferenced works.

For brevity, only the first author of a source is called out when a reference is cited in the text of the Caribbean Tsunami History. If two works for the same author were referenced for one event, one date will be identified parenthetically, and all other mention of an author's remarks will be the other date given. Cited page references may be for a specific reference, and may not indicate all pertinent information within that source. Quoted material may be somewhat condensed and not all statements have been directly referenced. In quotations, any clarifications or linear conversions are in brackets, while an author's own notations are in parentheses as written. For the most part ellipses [...] have been omitted. Grammar within quotations basically has been left intact, despite some lovely split infinitives and subtle confusion between which's and that's.

For each event, the likelihood of its occurrence has been assessed and a validity rating ascribed. Three reports were considered invalid. Of the remaining 124 tsunamis, 53 were evaluated as reliable occurrences. Another 8 probably occurred, 19 may or may not have occurred, and 44 are doubtful or improbable occurrences or the information is so meager that to ascertain if tsunamis actually occurred needs further research.

Sources have differing reliability levels. Some accounts appear to have biases that need to be taken into consideration or filtered. Occasionally, successive versions jump back and forth between languages, confusing the details. Descriptions by excited onlookers, not used to methodical reporting procedures, may be less credible than those of researchers, scientists, engineers, lighthouse keepers, ships' captains, or others who assess measurements of the sea—although this, too, could be misleading.

Often reports were requested by a church body to assess damage to their religious edifices and parishes. These were carried out by clergy who, overall, made excellent reports with multiple details of earthquake and tsunami effects.

Among noteworthy scientists known for meticulous research is the explorer Baron von Humboldt. Humboldt reported gathering information by listening to word-of-mouth traditions of the local people to obtain local tsunami and seismic histories. Prior to the codification of events and recollections set down in a written historical record, information about the past was recounted by word-of-mouth from generation to generation. These oral records were often ritually formulated and handed down to each new generation to preserve the memories, customs, and culture of a given people. Since no seismic instruments existed, the oral traditions and general knowledge of occurrences in the remembrance of a cultural group are invaluable for helping to inform the historical record.

Archaic spelling has sometimes been retained. In certain instances, *[sic]* is placed in the text to acknowledge such usage or missing words. The originality displayed in the spelling of words had little to do with ignorance, as many have assumed. Rather, prior to the codification of spelling with the advent of dictionaries, one had the freedom to spell as one chose.

Fathom (plural) and fathoms (plural) are both used. Suggested as more valid, Hawaii is spelled Hawai'i. Foreign words are shown *Italics* with the gist of the meaning following in brackets. Certain expressions were drawn on repeatedly to describe the observation of a tsunami. Terms often encountered were:

- Flux and reflux
- Ebb and flow
- Recession from the shoreline
- Agitation of the waters
- *Maremoto* [tsunami]
- *Raz-de-marée* [tsunami]
- *Mar de leva excepcionale* [exceptional rise or threat of the sea]

An instance of the latter was reported by Singer *et al.* (1983) for the event of March 1812 in Venezuela: A *mar de leva excepcionale* occurred at the La Guaira; "gigantic waves" broke stretches of the sea wall protecting the coast.

These expressions encompass a broad range of terms describing the action of the sea. Many were used interchangeably and their usage was sometimes obscure. Discerning distinctions among terms in earlier records was not always possible. Seismically induced waves were not well understood in times past. Prior to the instrumental period of gathering data and better methods of collecting information, these descriptions may be all that is available to indicate the possibility of tsunamigenic occurrences. When used with links to earthquakes and details of associated phenomena that often accompany tsunamis, the inference that tsunamis may have occurred is not unreasonable.

The sea may have been depicted as agitated by an oscillatory motion analogous to that felt on the shore. Grases (1971) cited an account of a severe earthquake affecting The Windwards, Surinam, and French Guiana, in 1767: "It seemed the turbulent sea would hurl itself on the shore," due to violent shocks at Surinam that were felt at the same time at Barbadoes and Martinique, where "the waters of the sea were very agitated" and "the sea ebbed and flowed in an unusual way." While "agitated waters" may represent a "seaquake" (not a true gravity wave) or other wave phenomena, it is likely that "ebbing and flowing" of the sea was descriptive of a tsunami.

Anecdotal material is incorporated to illustrate the actual experience of tsunami occurrences or to portray the circumstances surrounding the historical events. Extended descriptions, remarks, and quotations evoking particularly vivid images are cited. A number of remarkable narratives are recounted in the discussion chapters and repeated in the Caribbean Tsunami History. These noteworthy descriptions contain much of the best material to inform the present about the magnitude, distribution of effects, and aftermath of tsunamis in the historical record.

To render the material as authentically as possible, language taken directly from the historical texts may be presented in the chronicle, which may reflect "politically incorrect" beliefs or ideas of the period. No offense or disrespect is intended by any use of wording or views expressed in the historical record that represent prejudice, biases, or doctrines of historic times.

3.2. Waveheight

The runup height is the maximum height of the water above a reference sea level (Lander and Lockridge, 1989). It can be indicated by markings on walls, trees, or other features or traces of seaweed or other floating materials transported by a tsunami. The amplitude of a wave is the vertical rise or fall of the water from the mean sea level at the time. Marigram recordings indicate wave amplitudes. Runup height may include a wave surge as the tsunami meets the slope of the shore, resulting in a height somewhat greater than the amplitude. In general, runups can be considered more or less equivalent to the amplitude, or approximately half the full excursion of the wave from drop to rise. The full range from the lowest drop of the wave to the highest rise (trough-to-peak) is considered the waveheight.

Difficult to untangle are some uses of the words runup and waveheight. Almost impossible to differentiate as to whether the report meant amplitude or the full excursion of the wave, some observations in the older literature reported runup heights as waveheights. Early height data may have specified the rise or fall as "waveheight," therefore at about half the value of what would currently be reported as waveheight (trough-to-peak by today's definition). Thus, a 3-meter trough-to-peak wave may have been reported as a "waveheight" of 1.5 meters. When seemingly inconsistent height data were given, as with 25 or 30 feet as well as 50 or 60 feet, it may be inferred that the former was speaking of amplitude and latter of the full waveheight.

The meaning of terms can be perplexing when unknown or differing definitions are used. The vertical drop in height (often referred to as retiring, retreating, receding, or withdrawing) from mean sea level must be distinguished from the distance that the sea withdraws from the shoreline (often referred to as retiring, retreating, receding, or withdrawing).

Sometimes inundation, which is the farthest distance that the water enters inland, is confused with the height of runup flooding coastal shores. In some instances, information was so limited that it was difficult to tell precisely what was intended by the author. Distinctive details that could shed light on a particular usage have been included. Unraveling what actually happened during a past event may provide key input for forecasting potential waveheights, distribution of effects, and vicinities affected for disaster planning.

Mention should be made that nearly every statistic reported could have a qualifier added, such as approximately, up to, more than, at least, and so on. These have been kept to a minimum, although they are implicit throughout. Data are given as they were reported in the literature. Estimated measurements in the historical record, prior to better methods of collecting data and surveying the aftermath, particularly in the pre-instrumental period, must be met with mindful consideration. Nevertheless, it should not be assumed that data are exaggerated or erroneous if they differ from "accepted" thinking. Error may be perpetuated by discounting reported data that lie outside a recognized "norm."

Depths, heights, and distances are given in meters rounded to the nearest tenth. Maximum amplitudes specified in non-metric units of length are converted to meters using the linear equivalents specified in Appendix C along with a selection of archaic and foreign terms encountered in the historical record. The foot-to-meter conversion factor is 0.3048. If a measurement other than meters (m) kilometers (km) or centimeters (cm) was reported, the metric equivalent is given in brackets following the usage.

If multiple tsunamis were not explicitly reported as separate events, they were treated as a single entry in the text. For instance, word-of-mouth traditions preserved in the region of Cumaná toward the end of the 16th century held that "numerous" enormous earthquakes during the period resulted in the ocean rising

15 to 20 fathoms, with massive inundations that caused the populace to move further inland. (One fathom, equivalent to 6 feet, thus tallies a 27- to 37-meter wave.) Since the "numerous" tsunamis were not specified individually, they were grouped together as one event.

Somewhat differing waveheight estimations may be due to information from an earlier source being repeated in a latter period using the customary units of measurement of those nations. The interchange of old linear units can be somewhat confusing. Nevertheless, the intended point of reports—to describe great waves and far-reaching inundations—has come across the veil of time, language, custom, and units of measurement.

Considering that estimates were rounded off to roughly even numbers, some with a rather broad range, the reports of heights are remarkably similar. Herrera (1726) reported waves that engulfed the shores, "rearing up 90 to 100 *pieds* [Old French linear unit equivalent to 33 centimeters or roughly a foot, thus 30 to 33 meters]." Hoff (1840) similarly cited "frequent inundations by the sea over the low shore, even climbing as high as 15 to 20 *toises* [Old French linear unit equivalent to 1.95 meters, or 1.1 fathoms, thus totaling 29 to 39 meters]. These inundations induced inhabitants to build the part of town that stands on the declivity of the mountain." These estimates, in their differing units (fathoms, *pieds, toises*), range between 90 and 120 feet [total variance 27 to 39 meters]—impressive waves.

3.3. Validity rating

The validity ratings are assessed based on the judged accuracy of the reports. In evaluating the validity of tsunami occurrence, several criteria are taken into account. First, confirmation by means of associated earthquakes, marigraphic records, and ascribed tsunami and earthquake magnitudes is established. Whether there are multiple sources, observations, and reporting locations is then ascertained. For rating evaluations prior to the instrumental period, especially if reports are scant, it becomes necessary to consider modifying these criteria somewhat. Report authenticity and the judged reliability of the sources is evaluated. Also taken into account is whether contributors are thought to discern the difference between tsunamis and other wave phenomena. The Caribbean Tsunami History chronicles 127 reported tsunamis, comprising the known valid, probable, possible, doubtful, and invalid occurrences in the past 500 years. Validities are quantified on a scale from V0 to V4. See Table I.

- In 3 instances, it was judged that a tsunami did not occur or that it was an erroneous or invalid report. (V0)
- A total of 44 tsunamis were considered to be unlikely or improbable; details may have been doubtful or confused with other events or wave phenomena, or the paucity of information was such that an authentic tsunami occurrence could not be ascertained—although the possibility of a tsunami cannot be completely ruled out. (V1)

Table I. Tsunami Validity Parameters

Values quantifying validity rating parameters in the Caribbean Tsunami History.

Validity Rating	Quantifying Definition	Reported Tsunamis
V0	Tsunami did not occur; Erroneous or invalid information	3
V1	Unlikely or improbable tsunami occurrence; Information judged unreliable or doubtful authenticity, but a possible tsunami cannot be ruled out entirely	44
V2	Tsunami may or may not have occurred; Some question about the event or reports, data are unclear or insufficient to authenticate a tsunami occurrence, particularly in the pre-instrumental period	19
V3	Likely or probable tsunami occurrence; Reliable observations for corroboration are few	8
V4	Tsunami did occur; Data corroborated or deemed valid information	53
	Total Tsunamis reported for the Caribbean	127

Table II. Validity Rating Comparison

Validity ratings in *Tsunamis in the Mediterranean Sea 2000* B.C.*—2000* A.D., Soloviev *et al.* (2000), p. 105.

Validity Rating	Quantifying Definition	Approximate Probability
0	False entry	0%
1	Very doubtful tsunami	25%
2	Questionable tsunami	50%
3	Probable tsunami	75%
4	Definite tsunami	100%

Table III. Reliability Estimates

Reliability Estimates in accordance with a five-grade scale in from *Tsunamis in the Mediterranean Sea 2000* B.C.*—2000* A.D., Soloviev *et al.* (2000), p. 191.

Reliability Estimate	Quantifying Definition
R = Reliable	An event was recorded at least by one gauge or there exist many reliable observations
L = Likely	Few reliable observations
P = Possible	It is difficult to evaluate whether a tsunami or some other event (seiche, sea quake, wind surge, and so on) was observed
Q = Questionable	Probably some other event, different from a tsunami, took place, but the possibility of a tsunami should not be utterly excluded
W = Wrong	Erroneous indication of a tsunami

- In 19 cases, tsunamis may or may not have occurred; reports had little or no corroborating information to authenticate occurrence, especially in the pre-instrumental period, details were unclear or muddled, or there was some question about them. (V2)
- Based on one or more criteria, it was determined that 8 tsunamis were likely or probable occurrences. (V3)
- The remaining 53 instances were judged to be reliable or valid reports of tsunami occurrences. (V4)

Additional facts, confirmation, or corrections may impact the accuracy of any of the assigned validity ratings. Awareness of events judged to be invalid or unlikely or to have some question about them (V0s, V1s, and V2s) could bring to light information to help determine their authenticity. In one instance, a tsunami was discovered to have been on the Pacific coast. This report was expunged, and is not contained in the history.

Eliminating the 3 reports for the Caribbean determined to be invalid brings to 124 the tally of tsunamis evaluated as doubtful, possible, probable, and authentic occurrences. These are divided nearly equally between those that were judged improbable or possible (V1s and V2s)—totaling 63, and those that were judged probable or most definite occurrences (V3s and V4s)—totaling 61. The "numerous" undated tsunamis that occurred in the latter part of the 16th century at Cumaná, Venezuela, may never be disentangled or subject to evaluation, possibly adrift in the mists of time forever as undoubtedly so many others are in the prehistory of the Caribbean.

In certain instances, the paucity of information cannot be satisfactorily resolved without seeking still earlier source materials. When a reference is to an almost certain occurrence, a case for judging it as a possible or even probable tsunami has been made. Since so many data still need to be located, were not reported to any central authority, or were lost or destroyed—often by the destruction in these very events—a tsunami may have occurred even if there are only one or two reports (from sources deemed reliable).

Soloviev *et al.* (2000) included validity ratings comparable to those used here, with a single exception. In Table II, Validity 2 is termed "questionable." This casts a rather more negative or pejorative slant than may be desired for mid-range quantification and thereby gives quite similar meanings to Validity 1 and Validity 2.

The fine distinction in meaning is between a description interpreted as "doubtful" rather than as "having some question about it." Rather, for some instances of Validity 2, circumstances might need sorting out, details might be too meager; or corroboration might be needed, but the occurrence itself may not necessarily be doubtful. The Reliability Estimates of Soloviev *et al.* in Table III agree fairly well with the parameters used in this history. In all, except for the slight distinction made for Validity 2, there is general agreement among the tsunami validity ratings.

Similarity of phrasing among the sources indicates that the same basic source information was reiterated in successive reports. This holds true for many of the tsunami and earthquake compilations and, since it is important to preserve the rich detail from the early sources, it holds true as well for this history.

Another important point to recognize is that the repetition of the same account from report to report or catalog to catalog does not serve to confirm multiple observations or multiple sources. The sheer weight of so many repetitions can give a sense of certitude about an event that may not be warranted. On the other hand, repetition of accounts may not, in itself, be an indication of report duplication.

To discern ambiguities, uncertainties must be evaluated with care, particularly in evaluating the multiple sources, multiple observations, and multiple reporting locations. In a particularly perplexing example, in reports for events in 1775, the sources cited very few details, some with comparable wording, and with similar or uncertain dates and locations. Essentially, reports of as many as four events were muddled together, ostensibly: for the year 1775 [no date specified], for February 11, for March [no day], and for December 18.

Available source reports all had similar, yet slightly different wording, apparently more-or-less based on Southey (1827). Southey reported "three shocks felt at *Española* [Hispaniola], several storehouses thrown down, and great damage done by the sea." These details from Southey were reported by Taber (1922b) for the month of March, adding, "but he fails to give the precise location." Shepherd and Lynch (1992) quoted the report of three shocks at Haiti by Southey for February 11, and also cited for the same date the report by Chuy and Pino (1982) of earthquake damage at Santiago de Cuba that "destroyed houses and damaged churches from there to Santa Catalina." They assigned earthquake intensities for Haiti [MMI ≅ VI] and Santiago de Cuba [MMI ≅ VII]. Poëy (1857) reported three shocks at *Saint-Domingue* [the island of Haïti] and Martinique for March. Great damage from waves following three shocks was also reported for December 18 at Haiti and Cuba.

Rubio (1982) and Heck (1947) reported damage caused by waves at Haiti after three earthquakes for 1775 [year only given] with no location of effects specified. Rubio also listed a separate event for March 1775 [no day given] in which he reported that at the Dominican Republic an earthquake was "felt over the whole island" with "damage caused for most," and that there was no locality. For December [no day specified], Fiedler (1961) reported an MCS ≅ VIII+ earthquake, due south of Hispaniola at 9.3N and 70.4W, with the epicenter local to Trujillo, Venezuela. Centeno-Graü (1969) reported that Trujillo was ruined and gave details of a cathedral and a monastery that were damaged. Grases (1971) listed the Trujillo reports separately from those for "Haiti." As if this jumble were not complex enough, Rubio mistakenly used the year 1755 in his tsunami catalog, presumably for the 1775 wave damage at Haiti. However, it was located chronologically between 1770 and the separate report for March 1775 [no day given] at República Dominicana.

Such an instance of multiple reports with scant, ambiguous details needs careful scrutiny to straighten out what happened so as to assess validity. By piecing bits together, reckoning who used whom as sources, and reflecting on analogous events in the historical record, the details of muddled event reports can sometimes be unraveled.

3.4. Magnitude and intensity

The traditional *Qualitative Imamura-Iida Tsunami Magnitude Scale* (I–I) correlates waveheights, damage, and effects. See Table IV. These details are used in determining the values for tsunami magnitudes. An instance of this scale usage appears in the 1856 Honduras event report. Sutch Osieki (1981) indicated an Ms ≅ 7.0 to 8.0 shock that generated a noteworthy tsunami estimated at magnitude I–I ≅ 2.0, representing a wave of 4 to 6 meters. This event was cited by Cruz and Wyss (1983) as "one of the largest tsunamis to have affected the Atlantic Coast of Central America."

This scale, similar to earthquake scales, is especially convenient for applying to past tsunamis for which no instrumental records exist. Solov'ev (1990) indicated that it is more akin to an earthquake intensity scale than a magnitude scale. Solov'ev distinguished between the maximum tsunami height and the mean tsunami height. Solov'ev defined tsunami intensity i as

$$i = \log_2 \left(\sqrt{2} \times \bar{h} \right).$$

Comparison to the *Qualitative Imamura-Iida Tsunami Magnitude Scale,* where

$$m = \log_2 h$$

suggests that the mean tsunami height $\bar{h}$ be given as $1/\sqrt{2}$ times the maximum height h.

Earthquake magnitudes in the Caribbean Tsunami History are specified as they were reported in the source literature. The symbol ≅ is used to mean "approximately equal to" when a magnitude or intensity is reported. Arabic numbers [Ms ≅ 7.5] are used for magnitude values, as with the *Richter Earthquake Magnitude Scale* and the *Qualitative Imamura-Iida Tsunami Magnitude Scale.* Roman numerals [MMI ≅ VIII] are used for earthquake intensities.

To help with relative comparisons of destructiveness, magnitude and intensity scales had been projected back, in some sources, to evaluate events from long before any instrumental magnitude or intensity data were available. In 1902, the *Mercalli Earthquake Intensity Scale* was developed as a measure of the destructiveness of an earthquake. It was modified in 1930 by U.S. seismologists and became know as the *Modified Mercalli Earthquake Intensity Scale* (MMI). The MMI scale is characterized by incremental values for destructiveness from I to XII. See Appendix D.1. If more than one scale was stipulated, MMI values were selected; if not, the specified earthquake intensity scale is used.

Table IV. Qualitative Imamura–Iida Tsunami Magnitude Scale: I–I

Correlated tsunami magnitude parameters. Values are approximately equivalent to $m = \log_2 h$, where h is maximum runup height in meters, measured at a point 10 to 300 kilometers from the source zone (Iida *et al.,* 1967). Similar to earthquake intensity scales, the values can be applied to past tsunamis for which no instrumental records exist.

Magnitude	Height	Damage from the Tsunami
4	30 m	Considerable damage along more than 500 km of coastline
3	10–20 m	Considerable damage along more than 400 km of coastline
2	4–6 m	Damage and lives lost in certain landward areas
1	2 m	Coastal and ship damage
0	1 m	Very small damage
−1	50 cm	None

The Rossi-Forell Earthquake Intensity Scale (R–F) uses values from I to X. See Appendix D.2. Milne and Ramírez both outlined classes from I to III. Robson defined classes III, IV, and V. For values greater than these three classes, Robson replicated those of the MMI scale. Rockwood and Hall both used increments from I to VI. See Appendices D.3.1 through appendix D.3.5. Rudolph also defined incremental values from I to VIII; Rudolph's Earthquake Intensity Scale is reported in a German translation from Grases (1971). See Appendix D.3.6.

Although in terms of instrumental and modeling capabilities, modern technology has made the intensity scales obsolete, nevertheless, the increments of destructiveness are useful when applied to past earthquakes for which no instrumental records exist. The method is often used to "fill in the blanks" of the seismic historical record.

Mercalli's intensity scale is useful for getting a general idea of earthquake effects, although it lacks accuracy and does not measure the earthquake itself (Blocksma, 1989). Earthquake damage is influenced by many things such as type of ground involved, building structures, and population density to name a few. Blocksma makes the point that "seismologist Charles F. Richter was bothered by all this and devised a system, which he never intended to become so universally used, to measure the actual 'magnitude' of the quake, a term he used to take the emphasis off the effects (intensity) and place it on the physical activity—the earth's action—during the quake."

The *Richter Earthquake Magnitude Scale* is a logarithmic measurement of earthquake size. See Appendix E. Richter defined the magnitude as "the logarithm to base 10 of the maximum seismic-wave amplitude (in thousandths of a millimeter) recorded on a standard seismograph at a distance of 100 kilometers from the earthquake epicenter (Bolt, 1993)." A distance factor is added to normalize magnitude to take into account the diminishing amplitude with the distance from the source.

Richter measured earthquakes using a seismograph to record size and time of tremors. He measured an earthquake by relating the amplitude or size of the tremor to the time between the primary and secondary tremors, according to Blocksma (1989), "much like an obstetrician judging the stage of labor by timing contractions and analyzing their intensity."

Using the logarithmic scale, every time the magnitude goes up by 1 unit, the amplitude of the earthquake waves increases by 10 times (Bolt). The scale has no upper or lower limit. The difference between a magnitude 1.0 and a magnitude 9.0 on Richter's scale actually represents an increase of ground motion of 100 million times. Each successive whole number indicates a tenfold increase [in ground motion] from the level before. According to Blocksma, this is sometimes misleading, "making a 6.0 quake not twice as great as a 3.0, but a thousand times greater, and an 8.0 earthquake is 10 million times greater than a 1.0." See Appendix E.

Myles (1985) reported that both Richter and Gutenberg assigned a magnitude of 8.75 to 9.0 to the colossal 1755 Lisbon earthquake. Bolt stated, "This logarithmic relation indicates that an increase in magnitude of M_S (surface-wave magnitude) of 1 unit increases the amount of seismic energy E released by a factor of about 30 (not a factor of 10, as is sometimes quoted)." Myles described an 8.0 earthquake classification as containing the energy equivalent to the detonation of 5,643,000 metric tons of TNT.

Tazieff (1964) attempted to describe the awesome power of such immense earthquakes. He estimated the prodigious energy to be in the nature of 10^{26} ergs, the common metric unit of energy, or about a million, million kilowatt-hours, released in a few minutes. Interestingly, Bolt reported in 1993 that "the present total consumption of energy in the United States per annum is about 10^{26} ergs." From measurements of the seismic-wave energy produced by the sudden fracture, Bolt estimated that for each year the total energy released by earthquakes throughout the world is between 10^{25} and 10^{26} ergs.

Generally speaking, shallow earthquakes have to attain Richter magnitudes of more than 5.5 before significant damage occurs near the source of the waves, according to Bolt. When seismograms from various stations worldwide are used to calculate the energy of the waves, a Richter-magnitude 5.5 earthquake "turns out to have an energy of about 10^{20} ergs."

3.5. Event reports

It should be recognized that due to brief, ambiguous, or absent reports, at best, a limited picture of tsunamis in historical times can be achieved. The dimensions for many tsunamis in the historical record are unknown. Much information is yet to be located. The not infrequent changes in possession by colonial powers may account for lapses in availability of reports, to say nothing of the detriment from years of wars and rivalry among the European nations. The legacy of privateering, freebooting, and piracy during the rich and colorful history of the Caribbean also took its toll. Imagine the details lost just in ship's logs, and government dispatches from all those sunken ships.

Places lack reports for any number of reasons, especially because many coasts were sparsely populated and reports were not centrally collected. Population centers and larger towns generally had more fully reported information than sparsely settled areas or islands.

Areas reporting tsunamis may not have been located near the primary areas of impact. In some instances, accounts may not accurately represent the full impact of the occurrence because of the lessening of effects at varying distances from the site of occurrence. For instance, waves with amplitudes measuring 60 to 70 centimeters that were recorded at San Juan, Puerto Rico, Limetree, St. Croix, or Galveston, Texas, may have appeared minor simply because of the great distance from a tsunamigenic source.

During the period from 1777 to 1923, the number of shocks large enough to deserve mention in newspaper accounts and other records felt in the northern Virgin Islands tallied 32 earthquakes (Holcombe *et al.,* 1977). During the same period, only four earthquakes affecting St. Croix were mentioned in the records that were examined by Robson (1964). Holcombe *et al.* commented, "To what degree this might be due to fragmentary reporting is not known."

An obvious reason some areas have fewer reports is that they experience fewer tsunamis. Reefs or other formations may protect the coasts. The shoreline configuration can play a role in lessening the effects of tsunamis as well. "Along some coastal areas, fringing reefs provided protection against runup effects by damping tsunami waves. Co-seismic coastal uplift also helped minimize the impact of flooding," according to Nishenko *et al.* (1992) relating events of the 2- to 3-meter tsunamis from Limón, Costa Rica to Archipiélago de Bocas del Toro, Panamá, on April 22, 1991.

Occasionally, bizarre reasons have come to light for the lack of reports. Humboldt and Bonpland (1818) reported that no records existed at Cumaná from the end of the 16th century, due to "the continual devastations of the termites...the archives containing no document farther back than 150 years." The archives were thus besieged prior to the mid-17th century, since Humboldt's research on "equinoctial regions" was conducted during his travels from 1799 to 1804. No reports of tsunamis other than the "frequent inundations" of the late 16th century preserved in the memories and oral tradition of the inhabitants were included in the catalogs for Venezuela during this period. A gap of 183 years, from 1543 until 1726, remains devoid of any known written reports.

Accounts of tsunamis were entirely lacking from the Southern Antilles of Aruba, Bonaire, Curaçao, and many of the Dependencias Federales. Neither Colombia nor its Caribbean islands had any tsunami reports come to light, although several major earthquakes affecting the coast of Colombia at Santa Marta, Ciénaga, Barranquilla, and Cartagena were likely to have produced tsunamis.

During particularly severe earthquakes at Colombia, shocks were felt in the Antilles and produced undulatory ground movement at Martinique. Occurring in 1834, these earthquakes devastated coastal areas of Colombia. Meyer (1996) stated, "A preliminary analysis of information related to tsunami hazards for Colombia's Caribbean coast has shown a series of historical records of damage to the ramparts of Cartagena—due to 'heavy seas', but it remains to be found whether these were due to tsunamis or due to frequent (almost yearly) episodes of swell." Events in the historical record, which may be comparable to occurrences that have generated tsunamis in the past, are good candidates for further investigation.

Tsunami reports from the Islas de la Bahía, Honduras [Bay Islands] (Isla de Roatán, Isla de Guanaja, and Isla Utila), the Islas Santanilla [Swan Islands] off Honduras, or the Cayman Islands are almost completely lacking, as were any from Belize. Almost no reports were available from the coastal regions of Mexico's Península de Yucatán or its small islands.

Reporting on the region's relative seismicity, Taber (1922b) commented that few reports were included for Grand Cayman, 2.5 kilometers in length, or the Swan Islands (Islas del Cisne), "where the record covers a very brief period. The entire absence of epicenters from certain other localities, as between The Caymans and Santiago de Cuba, is probably due to lack of information rather than to low seismicity." Many of the tiny islets near the Caribbean subduction and fault zones simply have no reports available.

Far-flung reports from opposite ends of the Antillean archipelago, with no reports available from neighboring islands, exemplify absent accounts. In 1831, severe shocks were reported at St. Kitts and Trinidad. Violent agitation of the sea was reported at Trinidad. Effects were mentioned at Grenada, and shocks were felt at St. Vincent, Tobago, and Guyana, but reports are not known from islands in the considerable distance between St. Kitts and St. Vincent. Note also the great distance between reporting areas mentioned for the March 1868 event.

3.5.1. March 17, 1868, Puerto Rico and Bequia

Considered a culminating aftershock of the massive November 1867 earthquake, it was reported as heavy all over Puerto Rico [R–F ≅ IX]. The event was considered "nearly as strong as the earlier violent earthquake." Damage reports were available only from San Juan, where nearly all important buildings suffered considerable damage and ships in the harbor distinctly felt the shock. A small "flux and reflux" with flooding occurred at Arroyo and Naguabo. At Isla de Vieques, the sea receded and a small tsunami was observed. Two-foot [60-cm] waves were observed at Charlotte Amalie, St. Thomas, with recession and flooding. The sea dropped 1.4 meters, with a 10-minute period *(Sanct Thomæ Tidende,* 1868). The only other report for this event was by Browne (1869) for Bequia in The Grenadines, just south of St. Vincent:

> On the 17th of March at 8:00 [a.m. or p.m. not given], a steady strong wave was seen bearing down on Admiralty Bay; it had no perceptible crest, and

> was 3 feet [90 cm] in height; it encroached upon the land varying distances from 70 to 400 feet [21–122 m]. A second smaller wave followed. No shock of an earthquake was felt.

A country governing a colonial territory may have contributed, in a way, to the scattered reporting areas. If published sources were only available from a particular colonial government, then absent reports from islands between the country's colonial possessions may explain the "hop-scotch" nature of some of the event reports. In the following cases, French colonies figured significantly.

3.5.2. May 7, 1842, Haiti

The disastrous 1842 earthquake at Hispaniola killed 5,000 people at Cap-Haïtien alone. It was also felt aboard ships in the roadstead (anchorage). A 4.6-meter tsunami occurred at Port-de-Paix. Two-meter waves were observed along the northern coast and a 1.6-meter tsunami was reported at Santo Domingo. St. John experienced a 3.1-meter tsunami. The *raz-de-marée* [tsunami] was observed all along the Atlantic Coast as well. Destructive effects were reported at St. Thomas, Grand Turk, and throughout the Antilles.

While earthquake damage was reported in Puerto Rico, a tsunami was not. *Le Moniteur Belge* reported, "The effects extended to Puerto Rico and the other Antillean Islands, and as far as the *Montagues Roqueses* [Rocky Mountains], in the proximity of the Pacific Ocean." Poëy, who was French, may not have had access to reports from non-French islands, which consequently could account for the lack of reports from islands in between the colony at Haiti and the other French islands.

3.5.3. November 30, 1827, Martinique

At Martinique in 1827, "the most severe shock there remembered" was also felt at Antigua, Guadeloupe, Marie-Galante, Barbados, and at Guyana on *Terre Firma* (Colombia and Venezuela), according to Mallet (1851). Rudolph (1887) said it was widely reported "by ships *Martiniquais*," in the roadsteads of Saint-Pierre and Fort Royal (Fort-de-France). He reported that it was felt by "all vessels at Pointe-à-Pitre."

The breadth of the area affected can be difficult to define. Whether Dominica, which lies between Guadeloupe and Martinique within the area affected, was the island in the reports, or whether the effects encompassed a much broader area that included Haiti, is the source of some perplexity. Langrené (1752) reported *l'île de Sainte-Domingue.* Grases (1971) cited Poëy's (1857) report about *Saint-Domingue,* thence translated from the French as *Santo Domingo,* to designate the French colony of Haïti. Shepherd and Lynch (1992) postulated that *Saint-Domingue* might have been mixed up with *La Dominique* [Dominica]:

> The effects of this earthquake seem to be less than previously supposed. Poëy translates 'Dominica' to 'La Dominique', which Mallet translates back

> into English as 'Santo Domingo'. Subsequent catalogs take this to mean the Dominican Republic, increasing the area of perceptibility by a very large factor.

Saint-Domingue translated as Santo Domingo was common usage at the time, although in many earlier accounts place names such as Trou or Port-au-Prince were specified, serving to confirm Haiti as the location. No towns were mentioned in this 1827 occurrence to verify whether Dominica or Haiti was the location reported. For an earlier event in 1771 at "Santo Domingo" with similar ambiguity, Montadon (1962) queried: *"S'agit-il de la ville, du pays ou de l'île?"* [Is it about the town, the country, or the island?]

If effects reached Haiti, this considerably broadens the area of impact, otherwise described as outward from Martinique to Antigua, and also down through the Windward Islands as far as Guyana. Reports are lacking from islands between Haiti and Antigua. Discerning similarities among events that did propagate throughout this far-reaching area could shed light on such incongruities.

Information in the historical record can be augmented with the help of individuals, who can greatly assist by contributing information about their region and experiences. Individual observers should be encouraged to log any sightings of wave action and inform local authorities. Descriptions with as much detail as possible can contribute significantly. The appearance of the wave—scope, dimensions, color, if cresting, speed—height above and below mean sea level, full excursion of the wave from trough to peak, time of occurrence of both the seismic disturbance and the tsunami, sound, distance the ocean receded, interval between waves, height reached above mean sea level on land, how far inland the waters inundated, markings left by the seawater, topographical and geophysical characteristics of the shoreline, any benchmarks for indicating proportion, and other pertinent details about affected areas should be noted and reported. These data could be added to the store of information gathered for an area's unique form of the hazard.

3.6. Present United States Caribbean territories affected by tsunamis

Besides the mention of tsunamis noted above in 1842 and 1868 at the U.S. Virgin Islands (part of the Danish West Indies at the time) and Puerto Rico, twelve other tsunamis were reported that affected any present U.S. territories in the Caribbean. Appendix F contains brief summaries of these 14 events. Most of these were minor, with relatively few effects on U.S. in the Caribbean territories.

Many reports lack pertinent details and some events undoubtedly went unreported. No reports were available for Puerto Rico or the U.S. Virgin Islands for several major Caribbean-wide events. Noticeably absent are reports about the two teletsunamis of November 1755 and March 1761, which affected the entire Caribbean as well as the Atlantic Coast of North America.

No mention at all was made about the U.S. territory of Navassa Island situated in Canal de Jamaïque off the southwestern tip of Haiti, nor about the U.S. Guantanamo Bay Naval Station at Cuba. Navassa Island, which is a protected U.S. wildlife refuge currently used for marine research, and Bahía de Guantánamo were very likely to have been affected by many of the same earthquakes and tsunamis reported affecting Santiago de Cuba and western Haiti down through the years. Naval or other records could perhaps reveal details about any tsunamis that may have occurred in this region of the Caribbean.

Beginning with the report from Taylor (1888), which mentioned a recession of the sea at St. Thomas in 1690, up through the present, tsunamis affecting U.S. territories in the Caribbean have been quite infrequent and fatalities relatively few. With several significant exceptions, damage from the tsunamis was minor, insignificant, or unreported.

The two catastrophic exceptions were in 1867 in the U.S. Virgin Islands and Puerto Rico and in 1918 again in Puerto Rico. The 3.1-meter tsunami at St. John in 1842 was not insignificant despite the dearth of detail. Two formidable events, in 1690 and 1843, which strikingly paralleled each other, probably had significant widespread effects, although waveheight data are lacking. The tsunamis from these massive events probably wrought havoc throughout the U.S. territories, although there is scant detail.

津波

CHAPTER 4

TYPES AND EFFECTS OF TSUNAMIS

4.1. Teletsunamis

A teletsunami (TEL-leh-soo-NAH-mee) is a tsunami generated more than 1000 kilometers from the affected area, traveling over long distances to opposite shores, sometimes crossing entire oceans. Teletsunamis are usually associated with subduction zones. In a subduction zone, continental crustal plates, colliding with the heavier oceanic plates, force the underthrusting of the heavier plate under the continental plate creating ocean trenches and building great stress in the contact zone between the plates. The resulting earthquakes (stress releases) can cause tsunamis to propagate great distances. Teletsunamis are the principal type of tsunamis affecting Hawai'i and the U.S. West Coast. Most teletsunamis occur in the Pacific Ocean. Only two reports of teletsunamis affecting the Caribbean, in 1755 and 1761, are contained in the known history.

Teletsunamis, though infrequent, are significant for the Caribbean to be aware of because of the widespread devastation that could occur. The long distances, however, do give time to prepare for the arrival of a teletsunami. Since they originate at such distance, warning systems have time to detect the causative earthquake, verify the existence and magnitude of a tsunami, and warn the populations at risk. For this reason, communicated warning systems are very important for teletsunamis.

4.1.1. November 1, 1755, Lisbon, Portugal, and the West Indies

This violent earthquake was, in all likelihood, caused by the Eurasian Plate subducting the African Plate, which is the colossal crustal boundary that extends to the Azores–Gibraltar Ridge (Myles, 1985). The epicenter of this powerful seismic convulsion was probably near the Gorringe Bank off the Iberian Coast near Lisbon, Portugal. Myles stated that the 5-minute shock was later assigned a magnitude of 8.75 to 9.0 by both Gutenberg and Richter.

A teletsunami generated by this mammoth earthquake propagated across the North Atlantic Ocean, one of the most noteworthy tsunamis ever to affect the Caribbean. This teletsunami can serve to inform our understanding of long-distance wave propagation. Transatlantic velocities to the Antilles were calculated by Milne (1841) to Antigua of 528 feet per second [161 meters/second] and 642 feet per second [196 m/s] to Barbados. Milne also computed velocities for seven other paths: 148 m/s to Lisbon, Portugal; 134 m/s

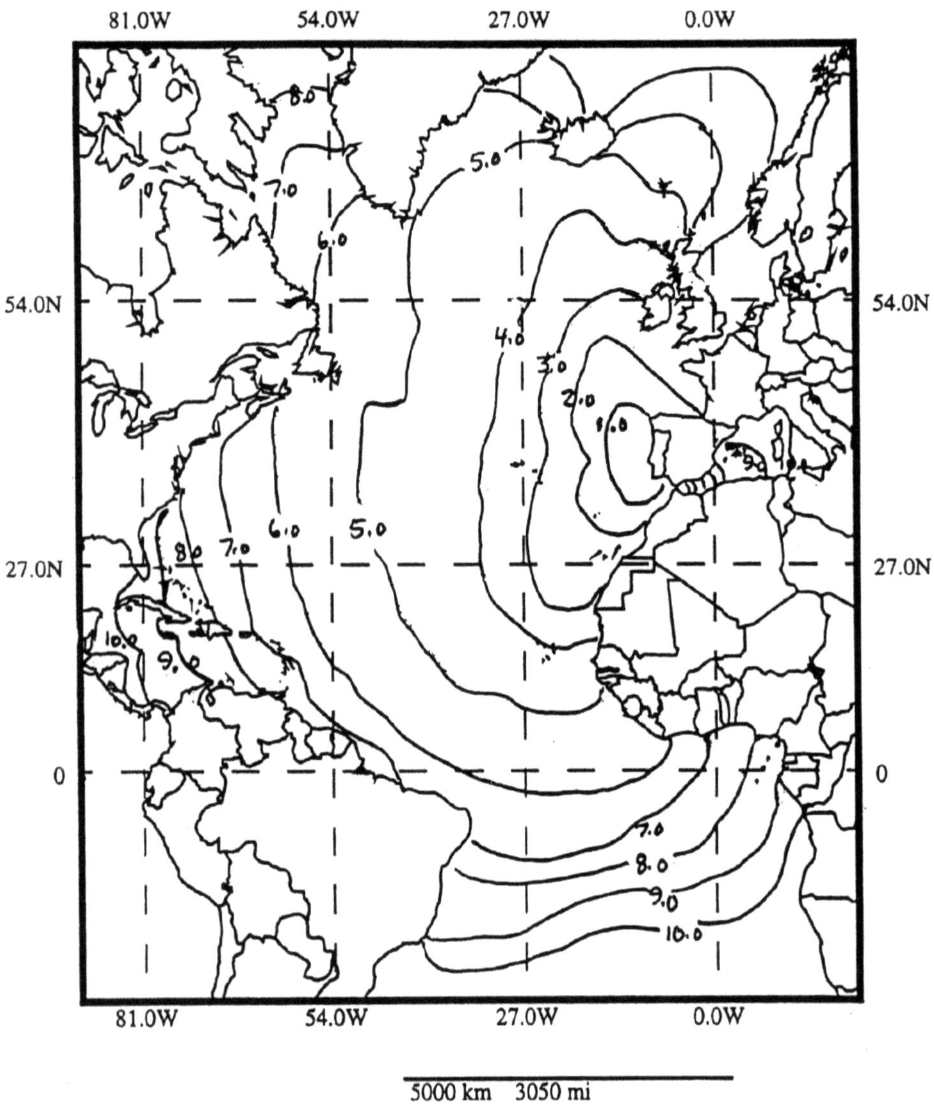

Figure 8. Travel times for the North Atlantic teletsunami of November 1, 1755. Connected lines indicate hourly isochrons. Time is indicated in hours after the originating earthquake on the isochrons. (Plot courtesy of Mader, 1997.)

to Ayamonte, Spain; 99 m/s to Funchal, Madeira; 97 m/s to Cádiz, Spain; 73 m/s to Kinsale, Ireland, and Mount's Bay, Cornwall; and 56 m/s to Plymouth, Cornwall. With runup heights of 5 to 10 meters in the Caribbean, the teletsunami reached Antigua 9.3 hours after the Lisbon earthquake, moving at an approximate speed of 600 kilometers per hour, as estimated by Schubert (1994). Figure 8 shows a plot of the travel times to the Caribbean in hourly isochrons

(Mader, 1997). According to Davison (1936), "The seawaves swept across the Atlantic and were observed in several of the West Indian Islands, where the usual rise of the tide is little more than two feet [61 cm]." Myles reported that dispatches from Saba that spoke of waves 21 to 25 feet [6.4–7.6 m]. Habitations on the extinct volcano were not affected because the capital of Saba, aptly named The Bottom, was built in the floor of the crater 244 meters above sea level.

At Martinique, a drop of 4.6 meters was observed. If the waveheight was proportional to its drop (the generally accepted approximation), the full excursion of the wave could have been as high as 9.2 meters. Samaná Bay, Dominican Republic, experienced 3.7-meter waves and the teletsunami almost completely inundated Santiago de Cuba.

Writing about the "Agitation of the sea at Antigua," Captain Affleck (1757) of the Man-of-War, *Advice*, related to his correspondent in Portsmouth, England:

> On the 1st of November last, I find you had a remarkable sudden flux and reflux of the sea, in like manner, at the same time, on the coast of America, and all these islands. The tide rose here 12 feet [3.7 m] perpendicular several times, and returned almost immediately: The same at Barbadoes. At Martinique, and most of the French Islands, it overflowed the lowland, and returned quickly to its former boundaries. The people at Barbadoes were never more astonished; the rising water in Carlisle Bay appearing as black as ink, instead of clear sea-green.

Humboldt and Bonpland (1818) reported that "At Carlisle Bay, which had an average tidal rise of 24 to 28 inches [60–70 cm], the water rose 20 feet [6.1 m]. It became black as ink, because, without doubt, it was mixed with petroleum or asphaltum, which abounds at the bottom of the sea." Affleck continued:

> Since I wrote this, I have taken a more particular account of the flux and reflux above-mentioned, from an observing man of this island, who remarked that "Here it began at 3:30 p.m., and flowed, every 5 minutes, 1.5 meters perpendicular, till as much after 6 p.m., without any violent disturbance on the surface of the water."

This could have been due to a local landslide tsunami or a seiche, since the wave interval of the teletsunami would have been greater than five minutes. The "flux and reflux above-mentioned" is ambiguous as to whether Antigua or Barbados was intended. It could refer to Antigua, the "flux" first mentioned and topic of the letter, or to Barbados, the topic of the immediately preceding sentence, which described the waves at Carlisle Bay. Earlier Affleck used "this island" to mean Antigua. Barbados was reported in some sources from then on with no mention of ambiguity. Thereafter, in many subsequent reports, Barbados was the island reported. Myles understood the effects to be at Antigua, citing that "at Antigua, the sea rose 12 feet [3.7 m] several times, and every 5 minutes afterwards until 10 p.m. (Lisbon, time), when it rose 5 feet [1.5 m] without any violent disturbance on the water."

The teletsunami also affected the eastern coast of North America. In an account of tsunami effects in Canada, Smallwood (1980) reported that the harbor at Cape Bonavista, Newfoundland, drained and overflowed, giving rise to a popular folk song entitled, "A Great Big Wave Hove in Long Beach." Humboldt and Bonpland reported that this massive earthquake was felt at almost the same moment on the coasts of Sweden, at Lake Ontario (bordering Canada and the U.S.), and at *Martinico* [Martinique].

For many islands in the Caribbean, including Puerto Rico and the Virgin Islands, reports are entirely lacking. Government dispatches and other news may have been carried from the West Indian colonies to Europe by ships that had been in the area. Upon their return, ships logs and eyewitness accounts may have been recorded with colonial authorities.

Reports about the local tsunami at Lisbon are instructive with regard to the amalgam disaster effects and because of the pivotal role one figure played in the history of seismology. Myles (1985) vividly described the horrific amalgam of events at *Lisboa* [Lisbon] on All Saints Day, 1755. While much of the populace was in candle-lit cathedrals, these edifices toppled onto the victims, setting the city ablaze. The fire induced terrific winds, creating a tornadic inferno that ravaged what was left of Lisbon's rubble. People, with no expectation of a tsunami, rushed to the waterfront to escape crashing buildings and the blistering furnace at their backs. Hundreds perished from the collapse of the quay where they fled for safety and the local tsunami that ensued.

The vast tectonic tsunami arrived 40 minutes to an hour later. "Oeirás Bay suddenly emptied itself to a distance of a mile [1.6 km]. The populace watched the sea withdraw in fascination, unaware of the imminent return," which drowned thousands more. The black waters, which rapidly crested in the shoaling waters, were computed at more than 50 feet [15.2 m] in height. The raging surge swept more than 1.6 kilometers inland, "its thunderous mass hurling ships and carrying all before it. Minutes later, many who had somehow survived the initial onslaught were caught in the all-embracing sweep of the backwash and drowned."

For the first time, an earthquake was intensively studied in Europe. Sebastião de Carvalho, Marquês de Pombal, chief minister to King Joseph of Portugal, played a pivotal role in the history of seismology by his detailed investigation of the seismic effects of this event. When asked what to do about the calamities, Pombal made his famous reply: "Sire, we must bury the dead and feed the living." His heroic efforts averted an outbreak of typhus through the expediency of rapid burial—by means of dumping thousands of corpses at sea, much to the wrath of the Jesuit clergy, since the urgency allowed no individual last rights. (Epidemics were and still are common after major disasters.)

According to Myles, Pombal's single-minded dedication instituted the lifesaving practice of seismic investigation and analysis in the Western Hemisphere. Studying the extent of the damage, Pombal personally surveyed and plotted the fissures, kept notes of observations, and even brought in Japanese experts as consultants. He labored for months to remove the ruble and to rebuild an "earthquake-resistant" metropolis, perhaps one of the reasons for the lesser effects from the next large Lisbon, event, which followed in 1761. Concern about a recurrence of such an event is the basis for the establishment of a Portuguese tsunami warning system. Those responsible for emergency response and warning systems in the Caribbean need to be prepared to act if a warning is ever received from Portugal.

4.1.2. March 31, 1761, Lisbon, Portugal, and the West Indies

The second historical teletsunami to reach the Caribbean was due to a powerful earthquake off the Iberian Coast, this time originating at the Galicia Bank near the same epicenter as in 1755. Davison (1936) reported, "Despite the violence of the great shock of similar 5-minute duration, *Lisboa* [Lisbon] was not as severely damaged." Perhaps this was due in part to Pombal's efforts. Mallet (1854) said the sea fell and rose 2.4 meters vertically, with a 6-minute period, until evening at Lisbon, 1¼ hours after the terrible shock at 12:30 p.m. local time.

Effects of the teletsunami, manifest far beyond the seismically disturbed area, were estimated by Davison as "a million or half-a-million square miles," but the effects were much less widely noticed than in 1755. The teletsunami traversed the North Atlantic, reaching Barbados, more than 1,200 leagues [5,800 km] from Portugal, in 8½ hours, arriving at 4:30 p.m. Barbados time. Seiches were set up in many places, including the waters of Loch Ness, Scotland, which rose 30 inches [76 cm], continuing for 30 to 45 minutes.

Throughout the North Atlantic effects were observed, such as at Cornwall, England, Ireland, and other places in the British Isles, Holland, France, Madeira, The Canaries, and at Ilha Terceira in the Azores. According to Davison, at Barbados" "the sea, then near low water, suddenly retired from the shore, and in 3 minutes returned, rising to the height of 4 feet [1.2 m]. The flux and reflux continued until 6 a.m. the next morning. Mallet reported "violent agitation at Barbadoes (with no land shock)." Schubert (1994) reported a 12-meter wave at Barbados, undoubtedly a decimal point error.

As with the 1755 teletsunami, reports are lacking for many Caribbean islands, including Puerto Rico and the U.S. Virgin Islands. Berninghausen (1968) may have cited a report for Puerto Rico in his remarks about places affected by the teletsunami. Berninghausen reported that, in the *Açores* [Azores], "The sea rose to great heights and fell so low the quays were left dry. All the boats hauled onto the beach at Portorico were swept into the bay and destroyed. Waves were noted throughout the British Isles and the Low Countries as well as on Barbadoes."

Enumerated by Berninghausen from sentence to sentence are the Azores, Portorico, British Isles, Low Countries, and Barbadoes. The mention of 'Portorico' is uncertain in its reference. Both Portorico and Porto Rico are antiquated spellings for Puerto Rico. This could be a report at another location in the Caribbean, besides Barbados," for this teletsunami. Researching older source material may help to elucidate any ambiguities. The older literature could also yield mention of effects from other historical events that may have produced waves that propagated across the Atlantic as teletsunamis.

4.2. Tectonic tsunamis

Tectonic tsunamis are caused by the movement of oceanic plates, with the subsequent uplift or drop of a portion of the sea bottom. An active subduction zone exists at the eastern margin of the Caribbean Plate where the North American Plate subducts beneath the Caribbean Plate at the island arc of the Lesser Antilles archipelago. The local tsunami hazard in those areas a near subduction zone is generally very severe. Throughout the northern boundary of the region, numerous strike-slip faults exist, indicating potential danger from tectonic events (Mercado Irizarry, personal communication, 1997a). The faulting systems that limit the Caribbean Plate to the south, along the coast of Venezuela, similarly present danger from tectonic events.

4.2.1. September 1, 1530, Cumaná, Venezuela

The earliest reported tsunami thought to be an authentic tectonic tsunami took place in 1530 in Venezuela, along the coasts of Cumaná, Península de Paria, and Isla Cubagua. "On a serene and tranquil day, in an instant, the sea rose up 4 Fathom [7.3 m] in height (Oviedo, 1547)." Las Casas (1876) commented, "*Grande y caudal* [large and voluminous], it covered the whole plain, encircling half-a-league [4.8 km] around the slope of the mountains." Silgado (1985) reported, "The sea rose like a miraculous thing to see, fell forcefully, and entered the land with a roaring that continued along with the temblors from 9 a.m. until noon."

The massive convulsion caused the rough waves to smash the natural dike between the coasts of Cumaná and Península de Araya, opening a connection to the sea. The Montañas de Cariaco were cleft, leaving an immense breach. Water entered 14 leagues [68 km] into the gap, forming the Golfo de Cariaco. Humboldt and Bonpland (1818) placed the mammoth rupture that opened up the Gulf somewhat earlier.

According to the remembrances of inhabitants, "The Gulf owes its existence to a rent of the continent attended by an irruption of the ocean." This formidable event had been referred to as "very recent at the time of Christopher Columbus' third voyage [1498]." This may be a reference to the earliest reported tsunami at Venezuela, in 1498. This event at Boca de la Sierpe, Pedernales, is another possible date for the opening of the Gulf of Cariaco.

4.2.2. November 18, 1867, Virgin Islands, Puerto Rico, and Antilles

Several phenomena occurred during this event, including a double tectonic tsunami generated by violent back-to-back earthquakes [Ms $\cong$ 7.5] that originated 15 to 20 kilometers southwest of St. Thomas, along the north scarp of the Anegada Trough. Characterized by initial vertical oscillations, probably due to a massive down-drop of the fault blocks, the two tremendous shocks separated by an interval of 10 to 15 minutes launched immense tsunamis that propagated in all directions, the second larger than the first. The leading trough caused the

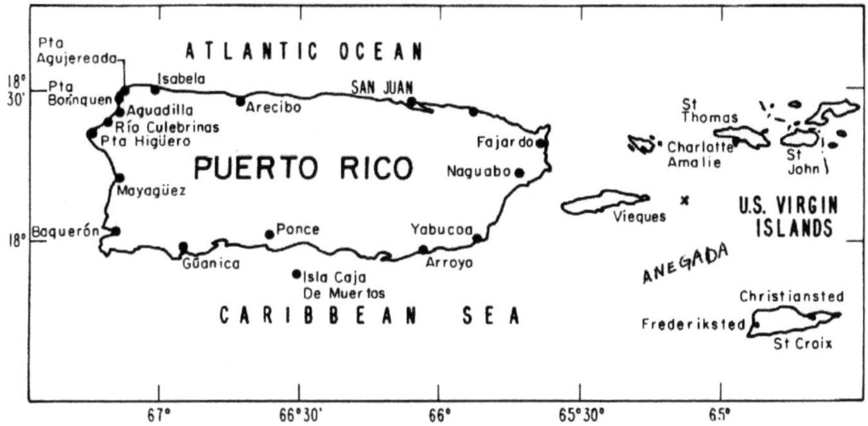

Figure 9. A triangle formed by Isla de Vieques, Puerto Rico; Saint Thomas; and Saint Croix encloses the tsunamigenic origin of the formidable November 18, 1867 earthquake, its epicenter placed 15 to 20 kilometers southwest of Saint Thomas. Near the middle of this triangle, it is 2,500 fathoms deep [4.6 kilometers], the deepest part of the Anegada Channel passing from the Caribbean Sea to the Atlantic Ocean between Saint Croix and Saint Thomas.

ocean to recede from shores on all sides of the seismic convulsion. Produced as well, were the landslide tsunami at Guadeloupe and possible volcanic activity at St. Thomas and at Grenada. The most momentous tsunami known to have affected the U.S. Virgin Islands, the Danish West Indies at the time, it was also significant because of its impact on much of the rest of the Caribbean. Waves were reported from Jamaica to Grenada.

Figure 9 shows the area where St. Thomas, St. Croix, and the Puerto Rico islas of Vieques and Culebra, more-or-less form a triangle surrounding the epicenter. Tsunamis of tremendous force deluged all these islands. The huge tsunami also struck St. John, 3.2 kilometers further east just minutes later, but details and waveheights are lacking.

Within a 12-hour period, 89 aftershocks jolted St. Thomas as well as the entire Virgin group, Puerto Rico, and the Antilles. Fuchs (1883) reported an earthquake at St. Thomas and "in the sea such a large wave; simultaneously a shock being felt at St. Croix and Porto Rico, and extending itself as far out as the French Islands; this same shock was felt at Cuba and all the Antilles, and even Havana; likewise *St. Jean, Hauptstadt von Porto Rico* [San Juan] similarly damaged; also seawaves at Jamaika." Reid and Taber (1919a) said the shocks could not have been felt west of Haiti.

Sainte-Claire Deville (1867) stated, "The earthquake was extremely violent in the Leeward Islands, from the Virgin Islands, and in all the Greater Antilles." In translation, the word "greater" *(grandes)* was ignored, rendering it: "and in all the Antilles." The islands of Puerto Rico, Hispaniola, Jamaica, and Cuba comprise *Les Grandes Antilles.* If Sainte-Claire Deville had meant: "in the Lesser Antilles, from the Virgin Islands [downward], and in all the Greater Antilles," a more precise translation, then shocks may have been felt considerably further west than Reid and Taber maintained. The *Saint Croix Avis* (1867) spoke of its far-reaching effects:

> While Guadeloupe, Dominica, Antigua, and other parts of the Spanish Main have escaped with minor shocks and little damage, Jamaica has suffered to a fearful extent. False or true, we cannot vouchsafe its correctness. News by telegram, via America, said the whole of Europe has been visited by the calamity, and Mexico has been dealt with severely.

"The same seismic wave may well have propagated in one direction as far as the coast of Mexico, and in the other direction up to about the Azores," according to Perrey (1870). Interesting also is a reference in Soloviev *et al.* (2000), citing Schmidt (1881), for November 19th at 21:00 UT: "Aegean Sea, Asia Minor, Izmir [Turkey]: The sea level was 1.1 meter lower than usually." This would be almost due east along the fault-line described by Myles (1985)—the huge crustal boundary between the Eurasian Plate and the African Plate "that extends to the Azores–Gibraltar Ridge." Since the trough preceded the giant tsunami, perhaps the lowering of the sea level at the far end of the Mediterranean was a distant effect of the general recession of the waters.

For the busy port of St. Thomas, shown in Figure 10, a wealth of descriptive reports has become available. Many of the reported details are contained in the *Virgin Islands Daily News,* "History Corner," a series of weekly articles on the calamitous events taken from many earlier sources by Paiewonsky (1979 and 1981). Watlington and Lincoln (1997) have brought together numerous accounts by ships' captains, passengers, and many source reports that vividly illustrate the incredible effects of this massive tectonic tsunami.

Extended anecdotal selections, somewhat condensed, are included from the exceptional collections compiled by Paiewonsky and Watlington and Lincoln. The enlightening recounting of the saga by Maskell (1868), a New Zealand passenger aboard the British Royal Mail packet, *La Plata,* is excerpted in the text because of its wealth of detail and astute commentary.

An article in *Harper's Weekly* (1868) reported, "The tidal wave was described by those who saw it as a monstrous breaker, or rather a sea-wall, variously estimated at from 30 to 60 feet [9.1–18.3 m] high, and flowing at a rate of at least 50 miles per hour [83 kph]." Demolishing dozens of ships and small craft, this was the most destructive tsunami to strike the U.S. Virgin Islands.

At least 23 people perished at St. Thomas and St. Croix. The actual death toll from the tsunami may have been much higher. The *Sanct Thomæ Tidende* reported, "Loss of life does not exceed 50. Other islands in the Virgin group felt the shocks as severely. St. John suffered enormously."

Maskell recounted from passengers who had previously gone ashore from the steamer, *La Plata*, and had not been aboard ship during the tsunami:

> From them we learned that the loss of life was probably great, as numbers of the island's inhabitants, after the first great earth shock, had put off in boats to escape they knew not whither. The tidal wave destroyed most of

Figure 10. Saint Thomas Harbor showing Hassel Island, Water Island, and Prince Rupert's Rock that broke the onrush of the massive wall of water on its course from the southwest. The height diminished several meters as the avalanche of water crashed onto them. The mighty wave engulfed the shores, surging inland, and carrying ships, small craft, and wreckage with it. At Charlotte Amalie, 8-foot [2.4-m] waves plunged over the piers, which today might be lined with cruise ships and hundreds of tourists.

> these frail craft, and the great majority of their occupants must have drowned. Also vessels in the harbor had been driven ashore with the loss of many seamen.

Maskell pointed out that, although the West India coal wharf on Hassel Island was protected by a jutting point of land, "Yet the tidal wave worked its destruction here, too. Even in that sheltered corner the storekeeper reported that the water had rushed over everything 16 feet deep [4.9 m] and more."

Striking the other entrances to St. Thomas *Havn* [haven] obliquely, the harbor experienced a tsunami somewhat smaller in height. The "points" of Water Island, Hassel Island, and Prince Rupert's Rock met the brunt of the wave. The mountainous wall of water, "not less than 23 feet [7 m], broke and retarded on Prince Rupert's Ledge, and caused a great rise in the waters of the harbor (Paiewonsky, 1979)."

The *Sanct Thomæ Tidende* (1867) reported that at Charlotte Amalie, "The majestic rollers hurled themselves upon the beaches and burst 8 feet [2.4 m] over the piers." The wave would have needed at least 4 meters of height to reach that high above the pier. Today the piers might be lined with cruise ships and hold hundreds of tourists and shoppers.

The main street of Charlotte Amalie, parallel to the beach, was estimated to be 30 feet [9 m] above the water's edge and 200 to 300 yards [183–274 m] away from shore at the time. The altitude of Charlotte Amalie is relevant because it means the surge would have traveled uphill as it "swept clear up into our principal street," according to the *Tidende,* "and no store from King's Wharf westward but has had 4 to 5 feet [1.2–1.5 m] of water within it, spoiling the goods and hauling sloops, schooners, and buoys down the main street." Runups reaching up to 1.5 meters in buildings in an area located as much as 9 meters above sea level are impressive indeed.

An imposing wall of water was seen looming 4 to 5 kilometers out to sea, spanning the entire island of St. Thomas 10 to 15 minutes after the first terrible jolt. The ocean receded so far that the bay was left "half dry with multitudes of large fishes jumping about on the bottom." Advancing toward the harbor with the trough in front of the crest of the wave, the immense tsunami crested at "fully 40 feet [12 m]."

Diminished after engulfing Little Saba, the wall of water surged into The Gregeries, cresting at 12 meters. It crashed like an avalanche as it penetrated the harbor, driving ships and everything in its path before it. Its fury broke against the wedge of Water Island, attenuating the enormous wall of water still further to a 9-meter swell as it penetrated the harbor. The tumultuous waves coursed inland 250 feet [76.2 m] at Altoona.

The *Tidende* also reported that "The sea entered not only at the fort points, but passed into The Gregeries from the west and east ends of Water Island, and then again into the harbor at the haulovers, so that the seashore and the streets at the western end of the town suffered equally with those exposed to the wave through the main entrance to the harbor."

Discrepancies in the various estimated waveheights may have had much to do with the location of the observer as well as which areas of the harbor were being viewed. Raupach (1868) described the scene:

> The foaming wave stood up like a straight white wall, about 15 to 20 feet [4.6–7 m], and advanced very fast into the Harbor, sweeping or upsetting small vessels before it, and raising large Men-of-War and steamers to its top. The appearance of this wave was like a white masonry wall, erect and straight, as if built with a rule; it had not the appearance of ordinary waves. The rising of the waves was repeated a second time after an interval of about 10 minutes, and the second appeared to be even a little larger than the first and went farther inland.

In his ship's log, Rear Admiral Palmer, Commodore of the American flagship, *USS Susquehanna,* reported sighting "the heavy wall of sea advancing towards the harbor with terrible power, with a skirmish line of tumultuous rollers in front. The sea was rushing in and receding with great violence."

The eventful experience reported by Commodore Boggs in the ship's log of the American corvette, *USS De Soto,* depicts the scene. The warship was deposited on the iron pilings of the demolished American screwpile jetty by the wave, which "swamped our boats at the booms; then receded as violently." Boggs continued:

> At the first swing of the current we had 6 Fathom [11 m] under our stern,

> the ebb was so rapid that in a few minutes, we took the bottom, having only 1¼ Fathom under us [2.3 m, a drop of 8.7 m]. The second wave, 10 minutes later, came in much greater in volume, and inundated even further onto the shores. The next roller swept her off again into deep water; this third wave was not as large as the first two, and the fourth rose not more than 12 to 15 inches [33–38 cm].
>
> The current, full of whirlpools, swung the warship against the wharf, breaching three holes in her hull. The cruiser made at least twenty full swings around the compass before being able to check her movement around the harbor. Waves came in at irregular intervals until long after dark, when the *De Soto* was finally able to anchor safely.

In the aftermath of the destruction at St. Thomas, evidence was "everywhere conspicuously apparent. Numbers of small houses were constructed partially with bulkheads of wrecked ships; and even in 1873, six years later, there were a few wrecks still on the shore (Thomson and Murray, 1885)."

The other islands of the Virgins felt the shocks as severely, according to the *Saint Croix Avis*. The greatest damage At St. Croix occurred at Frederiksted, "which suffered greatly, having no barrier or reef to stop the force the sea, which as far as the eye could reach to the north and to the south was a high threatening wall of green water. It seemed to pause, as if marshaling its strength, and then on it came in a majestic unbroken column." The huge rollers surged through the harbor as "a wall of water 25 to 30 feet high [7.6–9.1 m]."

"The U.S. ship, *Monongahela,* had a most thrilling experience in the harbor," according to Reid and Taber (1920). "Floating on top of the great wave on an apparent level with the tops of the houses. The warship landed in an open area of Bay Street, but did not stay there. The massive, powerful receding water deposited the Man-of-War with architectural precision on a ledge of reef on the shore's edge." Another observer recounted that crewmembers were said to be able to touch the leaves of the trees.

Commodore S. B. Bissell of the American frigate, *USS Monongahela,* "a vessel of 1400 tons burden, 361 feet [110 m] long, drawing from 16 to 20 feet [5–6 m] of water," gave an account, claimed as exaggerated by some, had many of its details corroborated by other reports. In the ship's log the saga is recounted:

> At 3 o'clock a terrible rumbling and vibration of the vessel was felt as if blowing off steam violently from the boilers: it was a very severe shock of an earthquake and lasted about a minute. Then a quiet occurred until about 2 minutes past 3 o'clock, when the water receded from the shore between Butler's Bay and Sandy Point with astonishing rapidity, and the island seemed to be approaching us.
>
> In a few seconds a reflux tide set in from seaward towards the land with such velocity as to carry the vessel and two others toward the beach. Such was her velocity that the anchor chain was torn from the deck and rushed entirely out the hawse hole.

Figure 11. Canted at 15°, the 110-meter-long U.S. warship, *USS Monongahela,* lies beached after being washed ashore in Frederiksted, Saint Croix, and stranded by the tsunami on November 18, 1867. The Man-of-War required 5 to 6 meters of water to be floated off the coral, which was not accomplished until the following May. (Illustration copied from *The Virgin Islands Daily News* "History Corner" article of November 2, 1981, by Isidor Paiewonsky.)

> Thus adrift, she was washed in by a heavy swell over the beach, bumping injuriously several times on the rocks; but more water coming in, she remained yet afloat. Sails were instantly hoisted to pay her around to seaward.
>
> At this moment, a curling wall of water about 15 feet [4.6 m] high came in from seaward, struck the port broadside, rolled her deeply to starboard, and, receding, left her high and dry on the ledge of the rocky wharf where she now lies, parallel to the shoreline, careened 15° to port, which side looks seaward, her port bilge resting on the very edge of the rocky wharf: this wharf descends perpendicularly 2 feet [0.61 m], from which the bottom slopes gradually seaward, so that if she were to fall over it would be completely on her beam ends.

Had the ship been carried out a few feet farther over the edge of the reef, it would have been a total loss. The *Monongahela* lies tilted 15° where she was stranded on the coral shoreline after her incredible journey, is illustrated in

Figure 11. A monograph in Scribner's Monthly (1878) by midshipman van Housel of the *Monongahela,* vividly described its trials as she struggled to survive the tsunami at St. Croix. Watlington and Lincoln (1997) commented, "Housel's non- scientific, anecdotal report unsuspectingly presents the historical and social circumstances of the people of the Danish West Indies at a significant turning point in their history."

> After the first severe shocks near 3 p.m., lasting 30 seconds, the receding sea quickly changed direction, and drove the *Monongahela,* lying at 7 Fathom [12.8 m], rapidly towards the beach, the shore approaching at railroad speed. Our vessel advanced and receded from the shore with the waters until, as if some great power had raised up the bottom of the bay, the sea rapidly closed in on the town, filling the houses and covering the street running along the beach to a depth of 24 feet [7.3 m]. Our ship, following the current, took a course toward the south end of the town, until over the edge of the street it swung her bow toward the north and was carried along, smashing a frame storehouse, and breaking down a row of shade-trees.

With the mighty rush of the backwash, they were carried back out 450 meters into the harbor. The vessel grounded in 4.6 meters of water and "careened over onto her port beam's ends." The bottom was visible for 800 meters beyond, and:

> That immense body of water [usually 55 to 73 meters deep], which had covered the bay and town, was reforming—with the whole Atlantic Ocean as an ally—for a tremendous charge upon us and the shore. This was the supreme moment of catastrophe. As far as the eye could reach to the north and to the south was a high threatening wall of green water. It seemed to pause, as if marshaling its strength, then on it came in a majestic unbroken column. The suspense was terrible.
>
> Our noble vessel seemed as a tiny nutshell to withstand the mighty rushing Niagara advancing upon us. Many a hasty prayer was muttered by lips unaccustomed to devotion. The tidal-wave struck the ship with gigantic force, making every timber shiver, but the decks were not engulfed.
>
> Being rather flat-bottomed [drawing from 5 to 6 meters], the first effect of the blow was to send her over on her starboard beam's ends, giving the water an opportunity of getting well under her before righting, when she was buoyed to the crest of the wave and carried broadside to the shore, finally landing on the edge of the street in a cradle of rocks that seemed prepared for her reception. Here she rested with her decks inclined at an angle of 15°. A small Spanish brig was carried bodily inland across the cane fields, landing in the midst of the King's highway.

Although Puerto Rico must have been severely affected, reports are brief in comparison to reports from the Danish West Indies and many of the other islands. The shocks were very severe at Vieques and Culebra, the islands closest to the tsunamigenic origin. The largest waves were observed at Vieques, where on the south coast, immediately after the shocks, "A high wave broke on the shore (Reid and Taber, 1920)." Later on waves washed the northern shore. Unfortunately, no precise waveheights are known for Vieques, where "the sea had the same characteristics as at St. Thomas."

A 6.1-meter tsunami was observed principally along the eastern and southern shores of Puerto Rico. The *Tidende* gave an account of "very high" waves at Salinas. According to Latimer (1868), "The same phenomena of earthquake and tsunami were produced" at Puerto Rico, causing damage throughout the island. "At Arroyo, the sea had the same characteristics as at St. Thomas," where the wave advanced 40 meters over the land. Citing the Alcalde of Yabucoa, Reid and Taber said the sea retired about 100 yards [137 m], and then returned an equal distance over the land.

Details about effects at Grenada were reported by Parsons (1868), Staff Commander of the West India Survey, who was assigned to survey the islands in the aftermath. Parsons said the chief effect was a wave that presented alarming features and frightened the inhabitants:

> The sea suddenly dropped 1.8 meters leaving much of the harbor dry, the reef opposite appearing to rise from the water, the steep sides and generally dark coral showing as a dark wall. The lagoon poured out its water, and all the retreating streams made a head at the entry to the Carenage. In a few minutes, this ridge quickly advanced to the shore, but did not overfall, and flowed over the wharves 1.5 meters. A sloop of 30,000 tons was placed on the wharf by this wave, and carried off with the next. The second wave, occurring one minute after the first, was of superior elevation; a lesser third wave followed, and then the waters of the Carenage obtained their normal condition, after a series of undulations and eddies, which lasted for an hour.

A wharf was demolished and its 90-kilogram stones displaced 3 meters by the backwash of the wave. Parsons noted that several small houses were moved inland 50 feet [15.2 m], together "with their inmates," and "all loose things on the shore were sent floating about the Carenage." Reid and Taber (1919a) reported the action of the wave being repeated six times, then all was quiet. Steamers at the West India coal wharf were left dry, a drop of 3.7 meters in that part of the Carenage. At Gouyave at about 5 p.m., the incoming wave that was "quite equal to that at the Carenage, St. George's" damaged boats and buildings. Trough-to-peak vertical oscillations of 20 feet [6.1 m] were reported.

4.3. Landslide tsunamis

Tsunamis that are generated by landslides are usually, though not always, associated with earthquakes. Since the source of landslides is near shore, the warning time for landslide tsunamis may be only a few minutes. Landslides can occur when a tectonic trigger causes the ocean floor to slump or when huge chunks of glaciers, volcanoes, or other debris slides crash into the sea. Whether they are submarine or on terra firma, landslides can cause massive tsunamis and have devastating local effects. The largest tsunami ever reported in the Caribbean was probably a landslide-generated tsunami, located off the northern coast of Guadeloupe, part of the far-reaching 1867 event.

4.3.1. November 18, 1867, Sainte-Rose, Guadeloupe

The enormous tsunami was observed by an eyewitness at Sainte-Rose, cited by Saint-Claire Deville (1867):

> The ocean suddenly withdrew 100 meters from the shore. This was preceded by earthquake oscillations of perhaps 5 to 6 seconds' duration. Then suddenly, the first wave, rising from the north, at least 60 *pieds* [feet] high, [1 *pied* = 0.33 centimeters, roughly a foot, thus 19.8 m] and 3 *milles* [nautical miles] [5.6 km] broad, rolled violently towards the land where it crashed, immersing the whole shore, and flooding the houses. Boats, fishing nets, wood, and everything on the beaches were left adrift.
>
> At Deshaies, on the northwest shore, the tsunami ravaged the town and smashed nearly all the houses. The difference in the extremes of the water levels was not less than 2 meters (trough-to-peak) at Basse-Terre. At Îles des Saintes, off the southwest coast, the violence of the displacement of the ocean resulted in Fond-du-Cure being submerged by 1 meter of water, the inhabitants running away before the looming sea. The wave at Pointe-à-Pitre, protected by the Petit Cul-de-Sac, was hardly more than a swell.

Given that the first motion was a withdrawal and that the waves arrived immediately after the shocks, the origin was probably a locally generated submarine-landslide tsunami. It was most likely triggered by the same great shocks that launched the double tectonic tsunami, considering the tremendous height and breadth of the tsunami at Sainte-Rose in comparison to other islands in the path of propagation as well as compared to other areas in Guadeloupe.

4.3.2. October 11, 1918, Puerto Rico

One of the most severe earthquakes ever to strike Puerto Rico originated in the Mona Canyon Fault Zone, 24 kilometers west of Punta Higüero (Jigüero). The probable trigger mechanism was thrusting into the trench near the Brownson Deep, the deepest trench in the area (Reid and Taber, 1919b). Generally reported as magnitude 7.5, Punta Borinquen, closest dry land to the epicenter and lying within isoseismal MMI ≅ IX, experienced an Ms ≅ 9 earthquake, according to the map in Reid and Taber's (1919a) report to the U.S. Congress. The 15-meter lighthouse solidly built on rock was shattered badly and would have fallen but for the support of the central iron stairway. The lighthouse keeper said waves reached 4.6 meters above sea level, "not counting the wash of the wave."

Although Añasco [MMI ≅ VIII], built on alluvial material, lay 2½ times as far from the epicenter as Rincón [MMI ≅ VI], built on rock, Añasco experienced considerably greater earthquake intensity. Reid and Taber said the injury to property was so much less at Rincón that it caused general comment. An eyewitness, 4 kilometer southwest of Rincón, noticed surface waves that moved along the ground as fast as a train with a 1-meter wavelength and an amplitude of 15 centimeters.

The *Denver Rocky Mountain News* (1996) cited U.S. Geological Survey scientist, William Dillon, "Our data indicate that the Puerto Rico tsunamis may

Figure 12. The tsunami that swept into Mayagüez, Puerto Rico, on October 11, 1918, uplifted this house and damaged many more. (Figure from Reid and Taber, 1919b.)

have been caused by landslides that appear on the nearby sea floor." Dillon based his conclusions on data reprocessed from a seismic line that was provided by Shell International: "It shows tilting of the small portion of the Earth's crust upon which Puerto Rico sits." Dillon indicated that:

> Rubbing together of the Puerto Rico and North American tectonic plates caused the sliding of giant limestone slabs underwater, some up to 50 miles [80 km] wide, that may have generated the tidal waves.

The convulsion launched a devastating tsunami that lashed the northwest coast, where waves were highest, reaching maximum runups of 6.1 meters. The "great sea wave," with the trough preceding the crest, was seen almost immediately following the shock. Reid and Taber (1919a) said observers reported, "Wherever the wave was seen along the coasts of Puerto Rico, the ocean first withdrew from the land, exposing reefs and stretches of sea bottom never visible during the lowest tides, and then the water returned reaching heights that were equally high above normal. At some places the great wave was followed by one or more smaller ones, and especially in sheltered bays, the water continued to ebb and

Figure 13. On October 11, 1918, at Mayagüez, Puerto Rico, a small house was carried seaward by the backwash of the 1.5-meter tsunami and left stranded a short distance from the shore. (Figure from Reid and Taber, 1919b.)

flow for some time." The tsunami produced extensive damage, especially to villages established in a flood plain. Near Punta Agujereada [MMI ≅ IX], with 5.5- to 6.1-meter runups along the base of limestone cliffs 100 to 120 meters in height, the tsunami drowned 8 people, turned a narrow strip of beach used for pasture into a sandy waste, and uprooted several hundred coconut palms. Two minutes after the sea receded from Punta Higüero [MMI ≅ IX], waves 5.5 meters above sea level uprooted coconut palms and stranded fish between the rails of railroad tracks that were 5.2 meters above sea level.

Although Mayagüez [MMI ≅ VIII] was not affected by the brunt of the waves, Figure 12 shows the now familiar photo of a house uplifted by the tsunami that flooded ashore there. Figure 13 pictures a small house carried seaward by the powerful backwash and left stranded near the shore. The retreating torrent is often as formidable as the initial upsurge of the wave and filled with all manner of debris. About 23 minutes after the earthquake, seawater entered the lower floors of buildings near the waterfront and destroyed a few native huts along the beach (Reid and Taber, 1999a). Watermarks left on houses indicated runups to 1.5 meters above sea level. Debris made many streets impassable, impeding fire trucks and rescue efforts. More than 100 people perished.

At Aguadilla [MMI ≅ VIII], where runups reached a maximum of 3.7 meters, 32 victims drowned and 300 huts built along the beach were destroyed by the tsunami. Evidence from stained wallpaper left long afterwards enabled fairly accurate measurements (Reid and Taber). Runup heights varied in different parts of the city, but "at no place were the measurements less than 2.4 meters above sea level, and near the head of the bay the crest of the wave must have been 3.4 meters in height."

The Columbus Memorial, which stood on the beach near the mouth of Río Culebrinas 4 kilometers southwest of Aguadilla, was demolished by the shock and 1000-kilogram limestone blocks from the wreckage were carried inland and slightly downhill by the 4-meter waves to distances of 45 and 75 meters. Reid and Taber said, "The water washed over dune sand that was 3.4 meters above sea level, and the effects on vegetation indicated that the wave could not have been less than 4 meters high."

At Isla Mona, the 4-meter tsunami washed away a pier and the receding water bared the reef. A small boat at in Bahía de Boquerón anchored 50 meters from shore, where the water was normally 1.5 meters, rested on the bottom for a few minutes. Tsunami and earthquake effects throughout Puerto Rico caused damage to property that exceeded $4,000,000.00. Submarine cables from San Juan to Kingston and St. Thomas to Puerto Plata were broken in several places.

At Krum Bay, St. Thomas, waves rose 1.2 meters during a period of half-an-hour, where oscillations continued for 3 hours. At Charlotte Amalie, the water rose and fell up to 45 centimeters several times. Disturbances of the sea 70 centimeters in height were reported at Tortola. Earthquake effects with lessening intensities were felt at St. Thomas, Tortola, St. Martin, Saba, and in the Dominican Republic at Santo Domingo, Puerto Plata, La Romana, Samaná, Santiago, and San Pedro de Macorís, and in Haiti at Port-au-Prince, Gonaïves, Cap-Haïtien, Bayeux, and Port-de-Paix. Submarine cables from San Juan to Kingston and from St. Thomas to Puerto Plata were broken in several places.

Wave simulations in 3-minute snapshots are shown in Figure 14. The seabed collapse caused the "hollow" of the initial sea surface deformation, seen in the first simulation. Mercado and McCann (1998) point out that the wave simulations show how the wavefront is deformed by the faster propagation speed along the Puerto Rico Trench.

Travel times matched observations at the time very well, according to Reid and Taber (1919b): "In passing along the coast toward the south and toward the

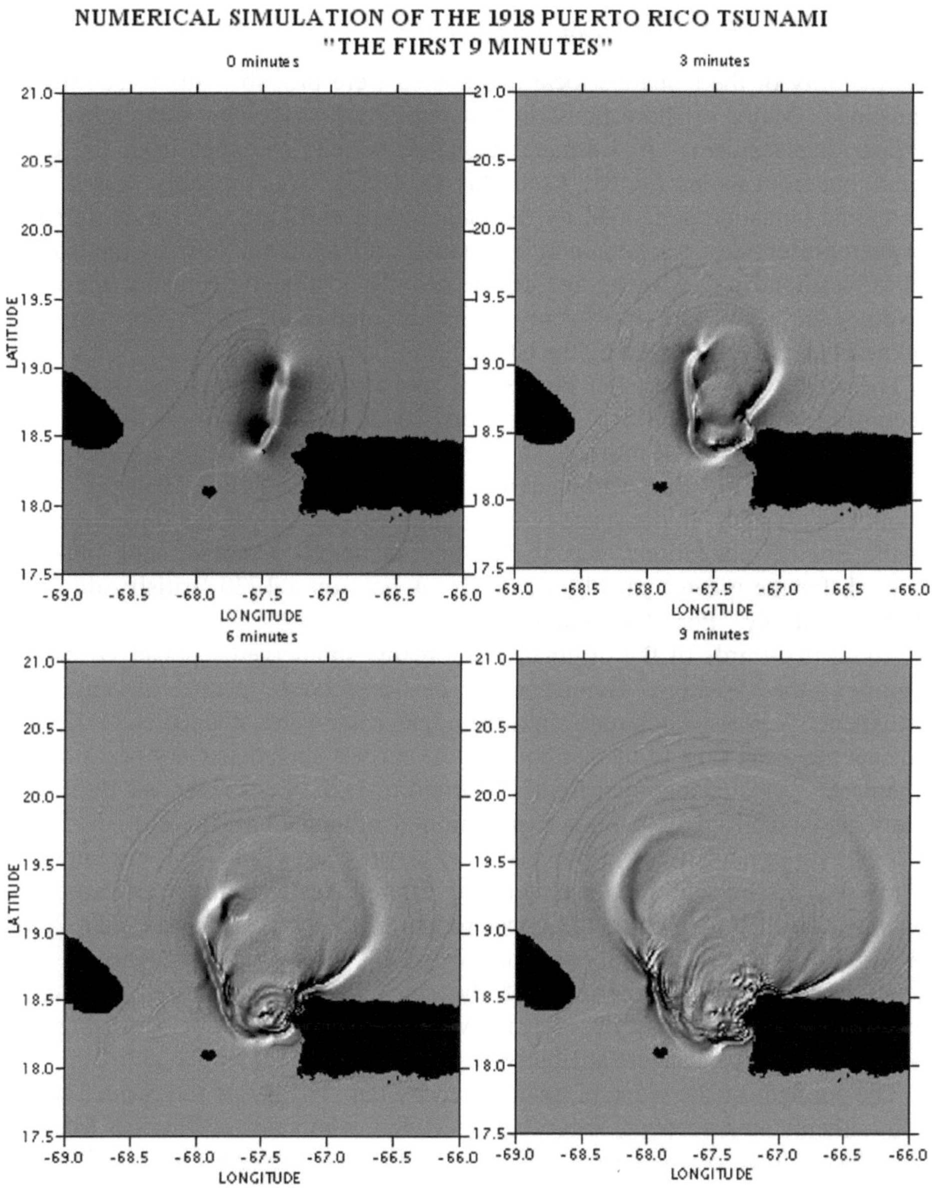

Figure 14. Wave-simulation snapshots of surface displacement at 0, 3, 6, and 9 minutes, for the October 11, 1918 tsunami. This model simulation used models from the Tsunami Inundation Modeling for Exchange (TIME) program. (Simulation courtesy of Mercado and McCann, 1998.)

east, the wave decreased in height, though not uniformly, and the time interval between the earthquake shock and the arrival of the sea wave gradually increased."

4.3.3. June 7, 1692, Port Royal, Jamaica

The majority of the tsunamis local to Jamaica were probably due to submarine landslides. Major offshore faults in that region are strike-slip with little or no vertical displacement. A 1.8-meter tsunami, which "rose not from the open ocean, but from within the bay itself (Myles, 1985)," was probably caused by a submarine landslide generated by the cataclysmic earthquake. The harbor lies between present-day Kingston and infamous Port Royal, enclosed by the narrow Palisadoes Sandspit. Moving against the heavily populated peninsula, the wave traversed the bay. More than twenty vessels moored in the harbor were capsized and nearly totally destroyed by the tsunami.

The violent subsidence left Port Royal submerged, overwhelmed by the sea. According to Leslie (1740), "Where Port Royal stood, the finest town in the West Indies, was almost intirely *[sic]* destroyed: so that in the place where rich merchants lived and fine warehouses stood, now is 20 Fathom Water [37 m]." Sloane (1694) said, "In the space of three minutes, Port Royal, the fairest town in all the English plantations, was shaken and shattered to pieces, sunk into and covered, for the greater part, by the sea." A note on a 1720 British Admiralty chart said simply, "Port Royall Sunk."

About two-thirds of the original city remains submerged, unchanged by the centuries—the Western Hemisphere's best-preserved record of an early settlement. A brass watch recovered from under the water stopped at 11:43 may indicate the exact time of the earthquake, the earliest such discovery of a stopped timepiece. (This exceptional find is credited to U.S. Navy diver, Al Barnesky, sent to assist the Link expedition exploration of sub-aqua Port Royal.)

What yet remained above water at Port Royal after it subsided was engulfed by the sea. Leslie said that the street that suffered least was "so overflowed that the water swelled as high as the upper rooms of the houses." More than 2,000 people drowned. Total fatalities reached 3,000. Given the subsidence of Port Royal and the great withdrawal of the sea along the shorelines, presaging powerful return waves, Myles incisively commented, "It is not surprising that most of these fatalities were attributed to drowning."

The strength of the tsunami, manifested by the "voyage of the *Swan,*" was a well-authenticated account, according to Myles, who said the English Man-of-War, "hove down on the beach broadside," was heeled over to caulk and tar the hull. "The sea's violent motion drove the Royal frigate, *Swan,* over the tops of many houses." Although it was not clear why, this was said to have been the reason more people were not injured when the *Swan* washed over the houses. Myles mentioned that Port Royal's many structures were two to four stories high. To surmount these would have needed an impressive wave.

The Ms ≅ 7.5 earthquake that originated off the northern coast of Jamaica was said to have caused even more damage in the countryside than at Port Royal. Waves of 2.5-meters struck all along the north coast. Although this event occurred over 300 years ago and the return period is not known, the return intervals for major earthquakes are on the order of 200 to 500 years in similar tectonic regimes. Hence, a repeat of this type of occurrence is possible at any time. Sloane reported an account of interest not only for its detail, but also because of the language of that earlier time. Replete with archaic spelling and strung together in one long sentence, the excerpt is set down precisely as it was in the text. Even the archaic letter 'ʃ' for the letter 's' has been retained:

> It will not be amiʃs likewiʃe to take Notice, That that part of *Port-Royal* which ʃunk was towards the Harbor Sandy, and the Sea adjoyning deep; and that Seamen who were then in the Port, aʃʃure me their Part and Sufferings in the Earthquake were very conʃiderable. One Captain had one of his Boys roul'd out of one of the Port-holes of his Ship in its motion, by which he was drowned; and moʃt of them loʃt their Anchors and Cables which were towards the Wharfs or Town, which I ʃuppoʃe came from Sand and Houʃes falling on them; and they after the Earthquake rode in fewer Fathoms Water than before, and one may believe that ʃome of the *Phænomena* may be accounted for from that.

4.3.4. January 14, 1907, Port Royal, Jamaica

A submarine landslide, which occurred at nearly the same location as in the great 1692 event, generated a tsunami that again overwhelmed the narrow Palisadoes Sandspit enclosing Kingston Harbor. Devastating earthquake damage was widespread and waves were reported at numerous places along the shores of Jamaica. At Annotto Bay, the sea was observed to withdraw 73 to 93 meters, 3 minutes after the shock, dropping 3 to 3.7 meters below its usual level. The returning wave raised the water level 1.8 to 2.4 meters above normal, sweeping into town and destroying houses. Similar details were reported at Port Maria and at Ocho Ríos, near St. Ann's Bay.

Again, the short span of time after the shock and the initial withdrawal of the sea suggest a local landslide as the tsunamigenic source. Hall (1907) reported that those who saw the sea immediately after the shock said it "was mountain high," and that the line of seaweed that remained visible confirmed the high wave. Unfortunately, no runup height was estimated.

A pilot and his crew who witnessed the occurrence reported that they saw "the huge wave strike Gun Cay completely cover it from view, and almost simultaneously the [Palisadoes] Point was struck and obscured from view." Then they "observed the same phenomenon when it struck Kingston."

According to Berninghausen (1968), seiches of 2.5 meters were set up in Kingston Harbor. Hall reported that "the great earthquake came without any warning, and in the space of 20 seconds, Kingston Town was shaken into a mass of shapeless ruins," which caused the immediate death of 1,000 people and injured 1,000 more, many of whom later died. Some 90,000 were left homeless. Mangrove bush that had been previously above water, at the eastern edge of the narrow Palisadoes Sandspit, was 4 fathoms [7.3 m] deep under water.

Hall placed the epicenter 5 kilometers south of the Palisadoes, between Plumb Pt. Lighthouse and Harbour Head. He determined that a second focus lay between Buff Bay and Enfield, on the north coast, which "formed below the geological fault in the St. George district, since the intervening mountains would have otherwise attenuated those shocks."

When the seismometer, disabled during the earthquake, was back in operation, recorded aftershock traces revealed the two foci. The first bearing cut through Harbor Head, and the second cut the northern shore between Annotto Bay and Buff Bay, according to Hall. Several other tsunamis were generated in ways that indicate a trigger by a submarine slump or a landslide from the precipitous trench walls in the areas of high seismicity.

4.3.5. September 7, 1882, Archipiélago de San Blas, Panamá

In 1882, in northeastern Panamá, a tsunami, "three meters high or more," swept the San Blas coast of Panamá (Camacho, 1994). Cables connecting Colón with the West Indies were broken, suggesting a submarine landslide originating from the massive Ms $\cong$ 7.9 earthquake. Sieberg (1932) reported, "The earthquake was accompanied by tsunami waves, which were observed from Panamá to Guayaquil [Ecuador]." Most at risk from such widely felt cataclysms are low-lying coastal areas. Molina (1997) reported that the tsunami washed out most of the islands of Archipiélago de San Blas, sweeping away villages built on the beaches as well as on the mainland. The islands remained submerged for several minutes.

Between 75 and 100 people perished, most of whom drowned. Mendoza and Nishenko (1989) said the outflow and inflow occurred four times, the greatest damage due to the third wave. A "seaquake" was felt aboard the *Honduras* in the roadstead 3½ kilometers from the coast, followed by rumbling. "Although no visible waves appeared on the surface, the sea began to move rapidly. The vessel, began to move, as if at full speed, and ran aground on a bed" (Montessus de Ballore, 1888). At Colón, the marigram of a distinct 62-centimeter train of waves from the French Canal Company tide gauge was lost.

4.3.6. April 6, 1690, Nevis and Saint Thomas

Landslides generated on Nevis Peak Volcano may have been responsible for the tsunami at St. Thomas in 1690. One of the highest-magnitude earthquakes in the eastern Caribbean (Dorel, 1981), it generated landslides on the Nevis Peak stratovolcano and the earth in the lower part of Charlestown was fissured and water was discharged. The sea withdrew a furlong [201 m] from Charlestown before returning in 2 minutes, according to Smith (1745).

Details of the sea retreating from land at Charlotte Amalie, St. Thomas, were reported by Taylor (1888). His description of the effects is noteworthy for being the earliest record of a tsunami affecting any present U.S. territory. Poul Olsen (personal communication, 1988), Royal Archivist for Denmark, translated from Danish a letter from the files of the Danish West Indian and Guinea Company that supported Taylor's account: In the *Breve og dokumenter fra Vestindien*

1674–1679, from the Governor of St. Thomas, dated April 8, 1690, it said, "Last Sunday, which was April 6, this island has been hit by a terrible earthquake, so that people feared their houses would come down. Some hours after that the sea receded so that one could walk about 9 to 10 Fathom [16–18 m] into the sea and pick out fishes on dry land." Greater detail was reported for the remarkably similar event in February 1843, following, which may help to shed light on this earlier 1690 event.

4.3.7. February 8, 1843, Guadeloupe

Notable as the strongest estimated intensity earthquake ever to occur in the Caribbean, the catastrophic Mw ≅ 8.3 earthquake devastated Pointe-à-Pitre, Guadeloupe, leaving it in total ruins. One-third of the inhabitants perished, reported as 1,500 fatalities by Montadon (1962). Losses amounted to an estimated 110 million francs. According to the *Antigua Weekly Register* (1843), "The sea rose 1.2 meters after the earthquake at St. John's, but sank again immediately remaining calm throughout." At English Harbour [MMI ≅ IX–X], the wharf built on "made-ground" sank and "afterwards had an undulating appearance like waves on the sea."

The affected area, notably similar to the April 1690 event, extended throughout the Antilles. Remarkable parallels between the two events were reflected in the magnitude, intensity, vast area, and other details. For instance, Grasses (1971) reported that in 1690, "at a rocky islet called Rockadunda (Redonda), a great part of it split and turned into the sea."

A nearly identical description, recounted by Rudolph (1887) about the rocky mass of *Felsinsel Redondo* (Redonda rock-island), came from the French warship, *Juvencelle,* near Redonda: "A mighty chunk of the rocky mass of Redonda broke off and fell into the sea, which resulted in a surging rise of the ocean." Oldmixon (1741) reported that "At a rocky islet called Rockadunda (Redonda), a great part of it split and turned into the sea." Montadon reported that a major earthquake such as this is likely to produce rockfalls and landslides as well as a vast array of other seismic effects.

Besides Redonda, Guadeloupe, Marie-Galante, and Antigua, other islands were ravaged as well. Islands felt the effects in lessening intensities across the Antilles. Mentioned in various reports were Haiti, St. Thomas, Tortola, St. Martin, St.-Barthélemy, Saba, Sint Eustatius, St. Kitts, Nevis, Barbuda, Antigua, Montserrat, Dominica, Martinique, St. Lucia, St. Vincent, Barbados, Grenada, Trinidad, Guyana, Surinam, Bermuda, and South Carolina on the U.S. East Coast. The terrible Ms ≅ 8+ earthquake, with a large mesoseismal area (Dorel), "laid some buildings in rubbish and some persons were killed" at Antigua.

Capadose (1845) said that at about the time of the February 8th earthquake, a spiral column of white smoke and unusual activity came from the crater fumarole of the Mount Misery Stratovolcano on St. Kitts. Robson and Tomblin (1996) said the volcanic activity at Mount Misery Volcano was due to an eruption in the central crater with either solfataric activity or a phreatic explosion.

4.4. Volcanic tsunamis

Volcanoes can create tsunamis in a number of ways, including explosions, both aerial and submarine, caldera collapse, submarine landslides, and landslides issuing into the sea. According to Sigurdsson (1996):

> About 5 percent of all tsunamis are believed to be caused by volcanic events; however, these are among the deadliest of all. Furthermore, about one-quarter of all casualties are killed by tsunamis. The most serious aspect of volcanic tsunamis is the very great distance over which this mechanism can transfer lethal volcanic energy. Volcanoes in the sea, either shallow submarine or volcanic islands, are therefore potential generators of volcanic tsunamis. This includes a great number of volcanoes in an island arc, such as the Lesser Antilles, as well as emerging hot spot regions, such as the Azores, Iceland, and Hawaii.

Their eruptions can transfer energy in three principal ways, according to Sigurdsson:

- By explosions that displace directly a large volume of water due to bubble expansion or bubble collapse—this is a reflection of thermal energy in the eruption magma
- By displacement of sea water due to the rapid entrance of pyroclastic flows or ash flows into the sea—kinetic energy
- By collapse of the volcanic structure of caldera, leading to sudden rush of sea water into the crustal cavities—potential energy

Submarine debris in the vicinity of Dominica, indicated a large-scale volcanic deposit dating to about 28,000 years ago (Sigurdsson *et al.,* 1980). This explosive eruption produced major pyroclastic flows, which advanced into the Grenada basin, west of Dominica. It flowed along the ocean floor and probably along the ocean surface for a distance of up to 100 kilometers and may have generated a significant tsunami.

4.4.1. May 7 to 9, 1902, Saint Vincent, Martinique, and Saint Lucia

Three days of intense volcanic explosions in May 1902 affected the seas near The Soufrière on St. Vincent, Mont Pelée on Martinique, and Hodder's Submarine Volcano near St. Lucia. From May 7 to 18 intermittent minor eruptions occurred. At Martinique, on May 20, another ensuing tsunami was reported due to continuing pyroclastic flows during the period.

4.4.1.1. The Soufrière, Saint Vincent, May 7 to 8

Beginning May 7th, mud and water from the crater lake were thrown out above the rim. An estimated 1.4 cubic kilometers of pyroclastic flow erupted from The Soufrière on St. Vincent. This was followed by floods of hot water and by vast mudflows descending the west flanks of the volcano along the Wallibou River and down to the mouth of the Rabacca River on the east coast (Robson and

Tomblin, 1966). Tsunamis were produced when these large mudflows entered the sea. Robson and Tomblin reported, "At its climax, pyroclast flows were emitted and swept down the major valleys leading to the west and east coasts, devastating all in their paths, and causing the death of 1,565 persons." The eruption column rose 5.5 miles (8.8 kilometers), and the explosions were heard at Trinidad, 280 kilometers to the south, and 672 kilometers to the northwest in the Virgin Islands. Preceding this, on May 6, the crater lake was steaming and discolored and some incandescent material was ejected onto the upper slopes of the volcano. Heilprin (1903) said portions of the sea bottom adjacent to St. Vincent, to the north, "sank incident to the eruption of The Soufrière."

Woodworth (personal communication, 2001a) cited articles for May 7 to 17, 1902, in the *Elizabeth* [New Jersey] *Daily Journal*, about the events from May 7th into the 8th on St. Vincent: "Wallibou and Richmond plantations and villages on the leeward coast were wrecked. Wallibou was partly under water, which had been swept in from the sea by a tidal wave. Five other plantations were gone." Richmond plantation remained under water. He commented that "the situation here sounds similar to Port Royal [1692], which, hit with a tsunami, was on land that was subsiding beneath sea level."

Reporting that the Carib Indian reservation was in this region, Woodworth noted, "They completely disappeared, with only a few Carib individuals left on St. Lucia and Dominica. It would be extremely important to see if we can find out if most, or all, of the Caribs were killed by the ocean, or other results of the eruption. If the ocean, this would be a very rare tsunami indeed; one that basically wiped out a branch of the human race."

Again citing from the *Elizabeth Daily Journal* (May 19-22, 1902), Woodworth (personal communication, 2001b) reported that "On the Richmond estate near the seashore, a new crater had formed, from which fire and ashes were being vomited." The Wallibou estate was destroyed, at least in part, by the wave. "The strange color and great depth of the new inlet of the sea on the Wallibou imply that an eruption occurred there and that a new crater has been formed which extends from the land into the sea."

Schubert (1994) mentioned this as one of the volcanism-produced tsunamis in the West Indies, taking place "in 1902 at Mont Pelée and St. Vincent." Either of Mont Pelée's 1902 events could have been meant—the May 8 event, since it occurred in concert with this eruption on The Soufrière, or the August 30 eruption, for which specific tsunami details were reported.

4.4.1.2. Mont Pelée, Martinique, May 8

The catastrophic Mont Pelée eruption occurred on May 8. When Mont Pelée's side blasted open, it sent a *nuée ardente* [blazing cloud; pyroclastic flow] down the mountain into Saint-Pierre. Saint-Pierre was destroyed and 38,000 people perished. The blast of pyroclastic debris was tornadic in the violence of its sweep, devastating an area of 52 square kilometers (Robson and Tomblin, 1966). Woodworth (2001a) noted, "All were killed except one prisoner in the depths of a jail. What did the pyroclastic flow do when it reached the harbor waters? It made a tidal wave!"

An eyewitness that saw the flow coming down the mountain said the fiery cloud reached his ship, "then another roar, and with it all the water in the harbor seemed to gather up and rush upon the shipping. Every craft heeled over to the great tidal wave and seemed to careen and sink." Afterwards, "The sea was rolling like the heaviest kind of ground swell." When this witness, who was severely injured, looked at Saint-Pierre, the town was gone. "All about us the ships were sunk or aflame, and between us and them and the shore dead bodies floated singly and in groups." Many ships were capsized by the tornadic blast of pyroclastic debris. A wall of fire perilously swept over the town and bay, striking the *Roddam* broadside with such force that she nearly capsized, according to her Captain. Woodworth (2001b) recounted the words of the *Roddam's* Captain Freeman, who described the blast as,

> An enormous black cloud, like a wall with patches of fire in it, approaching the sea from the land. With it came an immense tidal wave of boiling water, accompanied by a loud and terrible noise. Immediately the steamer was caught and tossed over on her side, almost capsizing.

Following the wave, Captain Freeman said, "The sea was a confused mass of boiling mud." From another ship, an eyewitness saw "the terrific tidal wave which had swept over the *Roddam* had parted her cable, and set the vessel adrift." Heilprin (1903) reported that the *Roddam* was "covered from stem to stern with tons of powdered lava, which retained its heat for hours after it had fallen. In many cases it was practically incandescent, and to move about the deck in this burning mass was not only difficult, but absolutely perilous." Chief-Officer Scott, of the less-fortunate *Roraima,* depicted the firestorm causing "a rain of burning ash that was directly responsible for burning most of the shipping in the roadstead."

Regarding the destroying blast itself, Heilprin said, "It can only be said in this place that it was tornadic in the violence of its sweep, of an intensely high degree of temperature, explosive in action, and necessarily gaseous in construction. In Captain Freeman's recital, according to Heilprin,

> It is said that a wall of fire swept over the town and bay, striking the Roddam broadside, and with such force as to nearly capsize her. A probably more correct interpretation of this phenomenon on would be that the swiftly descending volcanic cloud was surcharged with incandescent particles (and burning flames of gas?) and thus gave the appearance of a solid wall of fire.

According to Dudley and Min (1998), pyroclastic flows have been described as "incandescent, ground-hugging clouds, driven by gravity and fluidized by hot gasses." These flows can move rapidly off an island and into the ocean, their impact displacing seawater and producing a tsunami. In the eruption of Mont Pelée, "the flow's impact with the sea also generated a tsunami. The tsunami was observed as far as 15 miles [24 km] away at Fort-de-France (Dudley and Min)." Woodworth surmised, "we can assume all was because of the flow, but should check more sources. A tidal wave probably occurs with every pyroclastic flow that reaches a harbor. With such a force, the wave should propagate out to sea a bit before subsiding. A quartermaster on a ship well offshore suggests this did happen.

> On the morning of the 8th I was standing aft and watching the smoke pillar which rose high into the air. I saw what I thought was a pillar of water coming towards us astern. I waited to be sure before calling any one else, and it began to melt away and finally collapsed. I told the others about it but they laughed at me.

Continuing, Woodward said, "It seems to me that if there were another island or land mass near the place where a pyroclastic flow hits the harbor or ocean, the wave would hit that place before having had time to dissipate. Any flotilla of boats carrying people fleeing the eruption would probably not be safe, as the wave at least would reach them off island/land. If there were another harbor nearby to flee to, those who made it there safely, as well as the residents, might be killed in the pyroclast-flow tsunami, especially if they were in a direct line from where the flow initially hit ocean waters. I would imagine the closer the second land mass, the higher the tsunami."

4.4.1.3. Hodder's Submarine Volcano, near Saint Lucia, May 9

Another instance of volcanic activity that accompanied these events occurred on May 9, at the Hodder's Submarine Mount, lying 8 kilometers to the west of St. Lucia. Two bubbling white patches appeared and remained visible in the same position for 5 hours, from 8:30 a.m. to 1:30 p.m. The oval shapes, which were 8 meters apart, were approximately 91 × 55 meters and 91 × 137 meters. Then bubbling ceased and disappeared (Anderson and Flett, 1903).

Heilprin (1903) reported oceanic disturbances of greater or lesser magnitude noted about 11 meters westward of St. Lucia. "Sea captains remarked on a material change in the course of currents sweeping along the west and north coasts of Martinique. Heilprin concluded that, unfortunately, the observations which record these assumed disturbances still lack full confirmation.

4.4.2. May 20, 1902, Mont Pelée, Martinique

Robson and Tomblin (1966) reported that on May 20th, "a second large pyroclast flow struck the ruins of Saint-Pierre. Further pyroclast flows swept down the Rivière Blanche on May 26, June 6, and July 9." Still further pyroclast flows occurred in the later half of August. Woodworth (2001b) reported that the eruptions of Mont Pelée and The Soufrière continued, "and, yes, tsunamis were produced."

The *Elizabeth Daily Journal* article of May 21, 1902, "Tidal Wave Destroys A Village" was cited by Woodworth for Martinique, where "It was reported that a tidal wave had destroyed a portion of the village of Le Cabet." Woodworth noted that this tidal wave was on May 20, and that "It evidently extended to the capital of Martinique, Fort-de-France," where, as mentioned in the article on May 22nd, "Fort-de-France, on Tuesday, was covered with ashes, stones were falling, and a tidal wave added to the terror of the population."

To confirm whether the continuing pyroclastic flows generated during the period produced additional tsunamis needs further investigation. The *Elizabeth Daily Journal* article of May 22nd went on, "There has been a perfect calm in the air, yet the waters of the Caribbean are lashed to a fury, indicating that the same forces that cause the volcano to labor are working tremendous changes at the bottom of the sea." Woodworth concluded:

> Evidently, these 1902 volcanic tsunamis are only hiding because nobody looked at these events with tsunamis in mind. It appears we have major volcanic tsunami events here. Knowing now what a pyroclastic flow does when, or should it, hit harbor or ocean—Is this new knowledge, or just hidden knowledge? I guess we should make it Common Knowledge!

4.4.3. August 30, 1902, Mont Pelée, Martinique

In August 1902, a volcanic tsunami was observed in association with the eruption of Mont Pelée on Martinique (Heilprin, 1903). Heilprin reported that a violent eruption occurred with "intense lightning and electric manifestations," comparable to the May 8th eruption:

> A great flocculent volcanic cloud flowed from the crater of Mont Pelée, becoming dark gray in the center and turning white on the border, halting about half-way to Fort-de-France. With intense lightning and electric manifestations, it advanced rapidly almost to Fort-de-France, accompanied by a light fall of ashes and small stones.
>
> Simultaneous scintillant and zigzag discharges, producing an uninterrupted crepidation, from time to time effaced by a great glow, which illuminated the clouds. A strong odor of ozone was perceptible and the barometer depressed 3 mm and rose 4 mm in a 10-minute period. The sea retreated from Fort-de-France, followed by a more rapid rise of 1 meter, which covered the quays and inundated to the border of the *savane* [savanna].

Robson and Tomblin (1966) reported, "A number of pyroclast flows were emitted in the second half of August, including a particularly large one on August 30, which devastated an additional 56 square kilometers on the south-east flanks of the volcano." The destruction included the town of Morne Rouge, where 1,000 persons perished.

4.4.4. December 26, 1997, Montserrat

An example of caldera collapse generating a tsunami occurred at the Soufrière Hills Volcano on December 26, 1997, dubbed "The Boxing Day Collapse," the first recorded eruption since 1646 (Robson and Tomblin, 1966). The 3 a.m. slope failure, lasting about 15 minutes, caused a more than 44-million-cubic-meter caldera collapse, which "represents the highest-magnitude, and perhaps the most intense volcanic activity yet at the volcano," according to Calder *et al.* (1998). A volcanic debris slide of approximately 25 million cubic meters, slid into the ocean.

The wave was refracted around the coastline of Montserrat and inundated as far inland as 80 meters at Old Road Bay, 10 kilometers from the landslide site. A small tsunami moved inshore near the jetty, probably generated by the steep flow-front of the debris avalanche as it entered the sea at the mouth of the White River Valley (Calder *et al.,* 1998; Mangeney *et al.,* 1998).

The focusing of the wave at Old Road Bay can be attributed to the abrupt change of coastline at Old Road Bay. The wave flowing parallel to the shoreline would have met the shore head-on. The shallow offshore bathymetry and onshore topography in the area also aided the extended inundation (Calder *et al.,* 1998).

The tsunami was estimated to be 1 meter higher than the road, which lies 2 meters above mean sea level. Impact marks up to 1 meter were left on the side of palm trees facing the ocean. Models developed for the Montserrat Volcano Observatory determined that a debris slide entering the ocean at 50 meters/second would produce a source waveheight of 5 meters. The model conditions appeared consistent with estimates for the mouth of the White River soon after 3 a.m., according to Calder. "The coastal wave would decay at about the square root of the distance, so a wave should be a height of about 1 meter at 10.7 kilometers distance at Old Road Bay."

A waveheight discrepancy appears to exist in these descriptions of waveheights when the tsunami reached Old Road Bay. If the height was 1 meter higher than the road, and the road was 2 meters above the sea level, then the wave would have needed to be approximately 3 meters high instead of the wave 1-meter in height reported.

Since July 18, 1995, when the Soufrière Hills began erupting, several debris slides reached the ocean (Mangeney *et al.,* 1998). Unfortunately, no tsunami reports are known, although, depending of the strength of the ensuing eruptions, waves presumably would have been produced if debris landslides issued into the sea.

4.4.5. November 3, 1911, Trinidad

The explosion of a mud island volcano generated "extraordinary waves" that were observed south of Trinidad (Arnald and Macready, 1956). Singer *et al.* (1983) reported terrain uplift on the beach. Singer *et al.* described the phenomenon as an *isla de lodo efímera* [short-lived mud island], noted five days later on November 8th. Similar appearances of such mud island volcano explosions were reported, also to the south of Trinidad, in 1853 and 1874, prior to this event, and two more after it, in 1928 and 1934. The volcanic explosions

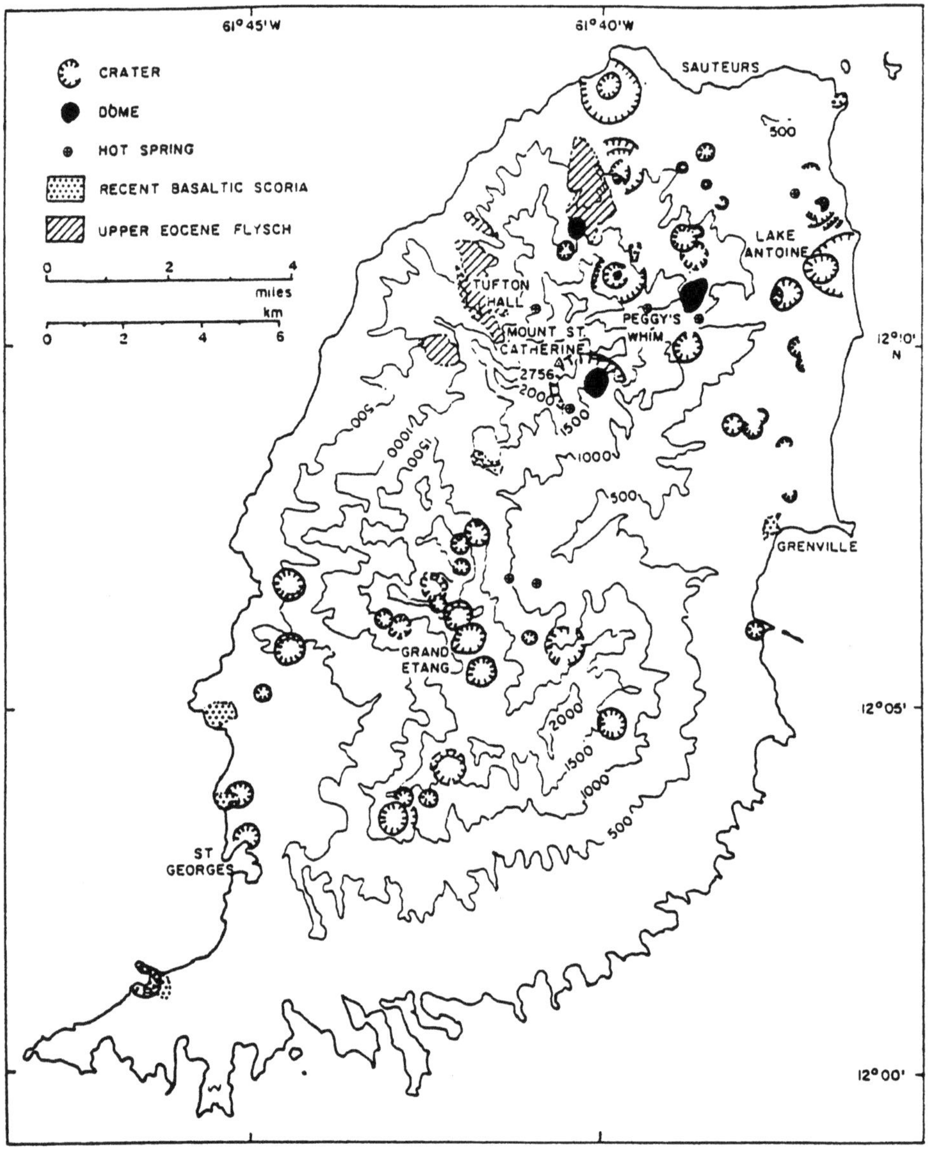

Figure 15. Volcanic map of Grenada showing an old crater located on the northern part of the Carenage at Saint George's Harbor on the southwest coast of Grenada. The crater lies near the location of a long-hidden volcanic crater, which was said to have come into action and discharged clouds of volcanic gases on November 18, 1867. Heights are in feet on the map from Robson and Tomblin (1966).

that formed the mud island(s) may have generated tsunamis for these events as well. Centeno-Graü (1940) reported that the island rose 8 meters higher than the level of the sea in 1928.

4.4.6. November 18, 1867, Grenada and Little Saba

Parsons (1868) reported that "On November 14th, the Central American volcanic mountains showed signs of more than usual activity, and discharges from them occurred constantly from that date until beyond that of the occurrences in the islands." Becker (1869) reported that a new volcano in Nicaragua [Cerro Negro] erupted for the first time on November 14. In addition, Mount Vesuvius erupted from November 14 to the 21st.

These phenomena may render plausible reports of the volcanic activity associated with the earth convulsions that occurred at Grenada and Little Saba. The thread of volcanic activity is continued with reports of volcanic eruption at St. George's, Grenada. Parsons reported, "Many inhabitants declared that a sulphurous *[sic]* stream was emitted from the middle or eastern arm of the Carenage." Skeptical about the occurrence of volcanic activity, Parsons refuted it: "The mud of the shallows being so disturbed that nasty vapors tainted the air for some time, but sulphur gases never existed at the moment."

However, according to the *Grenada Chronicle* (1867), a long hidden volcanic crater was reported to have come into action and "discharged clouds of gases, causing the waters of the Carenage to boil up and rise to a fearful height." An old crater located adjacent to the Carenage, near the area mentioned, can be seen on the volcanic map of Grenada, in Figure 15, from Robson and Tomblin (1966). Volcanic activity was correspondingly reported at Little Saba near St. Thomas Harbor.

Maskell (1868), aboard the West India Royal mail steamer, *La Plata,* during the massive earthquake and tsunami reported sighting a volcanic eruption on Little Saba Islet:

> The first glance thrown seaward showed, if not the cause, at least one principal accessory to the shock. Little Saba lying 7.2 kilometers to the southwest of our position, was in active eruption! A cloud of dark smoke hung over it. People on shore told us afterwards, that they saw a column of bright flame in our direction (in line with Little Saba), and thought the *La Plata* was on fire. Meanwhile, a strong smell of sulfur pervaded the atmosphere. A quarter of an hour passed, and at 2:35 p.m., a second cloud of smoke, this time white like steam, appeared over Little Saba. A moment more and the great wave reached Little Saba, and in an instant that rock was a mass of foam 100 feet [30.5 m] high.

The *La Plata* was situated between the islet and the town of Charlotte Amalie. Lying at anchor, 4 kilometers from the town, off the southern point of Water Island in the West Gregerie, the steamer would have been in a direct line-of-sight with Little Saba. Adding the reported distance of 7.2 kilometers to the islet from where she was moored to the 4-kilometer distance from town renders an approximate span of 11 kilometers with *La Plata* in between.

It is thus feasible that the flames from the volcanic explosion behind, in line with the ship, could have created the illusion that the *La Plata* was on fire. Sainte-Claire Deville (1867) cited the froth seen at the entrance to the *rade*

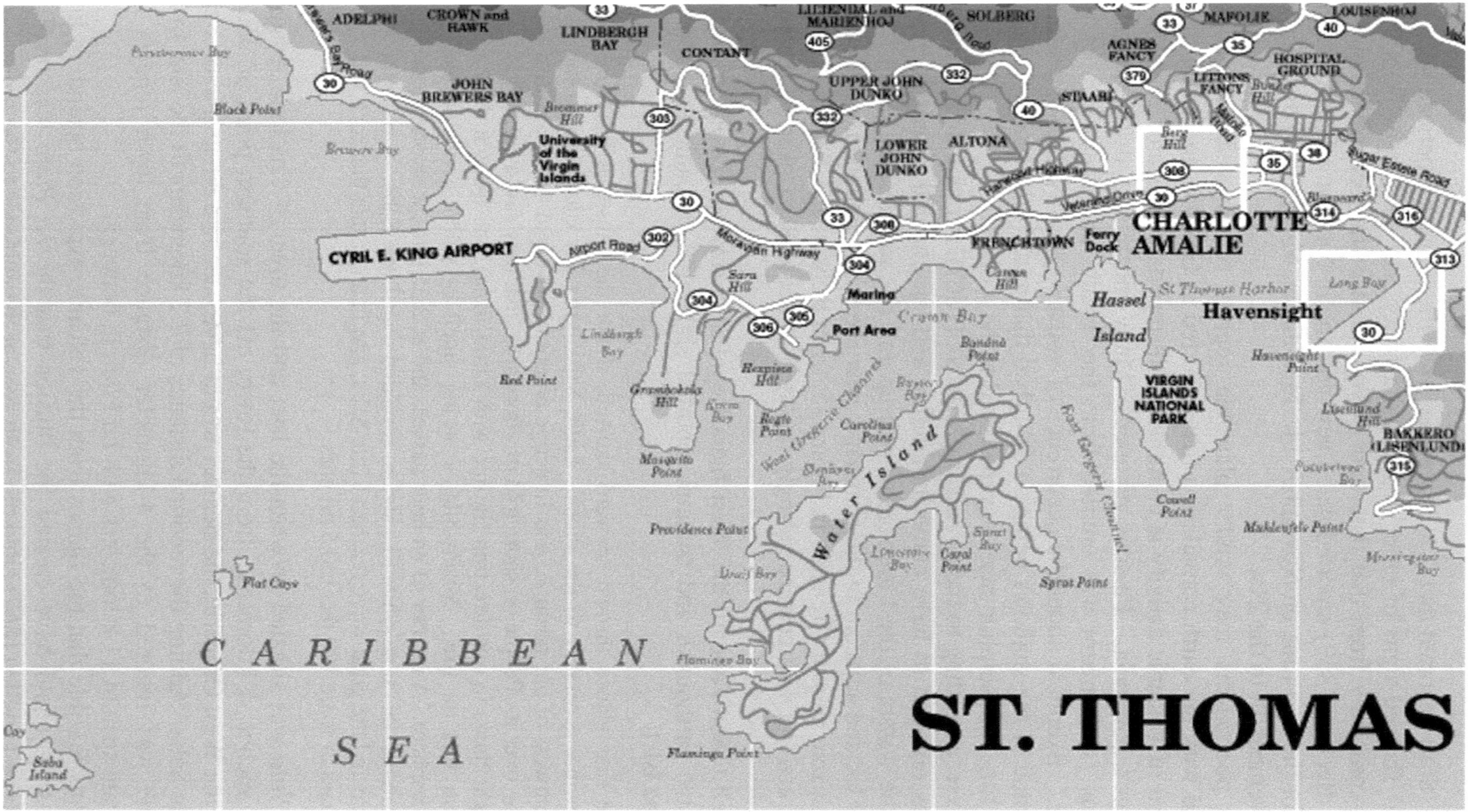

Figure 16. Saint Thomas Harbor including Little Saba Island. The immense tsunami entered from the southwest through the West Gregerie Channel, East Gregerie Channel, and the entrance to Saint Thomas Harbor. Little Saba lies 7.2 kilometers southwest of the entrance to the West Gregerie. A line passing from "Town" to Little Saba would pass through the region where the West Indian mail packet, *La Plata,* lay at anchor off the southwest coast of Water Island. *La Plata* was 4 kilometers from Charlotte Amalie in line with Little Saba.

[roadstead] as "a barrier of foam 100 metres high which deadened itself on the rocks in the middle, and surged on like an avalanche." Note height in meters from Sainte-Claire Deville versus feet from Maskell. The 100-meter height is not without feasibility—if the explosion of froth that would form when lava extruding from an old volcano was deluged by the great wave is taken into consideration.

Maskell concluded his recollections about the volcanic eruption at Little Saba with the account of passing Little Saba during their departure "two harrowing days later."

> Little Saba, itself, was rent and torn in all directions, its precipices almost shattered, and mounds of fallen rocks had piled up on its shoreline. In one place, a large crater gaped, and grey lava was still coming down from it to the sea. We were told by those who knew Little Saba well, that the whole shape of it was changed. It is probable that had not this small island acted as a safety valve and provided an escape for the imprisoned elements, the earth shocks at St. Thomas would have been far more severe. Not one stone would have been left on another, and few of us aboard the *La Plata* would have lived to tell the tale, were it not for this off-shore safety valve. Even now, one more earth shock came, as if it were a parting salute. It was severe enough to make the rocks roll furiously down the cliffs of the little island and to cause a fresh rush of lava to pour from its crater.

Moreover, in the *Saint Croix Avis* (1867) it was reported that at St. Thomas, "The sea rose 50 feet [15.2 m], and a volcano vomited forth lava." Watlington and Lincoln (1997) cited, "Little Saba was submerged for a short time, after being visited by earthquake and eruptions simultaneously." Latimer (1868) observed about Little Saba, "It is historical that some 75 years ago the small island of Saba was an active volcano, and on the 18th and 19th of November last, emitted smoke; thus showing that the volcanic action was not distant."

"Land slides on Little Saba Island, 7.2 kilometers southwest of the harbor, gave rise to a rumor of volcanic action there," according to Reid and Taber (1920). The map of St. Thomas Harbor, in Figure 16, shows Little Saba lying southwest beyond the entrance to the West Gregerie Channel.

Volcanic activity may have been occasionally confused with landslides, yet the distinct differentiation reported by Maskell (1868) between the rock falls and flowing lava seems to negate any confusion in this instance. It seems tolerably clear with a vivid eyewitness description of a large gaping crater with gray lava flowing down to the sea.

Disputes about volcanic activity occurring during earthquakes seem to stem from the idea that the volcanic upheaval could have been the tsunamigenic trigger. In a few events where possible volcanic activity occurred, there had been conjecture as to whether a volcano could have been the tsunamigenic source. Such accounts were dismissed out of hand as improbable or just plain wrong, and, furthermore, the same dismissal was apparently given to any mention of volcanic activity. However, as Örsted (1867) and Maskell recognized, an outlet can be provided for expansive gases in association with earth convulsions. An argument could be made in support of possible volcanic activity in conjunction with tsunamigenic earthquakes.

4.4.7. March 17, 1843, Marie-Galante, Guadeloupe

A submarine eruption occurred on March 17, 1843, subsequent to the disastrous February 8 event at Guadeloupe. Described by Blainville (1843) as erupting half-way between Guadeloupe and the eastern point Marie-Galante, "A large column of water, of blackish color, rose in a succession of jets, whilst steam was emitted around the column at the sea surface. The eruption continued for about half-an-hour." Waves may have resulted from successive jets spouting up for a half-hour from explosions beneath the surface. The shallowest point in the area was 7 meters below sea level (Robson and Tomblin, 1966). The surrounding sea floor was 550 meters deep, and then "the sea floor falls away abruptly to great depth."

Robson and Tomblin placed the point of eruption of the submarine vent as possibly the shallows named the Vaisseaux Bank, shown on British Admiralty chart No. 885 (1882). *The Barbadian* (1843) described, in similar detail, 100-foot [30.5-m] columns of water several feet thick [1–1.2 m] ejected from fissures in the ground, observed during the February 8 event. Robson and Tomblin put the date of the volcanic activity as February 17, whereas Poëy (1857) gave the date as March 17, 1843. The date discrepancy remains to be cleared up.

4.4.8. August 27, 1883, Saint Thomas

An unusual source for a tsunami in the Caribbean Sea was the major explosive event in the Sunda Straits near Java, of the Krakatoa Volcano. It lies approximately antipodal to St. Thomas, which may possibly have enhanced the affect there. The explosion generated waves as high as a 12-story building, according to González (1999a). Sigurdsson (1996) stated that the 26-meter eruptive column height implies a mass eruption rate on the order of 1.25×10^8 kilometers per second. The mammoth eruption (30 billion tons of magma), considered one of the world's worst modern disasters, caused 36,000 regional fatalities.

About 10 kilometers from the source in the Sunda Straits, the tsunami was estimated to have had waves 40 meters high. On the coasts, the wave crested at up to 36 meters (Sigurdsson). The tsunami arrived at Batavia (Jakarta), 155 kilometers away, 2½ hours later, with an amplitude of 5.5 meters (trough-to-peak height of about 11 meter). The volcanic explosion created air pressure waves that caused tsunamis in remote areas.

It has been debated whether changes in the barometric pressure from the explosive eruption or direct transfer of energy transmitted through the water propagated the tsunamis throughout the world's oceans. The waves were recorded at Australia, Japan, the South Sandwich Islands, Hawai‘i [Hawaii is used in the form preferred by the University of Hawai'i], California, and Alaska. At Table Bay, Cape of Good Hope, Africa. Berninghausen (1964) reported 14-inch [36-cm] waves; 36 centimeters was also recorded at the English Channel (Verbeek, 1886). According to Berninghausen, it could be assumed that similar small waves occurred all over the Atlantic Ocean.

A tsunami with three recessions from the shore occurred at St. Thomas, as reported in *Monthly Weather Review* (1883). Tsunamis were likely to have occurred at other locations in the Caribbean. Forel (1885) reported a subterranean noise heard at the time at *l'Îlot Caiman-Brac* [Cayman Brac Islet] in the Cayman Islands.

4.4.9. 1939 and 1990, Kick-'em-Jenny Submarine Volcano Mount

The Kick-'em-Jenny underwater Volcano Mount was first observed when a volcanic explosion broke the surface. A contour map from Robson and Tomblin (1966), showing the location of Kick-'em-Jenny appears in Figure 17. Lying in The Grenadines near Île de Ronde, Kick-'em-Jenny is close to The Sisters, north of Grenada. Kick-'em-Jenny was first observed from Sauteurs, Grenada, in active eruption on July 24, 1939.

Since then it has erupted at about 5-year intervals (Sigurdsson, 1996), but the vent has remained beneath the ocean surface. Robson and Tomblin reported effects of the submarine eruptions of 1939 and 1990. "The eruptions have been small," according to Sigurdsson,

> but have generated underwater acoustic signals, or T-phase, probably due to the collapse of steam bubbles due to reaction between the erupting magma at over 1000° C and the sea water above: a process variously known as hydro-eruption, fuel-coolant interaction, or phreatomagmatic explosion. The 1939 and 1990 events may have generated very small tsunamis of local extent.

In 1939, light shocks were felt from 8 a.m., at frequent intervals, in the northern part of Grenada, followed 4 hours later by 7 heavy shocks. The most prolonged 12-second shock "was accompanied by the discharge of a column of black smoke from a point in the sea estimated 5 miles [8 km] distant bearing 7½° true," according to Devas and Macadam-Sherwin (1939).

The eruption cloud rose 150 to 200 feet [45.7–61 m], and contained "solid, dark-colored bodies of considerable size, which fell back into the sea." At 6:04 p.m. the same day, a second discharge of brown smoke was projected rapidly to an estimated height of 900 feet [274 m] above sea level. All day from 8 a.m. to 7 p.m., *T*-phase tremors with a period of 1 to 3 seconds were recorded at the Montserrat Volcano Observatory.

Molard (1947) reported similar acoustic signals recorded in October 1943 and in October 1953, recorded at seismographs stations in Martinique, Guadeloupe, and Dominica. In October 1965, *T*-phase tremors were recorded from Puerto Rico to Trinidad, with ground tremors felt at Île de Ronde, Grenada, St. Vincent,

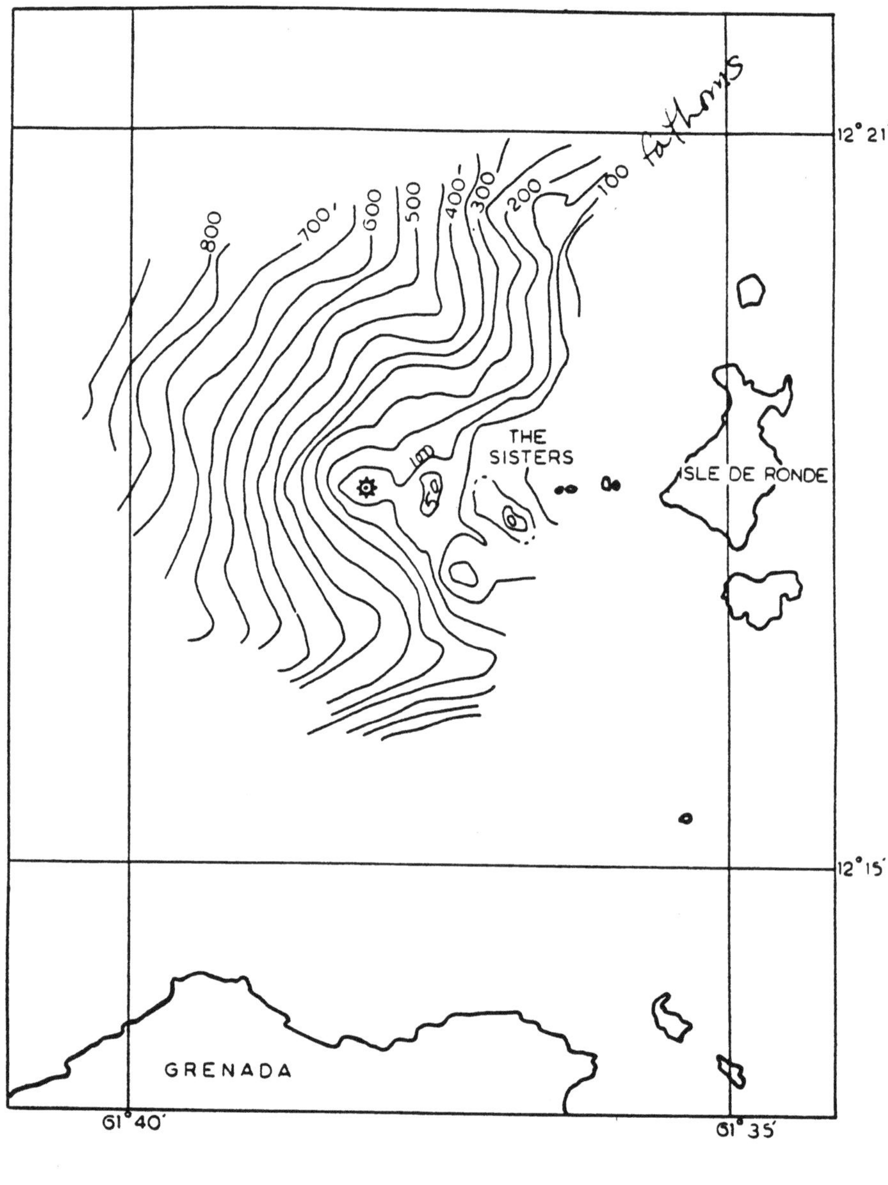

Figure 17. The Kick-'em-Jenny submarine volcano, first observed in 1939. The submarine mount lies situated near The Sisters, in The Grenadines, at 12.5° North Latitude and 61.5° West Longitude. Soundings in 1962, taken from the survey ship, *Vidal,* placed the depth of the vent between the contour lines of 50 and 100 fathoms [91 and 183 meters]. (Map from Robson and Tomblin, 1966.)

and Martinique. *T*-phase tremors originating near Kick-'em-Jenny were recorded generally in the eastern Caribbean in May 1966. From August 3 to 6, 1939, 168 *T*-phase events were recorded instrumentally, with two earthquakes felt at Sauteurs, Grenada, on the 3rd. Besides these, eruptive activity was also recorded in the years 1977, 1987, 1990, and 1997.

According to Sigurdsson (1996), Smith and Shepherd (1993) "have modeled hypothetical volcanic tsunamis from Kick-'em-Jenny in The Grenadines. This model relies on the thermal energy of eruption, which is thus fundamentally different from the kinetic energy of the Krakatoa 1883 tsunami, generated by pyroclastic flows." Smith and Shepherd proposed that the Kick-'em-Jenny tsunami hazard is:

> Due to the interaction of magma and water during eruption (hydro-explosions), which relies on the huge reservoir of thermal energy of erupting magma, and is strongly controlled by the water/magma ratio and confining pressure (water depth). In deep water (<300 meters), the confining pressure largely inhibits vapor bubble expansion. In the depth range of 300 meters to near-surface, the bubble expansion is large and energy yield is relatively efficient."

Sigurdsson stated that for a Krakatoa-sized eruption occurring at Kick-'em-Jenny the hypothetical tsunami generated by hydroexplosions, as modeled by Smith and Shepherd, "estimated wave amplitudes and runup heights throughout the Caribbean, yielding values that are truly terrifying." Sigurdsson also observed that:

> The runup height is of the order of 40 meters in Grenada, and up to 7 meters in the Virgin Islands. However, one must question whether this is a realistic approach. The latest Krakatoa-sized eruption in the Lesser Antilles took place in Dominica, about 28,000 years ago. Eruptions of this magnitude are extremely rare in the arc, and there is no indications that such an event is likely in the near future, either at Kick-'em-Jenny or elsewhere in the Lesser Antilles. The eruption would most likely occur in the subaerial environment and most of the thermal energy would be utilized in the convective rise of the eruption column. The most likely energy source for pyroclastic-flow generation would probably be from pyroclastic flows entering the sea.

Presently, there is concern about the currently active volcanoes of the Soufrière Hills on Montserrat and Kick-'em-Jenny (Sigurdsson, personal communication, 1997). González (personal communication, 1999b), citing Paul Abbott who related information from an article in *The Barbados Sun,* "concerning a sub-sea volcano, called Kick-'em-Jenny (locally referred to anyway), which my father has always talked about since I can remember, and the resultant tsunami that could be caused from an eruption of the volcano."

The Sun [February 13, 1999] article was headlined, "$3,000,000.00 Alarm Early Warning System To Be Set Up," by the Seismic Research Unit of the University of the West Indies, Trinidad. It was to "monitor the underwater volcano, Kick-'em-Jenny, situated north of Grenada." According to González, mention of Kick-'em-Jenny in the article was "one of the most under-publicized dangers that threaten the lives of thousands of Barbadians and Caribbean people." In the following excerpts, *The Sun* article reported on the hypothetical eruption of Kick-'em-Jenny:

> There could be an eruption any time now. The summit of the volcano is now around 130 to 140 meters below the surface and the pressure of the water dampens the explosions, but once it reaches 110 meters, Shepherd said eruptions would become more violent and a massive eruption will trigger a tsunami (tidal wave).
>
> Grenada is expected to be hardest hit, in the worst case scenario, with waves around 150 feet [45.7 m] in 15 minutes, while Barbados" is expected to be least affected—by waves around 12 feet [3.7 m] in 40 minutes. These waves will be traveling at 300 to 500 mph [483–805 kph]. We were extremely alarmed to read that such danger lay on our doorsteps, and was mentioned only as an aside to the warning system grant issue. Had there been no grant we would not hear of the seriousness of the matter.
>
> This deserves the utmost attention of all the people of Barbados, the government, and those in authority. There is need for a massive public sensitization program and the development of a national disaster plan to prepare, in the event that this volcano really erupts. The public needs to know which areas in Barbados are considered safe and which areas are most likely to be affected. Will people in town be able to seek safety? School children should be given emergency evacuation drills to get them to safety if this disaster happens. To realize the seriousness of the matter, all you have to do is think how Bridgetown would look if this happens at 11 a.m. on a normal busy weekday, or even worse, at 2 a.m., when all the people on the western coast in low-lying (flood-prone) St. Michael areas are sleeping. We must raise awareness of the real danger and seriousness of the situation and cause the authorities to act.

The much smaller, though not inconsequential, alternative eruption scenario by Smith and Shepherd (1993) was said to be more realistic, given the rate of growth of Kick-'em-Jenny since 1939. Sigurdsson noted that Smith and Shepherd were aware of the likelihood that they had overestimated the potential eruption magnitude. "The theoretical tsunami runup height from this size eruption ranges from 8 meters on Grenada to 1 meter in the Virgin Islands." Commenting about approaching activity possible for the rising seamount, Sigurdsson said: "Of great importance for Kick-'em-Jenny, is the water depth above the crater, which is now approaching maximum efficiency for tsunami generation in terms of water depth." As the water depth above the vent decreases further:

> The explosive efficiency for tsunami generation will be increasingly more partitioned into the atmosphere. Only time will tell if the shallowing of the crater in the next decades will result in larger tsunami-generation potential. With regard to the tsunami hazard from Kick-'em-Jenny, the most important factor at the present time is the probability of eruptions of larger magnitude

than observed in the historical past. Typically, volcanoes that are frequently active generally produce small eruptions; the eruption magnitude varies inversely with eruption frequency. On the other hand, volcanoes that experience a long period of repose (hundreds or thousands of years of dormancy) are likely to have a catastrophic and large-magnitude eruption at the initiation of an eruptive cycle.

4.4.10. La Palma, Islas Canarias

The *Denver Rocky Mountain News* (2000) reported that scientists from University College, London, issued tsunami warnings about a potential threat from a volcanic island in the Islas Canarias (Canary Islands). A huge chunk of the now unstable western flank of La Palma, the most volcanically active island in The Canaries, "is at risk of collapsing and generating a wave up to 160 feet [48.8 m] high that could swamp the coastlines of the Caribbean Sea and Florida."

Lead scientist, Simon Day (2000), reported in *New Scientist*, "If the flank of the volcano slides into the ocean, the mass of rock will push the water in front of it, creating a tsunami wave far larger than any seen in history. The wave would spread out across the Atlantic at the speed of a jet airliner until it strikes coastal areas all around the North Atlantic." Day identified scores of volcanic vents, configured in the shape of a three-pointed star, in the 7,957-foot [2,425-m] Cumbre Vieja Volcano, comprising the southern half of the island.

Day concluded that the western flank of the mountain, a mass of approximately 500 billion tons of rock, is slowly detaching itself as volcanic activity forces magma to the surface. A major volcanic explosion could cause the flank to detach and fall into the sea with the utmost catastrophic effects.

4.5. Triggering effect of hurricanes

Hurricanes appear to have an effect as a triggering mechanism on tsunamigenic events, a relatively new realization, although as early as 1852, Piddington had reported the effect in the *Sailor's Horn-Book for the Law of Storms*. These early insights have been shown to be well founded in a study of earthquake occurrence associated with hurricanes in the Caribbean Sea. A possible source of seismically induced waves may be triggering by hurricanes or low pressure centers.

A discernible increase in the number of reported earthquakes has been shown during the four days after the closest hurricane approach. Figure 18 illustrates this triggering effect, showing the number of earthquakes between 1982 and 1992, within 10 days and 1000 kilometers of the closest approach of the eye of a hurricane during its passage through the Caribbean. Data ascertained in a study with typhoons in Guam were found to be similar (Lander *et al.,* 1999).

In October 1990, Hurricane *Klaus* sideswiped the West Indies arc for four days, from the 6th to the 9th of October. As it moved westward, the path of *Klaus* curving above the arc of the Antilles and through the region near Puerto Rico is shown in Figure 19. During the hurricane, barometric pressures dropped to

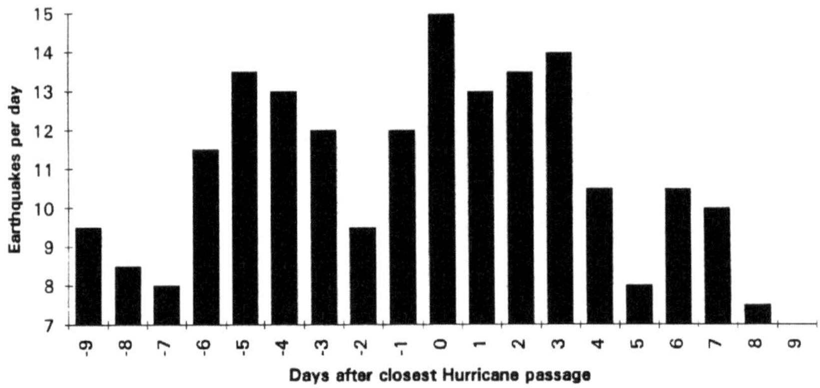

Figure 18. Illustrating the triggering effect of hurricanes is the number of earthquakes reported within 1000 kilometers of the eye of a hurricane and occurring within 10 days of the passage of the hurricanes through the Caribbean between 1982 and 1992.

985 Millibars. Earthquakes were reported along the path in close association with the passage of Hurricane *Klaus*. Asterisks indicate the locations of the eye of *Klaus* for the dates shown. Earthquake origin times (UT) are given next to circles of incremental size that are proportional to the earthquake magnitude. Two earthquakes [Ms ≅ 4.5 and Ms ≅ 5.0] occurred on October 7, within 12 hours of the maximum peak of the passage of *Klaus*, both within 50 kilometers of the eye. The 50-kilometer boundaries, drawn on either side, indicate a 100-kilometer swath bisected by the path of *Klaus*. As many as 6 additional earthquakes occurred in conjunction with its westward progression.

Hurricane *Klaus* illustrates the triggering of earthquakes by a specific hurricane, but many other examples show that this is not uncommon in the Caribbean (Dunbar and Whiteside, 1994). Hurricanes with paths similar to *Klaus* were *Frederic,* in 1979, *Faith,* in 1966, and *Donna,* in 1960 (Neumann *et al.,* 1988). Within 10 days and a radius of 200 kilometers (18° North Latitude and 63° West Longitude), each of these hurricanes was followed by an earthquake of Mb ≅ 5.0. The mechanism of this triggering is probably related to atmospheric loading, winds, or sea surges (Lander *et al.,* 1999). Destructive volcanic activity may occur preferentially during hurricane season in the Caribbean region as well (Simkin *et al.,* 1981). As with earthquake distribution, increased eruptive activity occurs during the hurricane season, especially during the months of September and October, the height of the season.

This suggests triggering effects due to atmospheric changes. Evidence suggests that volcanic eruptions occasionally show periodic tidal behavior. For example, well-defined 12-hour eruptive behavior was shown during the July 1997 eruption of the Soufrière Hills, Montserrat (*Bulletin of the Global Volcanism Network*, 1997), and eruptions at Kilauea, Hawai'i, have been tied to tidal periods (Wood, 1917). Hall (1907) made the following meteorological

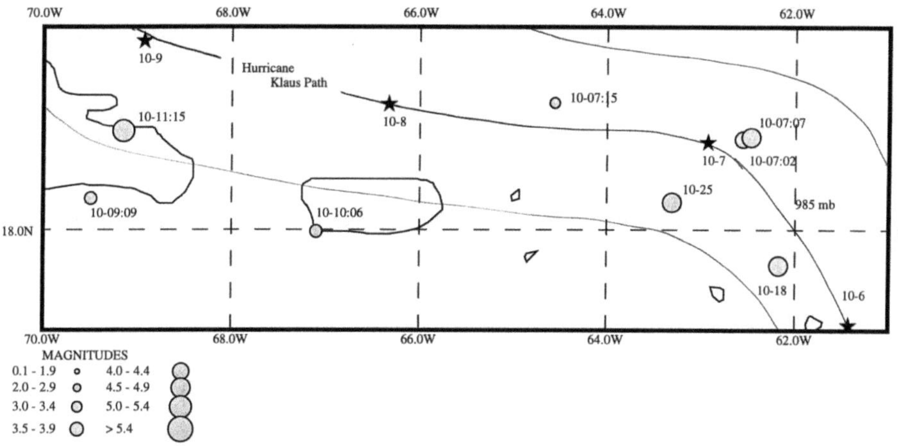

Figure 19. Path of *Klaus* from the 6th to the 9th of October, as it moved westward above and curving with the arc of the Antilles, and through the Puerto Rico region in 1990. Earthquakes were reported along the path in close association with the passage of the hurricane. Asterisks represent the locations of the eye of Hurricane *Klaus* for the dates indicated. Two lines indicate a boundary of 50 kilometers on either side of the path of *Klaus*. Circles plot earthquakes greater than Ms ≅ 3.0 in proportional increments. Date and time (UT) for earthquake origin times are given adjacent to the circles.

observations about atmospheric pressure waves and earthquake occurrence in his report on the 1907 tsunami at Port Royal. Based on barometric measurements:

> It thus appears that in some cases, two small atmospheric waves pass over Kingston about the time of an earthquake: that the first wave passes about 8 hours before the shock: that the second wave passes about 16 hours after the shock: and that the shock itself takes place during the small depression between them. For this event a wave passed at 7 a.m., or 8 hours before the shock, and the second wave passed at 3 p.m. on the 15th, or 24 hours, after the shock. The depression at the time of the shock was much larger than with the small shocks usually experienced in Jamaica.

Details noted by Watlington and Lincoln (1997) from the *Sailor's Horn-Book* (1852) included six West Indian hurricanes from 1766 to 1848 that were accompanied by earthquakes.

- It is stated that in the great Martinique hurricane of 13th August 1766, a shock of earthquake was experienced during the night of the Cyclone. (Chronological list of the West India Hurricanes)
- In the account of the great hurricane in Barbados in 1780, it said, "several notices to prove that there undoubtedly must have been an earthquake during its continuance." At St. Lucia, it was said to have happened "some hours after the greater severity of the gale." (Bland, Wernekian Society)

- An article on the climate of Port Antonio, stated that during a hurricane on 18th October 1813, two shocks of an earthquake were felt all over the island of Jamaica. They were mentioned as "remarkable," from which "it may be inferred that it was considered so by the inhabitants." (Arnold, Wernekian Society, *Edinburgh Philosophical Journal,* Volume VII)
- In the log of the H.M.S. packet, *Spay,* for the hurricane of Antigua in August, 1837, it stated, "At St. Thomas, the hurricane appeared to have concentrated all its power, force, and fury" and "in the midst of the hurricane, shocks of earthquakes were felt."
- During the Tobago hurricane of October 1847, a severe earthquake is said to have preceded "the first out-break," which was at about 10 p.m. at night; and that abundant proofs of it were apparent. *(Globe Newspaper)*
- For the Antigua hurricane of August 1848, it was stated that "at midnight, the wind raged furiously; lightning and thunder were incessant, accompanied by floods of rain. At this time a severe shock of an earthquake was felt, attended by very heavy gusts; 1:30 p.m., the mercury had fallen 4-__ *[sic]* of an inch, and the storm at this time __. By 2 a.m., it had abated."

The supposition that earthquakes may be felt during the passage of hurricanes through the Caribbean Sea appears to be warranted. C. F. Brooks (1919), a noted meteorologist, reported on the stresses placed on the Earth's crust by the violence of a storm in *Monthly Weather Review.* He stated that the immense pressure changes affecting the ocean floor from a hurricane could trigger earthquakes:

> A 10-foot [3-m] rise of water caused by a hurricane adds about a million tons of weight to each square mile of Earth's crust. This load effect upon an unstable sea bottom often provides the necessary trigger-action to set off an earthquake and the tidal wave that follows.

One of the most notable examples of a tsunami probably produced during the passage of a hurricane occurred in 1780 at Savanna-la-Mar. The eye witness account reported by Southey (1827) seems to support this: the shock at noon, followed by the 3.2-meter wave, after that the 7-hour hurricane, and then a subsequent aftershock and subsidence of the waters. The "great hurricane in Barbados," reported by Watlington and Lincoln (1997) from the *Sailor's Horn-Book,* may have been the same event as in Southey's report.

4.5.1. October 3, 1780, Savanna-la-Mar, Jamaica

At Savanna-la-Mar, Jamaica, a submarine earthquake devastated Savanna-la-Mar and Fort-Saint-Georges, and generated the 3.2-meter wave that "swept away the town of Savanna-la-Mar," following the several shocks. Taber (1920) reported that this wave might have been seismic in origin. He reports the dramatic events:

> The inhabitants of Savanna-la-Mar watched as the sea, swelling as it never had before, on a sudden, bursting through all bounds, and surmounting all obstacles, it overwhelmed the town, and swept away so completely on its retreat, as not to leave the smallest vestige of man, beast, or habitation behind. Where the sea did not reach, the destruction was nearly as effectual by the succeeding earthquake and hurricane. The gale began from the southeast at 1 p.m., at 4 p.m., it veered to the south and became a perfect tempest, which lasted at full force until near 8 p.m.: it then abated. At 10 p.m., a smart shock followed, and the waters subsided. All vessels in the bay were dashed to pieces or driven ashore.

Since an earthquake was distinctly reported before the huge wave, the tsunami was probably generated seismically, even if a hurricane was present. Beckford (1790), in describing the hurricane concluded: "The sudden swelling of the sea we may reasonably attribute to the heavings of the earthquake to which likewise the general ruin of our houses may be in some measure attributed." Hurricane-produced undersea landslides that may have produced tsunamis were reported by Woodworth (personal communication, 2001c), who offered the possibility in a letter about the CAT–5 Hurricane *Camille* in 1969:

> The hurricanes of 1938 and 1944 could have triggered submarine landslides, making the 50-foot [15.2-m] wave attacks on New Jersey "tsunamis," according to current definitions. Camille's slide destroyed an offshore drilling rig and damaged two others off the southern-most tip of New Jersey. Wave-height recording instruments [at South Pass] measured waves 20 to 23 meters before the instruments were destroyed. Maybe the South Pass recording instruments relayed the information and the data is extant. If undersea slides cause tsunamis and Camille has a documented undersea slide, then we may have a documented tsunami. The problem would be trying to separate out the details of this "event within an event."

The Caribbean Tsunami History may contain reports of non-seismic waves, despite efforts to ascertain which were true tsunamis. The greater number of tsunamis reported during hurricane season could represent incorrect reporting in the historical record of storm waves. Nevertheless, certain waves may be due to the atmospheric pressure changes of hurricanes triggering earthquakes, landslides, or volcanic activity. Woodworth declared, "It probably doesn't matter if a hurricane, storm, or earthquake produces an undersea slide. The end result is tsunami!" The application of computer simulation models and Geo-positioning satellite (GPS) data as well as greater understanding of the complex processes of oceanic and atmospheric physics can be valuable to help clarify details of historical events. If a definitive assessment is ever to be gained, data and reports must be purposefully investigated with tsunamis in mind.

津波

CHAPTER 5

EDUCATION ABOUT TSUNAMI RISKS

5.1. Tsunami fatalities

The Caribbean Tsunami History contains 11 events with deaths reported due to tsunamis. Many events of catastrophic proportions make no mention of fatalities. Most reports have estimates for earthquake fatalities, and deaths from tsunamis are lumped in with them or no account is taken of them. Countless others were injured, unaccounted for, or left homeless. In the 1780 Savanna-la-Mar account, about 300 people perished from a dreadful wave in Jamaica. The 3.2-meter wave surged 805 meters beyond the shore, and "so sudden and unavoidable was the destruction, although it took place at noon-day, that at one gentleman's house, consisting of 10 whites and 40 Negroes, not a soul escaped." Elsewhere this was reported as 10 deaths.

Unfortunately, in that era, oftentimes slaves, natives, and indentured servants were not counted in the death tolls. In historical times, frequently only the deaths of European colonists were reported. The 500 deaths reported for 1906 in Venezuela probably do represent a full count of tsunami victims.

The difficulty arises in distinguishing differences from the reliable counts of fatalities that were reported. Consequently, the fatality counts may be much misrepresented. Earlier data were sketchily reported, and in many instances, no details were available. In two instances in this century, fatality statistics were revised upwards when new information came to light about the 1918 Puerto Rico and the 1946 Dominican Republic events. These data have subsequently been included in the fatality statistics.

5.1.1. October 11, 1918, Puerto Rico

In the disastrous tsunami that lashed the northwest coast at Aguadilla in 1918, 30 people were said to have drowned at Aguadilla alone. This was later revised to 32 victims. The eight deaths of victims drowned at Punta Agujereada by the waves 5.5 to 6 meters in height brought the initial tally of people drowned for these two locations to 40. The greatest number of fatalities was said to be from the tsunami. The official death toll for this event was placed at 116. This total must have been earthquake fatalities only, since at Mayagüez, "more than 100" deaths were reported, plus the 4 deaths at Ponce (totaling 104). When added to the tsunami victim count of 40, this would tally 144. In 1997, Mercado Irizarry (personal communication, 1997a) updated the tsunami death toll, with the report of an account by an "old, very well known gentleman in Aguadilla." He said, "deaths should have exceeded 100. The reason was that many persons were unaccounted for and, therefore, not included in the official toll." Even more recently, Dillon *et al.* (1999) reported that 91 people were killed by the tsunami in northwestern Puerto Rico. With the significantly higher totals of those who perished from the waves, the total fatalities for the disaster from both the earthquake and tsunami approaches a number greater than 200.

5.1.2. August 4, 1946, Dominican Republic

An even greater increase in the number of deaths from tsunamis came to light from Christine Herridge de Guerrero, Coordinator of the Dominican Disaster Mitigation Committee (personal communication, 1998). In 1946, the Dominican Republic was devastated by a violent Ms ≅ 8.1 earthquake. Small (1948) reported that great rockslides collapsed into the ocean down the northward-facing escarpment of the Península de Samaná. A tsunami generated by the sudden disturbance of the ocean floor near the Mona Passage pounded the northeast coast. Lynch and Bodle (1948), who had been accompanied by Small, reported:

> Matanzas was so much damaged by the sea that it was abandoned. Loss of life, approaching a total of 100, was greater than at any other place in the [Dominican] Republic. The intensity of the earthquake was great, but most if not all of the deaths were from drowning. Reports indicated that the rise of water probably did not exceed 2.5 meters. When one considers that the beach is very low at this point, it is not difficult to understand why so much damage was caused by the waves, which moved inland for several kilometers.

Herridge de Guerrero reported, "The ocean had receded from the Matancitas coast and the people went to collect the fish exposed. The tidal wave entered almost a kilometer inland and swept the city and several villages into the ocean, killing approximately 1,790." This substantial addition significantly increases the number of fatalities in the Caribbean. She continued, "The great tragedy is that the Dictator, Trujillo, silenced the story for political reasons, and the population was never taught to be ready for earthquakes and tsunamis." Lynch and Bodle mentioned that Trujillo had commissioned the investigation into the events at the time to allay people's fears about the continuing earthquakes.

This significant increase places the total number of fatalities at 2,556 fatalities. Table V shows the comparison of fatalities due to tsunamis during the past 150-year period. The most tsunami-prone regions, those of the U.S. West Coast, Hawai'i, and Alaska, are compared to the Caribbean region. These data do not include fatalities that are not differentiated as to whether they are due to earthquake, tsunami, or both. For example, 113 deaths reported at Cumaná, Venezuela, in 1853 or total death tolls (probably mostly from the earthquake) for that wide-ranging event variously estimated as 600, 800, 1,000, and 4,000, were not included in the tallies.

Table V. Four-region 150-year fatality comparison.

Total fatalities for the past 150 years in the Caribbean are almost five times the total of Hawai'i, Alaska, and the U.S. West Coast combined. Table V is divided at the 150-year interval and the previous data are not counted. Data that were unclear as to whether due to earthquakes, tsunamis, or both were not included in the tally. If a range of numbers was reported, the lowest number in the range was used for tallying the sum of tsunami fatalities.

Four-Region Comparison of Tsunami Fatalities
in the Past 150-Year Interval

Hawai'i		Alaska		U.S. West Coast		Caribbean	
Date	*Fatalities*	*Date*	*Fatalities*	*Date*	*Fatalities*	*Date*	*Fatalities*
		1853	?100			1853	600–4,000**
						1867	23
1868	47						
1875	5						
						1882	75–100
		1900	5				
						1906	500
						1918	91–100
1923	1						
1946*	159	1946	5	1946	1	1946	1,790
						1946	75
		1958	5				
1960	61						
		1964*	106	1964*	17		
1975	2						
						1991	2
Totals	275		221		18		2,556

Total Three Regions Combined: 514 Total Caribbean: 2,556

* Triggered creation of a regional warning system.
** Includes data from both earthquakes and tsunamis—*not* included in the tally.

Moreover, the fatalities in 1842, just 4 years prior to this 150-year period, of 200 to 300 from the tsunami at Port-de-Paix, Haiti, are not included in these data. Earlier than this period, 2,200 fatalities from tsunamis were reported, using the most conservative estimates—there were undoubtedly more—principally from:

- Jamaica, 1692, 2,000 from drowning of the 3,000 fatalities at Port Royal
- Haiti, 1770, 200 from both the earthquake and tsunami at Port-au-Prince

- Jamaica, 1780, 300 from the "irruption" of the sea at Savanna-la-Mar
- Haiti, 1842, 200 to 300 drowned at Port-de-Paix; earthquake-related deaths totaled many more than the 5,000 reported for Cap-Haïtien alone
- Antigua, 1843, 12 to 40 perished at St. John's; 1,500 earthquake-related fatalities at Pointe-à-Pitre, Guadeloupe

The tsunami that devastated Hilo, Hawai'i, in 1946 prompted the establishment of a warning system for Hawai'i. The disastrous effects of the 1964 tsunami at Skagway, Alaska, gave the same impetus for establishing a warning system for Alaska and the U.S. West Coast.

The 2,556 Caribbean fatalities amount to almost five times greater than the three other regions combined for the same 150-year period. A much lower number of fatalities was estimated by Luis Odonel Gómez (personal communication, 1998). Doubting the 1,790 figure, Odonel Gómez did not think the number of tsunami fatalities could have been so high. Father Lynch's official estimate that was "greater [at Matancitas] than at any other place," had put the death toll as "approaching a total of 100," with "most if not all of the deaths" from drowning." The 2.5 meter tsunami "moved inland for several kilometers," at Matancitas where "the beach is very low." Despite Lynch and Bodle's report of the circumstances at Matancitas, Odonel Gómez put the death toll from the tsunami at no more than 25. These significant differences in the estimates remain to be investigated and set to rights.

Even if the much lower numbers were used in lieu of the 1,790 reported by Herridge de Guerro, the tally of Caribbean fatalities would still be of great consequence. Using the lowest estimate, 25 tsunami deaths, the total number of deaths from tsunamis in the Caribbean would be 719 for the 150-year period, one-and-a-half times greater than the combined total of the other regions—even so, a significant loss of life.

Approximations of fatalities for natural disasters in the historical record merely provide a representation of the nature of the disaster. The exact numbers (never to be precisely known) are not as important as their portrayal of the enormity of the loss of life. Over the centuries, hundreds of thousands have been left homeless. Countless non-European colonials have been unreported as casualties, due to the methods of record keeping at the time. Untold thousands have been injured or unaccounted for, a legacy that remains little known in modern times. Considering the populace at risk, the Caribbean is vulnerable to a significant threat from an impending tsunami.

This astonishing number of fatalities must be accounted for in establishing Caribbean mitigation plans. The concern of knowledgeable people is substantial. In the *Maine Sunday Telegram,* Anderson (1997) quoted George Maul, an environmental scientist and oceanography professor, who asserted emphatically, "The potential for a catastrophe is unimaginable. We need to get the attention of heads of state and ring their bells and tell them: This is something you need to pay attention to." Unfortunately, lack of general awareness may prevail—until calamity occurs.

5.2. Tsunami descriptions

Descriptions of tsunamis and their effects can offer an impressive picture of a powerful natural hazard. Perhaps even more than statistics, the descriptions found in the historical record give an understanding of what it must have been like to behold such a force of nature. The following description vividly portrays the impact of the imposing 1856 tsunami in Honduras.

5.2.1. August 9, 1856, Criba Lagoon, Honduras

Feldman (1993), citing Anthony [1856:167–171], tells of an account by an eyewitness who powerfully depicts the circumstances of the tsunami at the eight-kilometer-wide Criba (La Ceiba?) Lagoon:

> Then came a low rumbling sound, increasing in force as it came sweeping over the land from the south, so unlike thunder that we could not mistake it. The earth commenced trembling violently. Suddenly this changed, and a positive undulatory motion succeeded that threw us to the ground. The earth rocked like waves of the sea. The water receded from the 5-mile-broad Criba Lagoon, leaving the bottom dry. Away it went far toward the sea. The little island rose in the centre like a conical hill in some vast plain; its rough sides in dark relief against the retreating water. Upon returning, from every direction, the waters rushed to a common centre, where, meeting, they rose in the air in an immense column, filling the heavens with an ocean of spray. Then falling, they came toward the land with the force of an avalanche.
>
> On came the flood, carrying rocks from the bottom, now all foam as far as the eye could reach, and now an unbroken mass as solid as a rock. The destroyer tore off masses of earth in its passage, boiling up over the hill in a perfect whirlpool, in which whole trees, branches, and even stones were tossed about like feathers. I had been there about two hours—clinging to a tree, though I recollect when the water slowly turned toward its proper bed with but little less violence than when rushing upon the land. We were then knee-deep in the angrily moving flood. Natives [Sambo Indians] informed us that the water had swept back into the interior about 15 miles [24 km]. Very little of their property was destroyed, as most of the natives build their towns on rising ground. A respectable old Spaniard reported having seen a similar occurrence on the coast some years before. Then it was witnessed by many people, several of whom were carried out to sea by the retreating current.

5.2.2. June 3, 1770, Haiti

A striking description of a tsunamigenic event came from the 1770 earthquake and tsunami that demolished Port-au-Prince, Haiti. Scherer (1912) depicted the chaotic scene, "The earth boiled and became liquid; its movement resembled waves of the ocean. Stifling dust obscured the air and made it hard to breathe." Hot springs appeared and "earth fissures emitted noxious vapors," which Vivenzio (1788) said resulted in a volcanic eruption. Scherer commented, "The misfortunes in the country were no less than those of the city. All buildings were thrown down; it was believed that the whole island had reached its hour of destruction."

The base of La Selle Massif was partly submerged, at Grand Goâve. At Léogâne, "well-built public buildings surviving the 1751 earthquake, fell to the ground. The rest presented only ruins, as at the Port-au-Prince catastrophe, and 50 died." All houses from Lake Miragoâne to Petit Goâve were razed. Only one house remained standing at Petit Goâve. It collapsed from an aftershock the next day. At Les Cayes, part of the shore sank 6 inches [15 cm], and emitted sulfurous water. The Rivière Des Carpes rose 9 feet [2.7 m] and overflowed. Landslips and rockfalls made roads impassable, and severe shocks were felt at Croix-des-Bouquets, Môle Saint-Nicolas, Port-de-Paix, Cap-Haïtien, and Île Pierre-Joseph. Waves that were reported along the shore of Golfe de la Gonâve and Arcahaie prompted Taber (1922b) to place "the sudden vertical displacement of the sea floor" in Golfe de la Gonâve. He said the convulsion might have extended into Plaine du Cul-de-Sac, where the sea inundated the 1.5 leagues [7.2 km] from the shore.

Moreau de Saint-Méry (1797), deeply concerned that people would again forget the terrible lesson, gave this cogent reminder about the urgency of learning from such cataclysms. He said that authorities forbade any construction except with wood, "In fear that the lesson of the frightful earthquake would be forgotten, as was that of 1751, after which more than half the houses were rebuilt with masonry." Moreau de Saint-Méry declared that no foundation of masonry could be built more than 2 feet [61 cm] in height. Enclosures had to be of live hedges or sufficiently isolated so that "in their fall they could not touch any structure, or else open spaces were to be left."

5.2.3. February 22, 1798, Barra de Matina, Costa Rica

Occasionally, amazement mixed with indignant outrage felt by some eyewitnesses is portrayed in their comments, as happened with a tsunami at Barra de Matina, Costa Rica. According to González Víquez (1910):

> After an earth shock, between 7 and 8 p.m., the agitated sea was in a state of roaring that left all the inhabitants completely astonished. Following in 6 to 8 minutes, the ground began a trembling, which lasted for 15 minutes, and continued with slight oscillations until 8 a.m. the next morning, causing people to expect death at any moment. At once the peak of the sea, rising high up, looked as if it would come upon us, making us realize the tendency for this place to be inundated by the sea, and consequently considered an appalling place by all the retreating inhabitants seeking high ground.

5.2.4. January 1, 1907, Port Royal, Jamaica

A vivid image of the effects of the earth spasm rapidly progressing through the open ocean is described in the following account reported by the harbor master of Port Royal. A pilot sailing his sloop of 5 tons 3 kilometers southeast of Port

Royal, Jamaica, reported that, along with his crew, he observed: "A sort of half misty cloud coming from south-southwest over the surface of the water, resembling a spray of steam or a sort of boiling appearance." They felt the boat rock and toss as if it was bumping on the ground. In a few seconds, they saw a huge wave strike Gun Cay "completely covering it from view, and almost simultaneously the Palisadoes Point was struck by the tsunami and obscured from view." An enormous cloud of dust was seen to rise from town. The same phenomenon was observed when it struck Kingston.

The feeling of hitting bottom while in the open sea noted in so many ship reports might be akin to that reported by the sloop pilot and his crew. This well-articulated description can also give insight into whether tsunamis were possibly produced accompanying a distinct progression of temblors as reported in other accounts. In 1687, another Port Royal event, Mallet (1853) reported land undulating and that vessels in the harbor felt the earthquake shock. He said the ground had been observed "to rise like the sea in a wave, as the earthquake passed along." Terrain undulation and damage to ships may indicate the possibility of a wave being generated.

5.2.5. November 18, 1867, Saint Thomas

An extraordinarily enlightening representation of the actual experience of a tsunami—and its devastation—was published in a pamphlet by William Miles Maskell (1868), a passenger hailing from New Zealand aboard the Royal Mail packet, *La Plata,* in St. Thomas Harbor. Excerpted by Paiewonsky (1979) Maskell's recollections vividly depict his never-to-be-forgotten experience. Although somewhat further condensed, an effort has been made to preserve the rich detail of Maskell's narration. An illustration from *Harper's Weekly* (1868), depicting the *La Plata* upon the huge tsunami rolling into St. Thomas Harbor is shown in Figure 20.

> We were leaning over the rail at a quarter past 2 in the afternoon, watching the antics of the coal workers, 30 or 40 Negroes from three scows alongside, briskly carrying the coal and making the surroundings ring with their songs. Had we been alongside the coaling dock instead of being in the middle of the bay, women with baskets on their heads would have done the coaling, as at Jamaica; here only two women were working with the men. These women, by the way, work twice as well as the men, whose practices appeared to be a quarter of an hour's work to half-an-hour's rest.
>
> Five minutes later, without warning, came such a shaking of the timbers of our ship as few mariners or passengers before had ever experienced! It was not a quivering of our ship, but a downright shaking which threatened to leave not one bolt in place, nor two planks together in the ship. The shock was of no less than 2½ minute's duration. It was not one continuous shock. For some 4 or 5 seconds in the middle, there was an appreciable lull. Then it came again, if anything, with redoubled force.
>
> Our Captain, who ran on deck from his cabin at the initial shock, said that the first things that met his eye were stones shot from the cliffs at Water

Island like balls from a cannon! It was some minutes before we recovered our composure so as to be able to look around. When we did so, the first glance thrown seaward showed, if not the cause, at least one principal accessory to the shock. Little Saba, an island of perhaps a mile in length, and a few hundred yards wide [1.6 km × 275 m], lying perhaps 4½ miles to the southwest [7.2 km] of our position, was in active eruption! Meanwhile precautions were being taken in the event of further earth-shocks. The ship still trembled—a slight quivering, with every now and then a stronger shake. A quarter of an hour passed. Many on board thought the danger was over. Others knew better. Anxious looks were thrown by some of us to seaward.

Then it was we saw, looming behind that island, like a gigantic wall, a tremendous tidal wave rushing towards us! A small steamer happened to be passing outside. She turned her head to sea and rode over the wave; probably because she was in deep water. A moment more and the great wave reached Little Saba, and in an instant that rock was a mass of foam 100 feet [30.5 m] high.

The confusion on board was indescribable. We all seemed to make up our minds that our last hour had come, and resigned ourselves to our fate. Some passengers climbed into the rigging, so as, if possible, to float free clear of the ship; some went forward to the bows or on the paddle boxes with the same intention. Half wild with terror, some of the workers fell to their knees and broke into frantic prayer; others raved and tore their clothes with the intensity of their fear. About twenty of the Negro workmen, blind with fear, rushed headlong into one of the lifeboats. Their action induced two or three lady passengers and children to follow them. One instant before the great wave struck us, the forward tackle of the lifeboat broke, and it fell down nearly perpendicular. In the meantime, the terrible mountain of water that rolled swiftly upon us.

We had five minutes to prepare for the shock, though it is impossible to describe how those minutes went. Down on our ship came the roaring sea, literally piled up like a wall, rolling up on us at the rate of 50 mph [83 kph], with a perpendicular face of fully 40 feet [12 m].

Fortunately for us, about half-a-mile [805 m] outside our position, Water Island, ended in a low point. This point, exposed to the full force of the wave, met it like a wedge. It was observed that the moment the wave reached the Water Island point, it broke and dropped, so that when the wave struck our ship, it was perhaps not more than 10 feet [3 m] higher than the taffrail [a guardrail around the upper part of the stern].

The water, receding from the shore like a mill race to meet the wave, turned our ship around, so that we took the onslaught on our stern. When the first rush of water came on us, it turned our ship even more, and the main force of the wave fell on our starboard quarter. Three waves in one it came, with the roar of 100 Niagaras; and contrary to all expectations, the *La Plata* rose over it like a duck and was saved!

The sea smashed the starboard quarter-deck to ribbons, twisting davits like pinwire, breaking in pieces all on that side, and flooding all the saloons and after cabins. The coaling scows, fortunately, had been cast loose; two drifted straight away and the third sank like a stone. It is a mystery how the

Figure 20. Illustration of the West India Royal Mail steamer, *La Plata*, rising upon a mighty roller together with a coal scow. The *La Plata*, anchored in the West Gregerie Channel, off Water Island, about 4 kilometers southwest of Saint Thomas Harbor, rose over the wave and survived, sustaining damage only to the upper decks. Lithograph credit, *Harper's Weekly,* 1868.

lady passengers and children hung on in safety to the life-boat with so many workers in it that had fallen just before the wave struck us. Almost all the Negro workers were thrown out instantaneously. Only three or four of them were recovered afterwards. Probably ten or eleven of them were drowned, amongst them one woman worker.

Had it not been for the point of Water Island, which broke the full force of the wave, it is more than likely that the *La Plata* must have gone, and the chances of saving life in that sea would have been almost none. It was impossible at the time of the shock to catch a word. The low rumble heard on the first appearance of the wave had increased to such an extent that no human voice could have been audible. So the wave passed and in a short time beat in an awful fury on the shore, its sound mingling with its echoes from the hills.

After the great wave passed, one of our boats was lowered immediately to pick up the Negro survivors in the water. They were taken to Water Island. In view of the possible consequences of another earth convulsion, the Captain considered it advisable to send all the passengers ashore at St. Thomas. It was approximately 5:30 p.m. when we landed on St. Thomas. Before us was every imaginable kind of wreck and debris, flooded, overturned, or smashed by the great wave!

We were told that we had a mile and a half [2.4 km] of mud and mire to trudge across before gaining the town. The usual roads and paths were entirely washed away. The terror stricken population, wandering listlessly about, seemed utterly prostrated even to answer when spoken to; darkness was closing fast when we came, wet, weary, and entirely ignorant of our future movements, to the Market Square of St. Thomas. Later, we found several of our fellow passengers, who had gone ashore early that morning. From them we learned that the loss of life was probably great, as numbers of the island's inhabitants, after the first great earth shock, had put off in boats to escape they knew not whither. The tidal wave destroyed most of these frail craft, and the great majority of their occupants must have drowned. Also vessels in the harbor had been driven ashore with the loss of many seamen.

During the whole time, the earth trembled continuously. Often a shake, stronger than usual would send us into the street. Unless the earth opened up we were tolerably safe. The earth did open in one place on Main Street! We could hear the cries of terrified inhabitants and the crash of falling walls. The utter ignorance of what was to come next, formed not the least portion of our miseries.

Not for one instant did the subterraneous rumbling cease; not once was the trembling earth still, as if one continuous shock all night long. At intervals, sometimes of 2 or 3 minutes, sometimes of half-an-hour, a louder roll announced a heavier shock, and the whole place vibrated beneath our feet. However, we felt nothing equal to that terrible shake that started the convulsion. No less than 70 distinct shocks were felt by 8 a.m., without counting the lesser vibrations, the most severe at 10 p.m. and 4 a.m. Long, indeed, and terrible was that night; deep and sincere our gratitude to God when it had all passed, leaving but its terrible memories behind.

The heat, fierce, stagnant, pitiless; the shore covered for miles with a mass of sweltering filth, exhaling under that scorching sun horrible incentives to fever and disease. Added to all was the uncertainty about our ship, which we had left leaky and perhaps in great danger. With daylight, the force of the earthquake appeared to abate. The shocks, occurring at intervals of an hour, seemed not so severe.

The main street, running parallel to the beach, presented an extraordinary sight. Though fully 30 feet [9 m] above the water's edge, and 200 to 300 yards [183–274 m] away, the whole space had been flooded several feet [1–1.2 m] deep by the wave. The agglomeration of wreckage was astounding.

Looking strangely out of place among the ruins of shops and houses, a lighter, fishing boat, or a schooner lay among the wreckage. Mixed up with the ruins were many cases of cargo washed out of ships, together with the masts and riggings from the unfortunate vessels. Under the verandah of the hotel, a huge red buoy lay atop a mass of wreckage, heaped in almost inextricable confusion.

Later, as we rowed along on our way across to the *La Plata*, fresh proof of the horrors of those two awful visitations, hurricane and tidal wave, met the eye, with masses of wreckage strewn in all directions. Here was the steamer, *Conway,* rent and torn in all her frame. There was the *Derwent*, sunk by the hurricane and raised again by the earthquake wave. Then a forest of masts and spars of sunken vessels; and that strange conglomeration of wreckage at the sunken dry dock, with the *British Empire* on the top of the heap. We looked in vain for the massive American screwpile jetty that had seemed so firm and strong the day before, and of which, now, not a vestige remained.

Finally, the *La Plata* moored alongside the remains of the Royal Mail Company's coal-wharf. We were surrounded by black, stagnant, almost putrid water, charged with all kinds of festering abominations. What remained untouched was the mountain of coal, sweltering—one might almost say fermenting, and all around us, were ten thousand horrible odours, enough to generate the most fearful epidemics. We were glad that only a few hours would be needed to draw our supply, and we would leave that fetid den for Jamaica that evening, but again we were doomed to disappointment.

The workers on shore, though offered exorbitant wages, refused to work. All energy seemed to have deserted them. We had to wait until the cooler part of the evening allowed our own crew to get the coal into the ship themselves. Night brought no cooling and as darkness closed in, the oppression seemed to increase. The earthquake continued. Night seemed to have renewed its strength.

From dark to daylight the unceasing trembling continued, as during the night before. At intervals came the shocks, and as each rolled like thunder beneath us, its terrors were supplemented by the shrieks of the wretched inhabitants on shore reaching us like the lamentations of souls in agony. With the steady fortitude of British seamen in an emergency, throughout the night our crewmen labored. Sometimes as a stronger shock made the whole ship vibrate, it required all their courage to continue the work. But they persevered.

The next day as we saw the last hawser cast off, a sense of security came over us for the first time in 50 hours. The crewmen used their sounding lines carefully, lest a new reef had been thrown up by the convulsion. All was done quietly. The experience of the last two days had been too great for any noisy demonstrations of joy now.

In silence we slipped past an American corvette [*USS De Soto*] that was lying a few cables' length [650–875 m] from us. She too had suffered severely. With great difficulty, she had been kept afloat, though with three ugly holes in her. It was doubtful when we left whether they would have to run her ashore and abandon her. Her consort, a large frigate [*USS Susquehanna*], had sustained no damage, and had left in the night for Santa Cruz, another Danish island, 40 miles [65 km] away.

Passing Little Saba about an hour later, we discovered and picked up one of our boats. She had been torn away by the waves and dashed on that erupting island. Little Saba, itself, was rent and torn, and a large crater gaped; grey lava was still coming down from it to the sea. Had not this small island acted as an off shore safety valve, few of us aboard the *La Plata* would have lived to tell the tale.

5.3. Historical insight

Other enlightening anecdotes and excerpts are included in the Caribbean Tsunami History to give insight into the rare, but often formidable, tsunamis that have been experienced in the Caribbean. One benefit of using a detailed history of tsunamis and their effects is that observations from one event, if similar circumstances apply, can be extrapolated to fill in scant or missing details. Valuable clues to avert the repetition of disasters at many places as before in historic times can be gleaned through an overview of circumstances in the Caribbean Tsunami History. These can be used by scientists, researchers, engineers, architects, land use planners, and persons drafting emergency response or mitigation plans. When tectonic events originating on a particular fault are found or reported, anticipated effects may be applied to:

- Researching historical literature or marigraph records
- Discerning which references to pursue for further information
- Obtaining further geophysical data or measurements
- Seeking locations with similar conditions or tectonic regimes
- Developing models to replicate the conditions
- Discerning when or where to effect evacuation plans

Strength and nature of the effects or geophysical conditions can be gathered for consideration in future development by:

- Discerning similar tectonic conditions in comparable locations
- Identifying other events that might have included waves
- Ascertaining locations of potential future hazardous events
- Not repeating the mistakes of history

Research can be better focused through credible extrapolation from tsunamis generated and tectonic features of a particular area, whether in historical or current times. Whether the researcher is seeking details on crustal movement, depth, likelihood for damage, and so on, hints can be found in the anecdotal documentation.

5.3.1. October 18, 1751, Azua, Dominican Republic

An example is illustrated by the devastations at Azua, in the Dominican Republic. In 1751, Azua was dealt a fatal blow. It was devastated by a powerful earthquake and overwhelmed by the resulting *raz-de-marée* [tsunami] that completely covered the city and destroyed all the houses that had been rebuilt since 1691. According to Taber (1922b), Azua's situation at the head of Bahía de Ocoa increased the tsunami's destructiveness. Azua was finally abandoned, and rebuilt farther inland. Santa-Cruz del Seybó also "underwent the same fate as Azua, and they were also left to choose a new location."

The same geophysical and tsunamigenic conditions that had increased the destructiveness of the tsunami and brought about such a catastrophe at Ocoa Bay and Azua, would pertain to the complete destruction of Azua in 1691, and perhaps also to the earlier ruin of Azua reported by Scherer (1912) for 1673 and 1684. Little information is available to confirm the possibility of any tsunamis for these events.

If similar intensities and circumstances accompanied previous events, it may be possible to extrapolate to other events with comparable features that a greater potential existed for a tsunami to have taken place. The recurrence of information about horizontal and vertical shocks in the historical record provides a picture through reiteration of circumstances. When church bells were heard ringing, earthquakes were usually associated with horizontal shocks, and a tsunami was rarely reported. Thus, when cathedral bells rang, it could be an indicator of horizontal shocks. On the other hand, when vertical shocks were reported, tsunamis were more likely to have occurred. Further inquiry could be directed to these instances.

Comparing intensities and circumstances that accompanied previous events, in the case of Cumaná, Venezuela, reveal a number of events having relatively similar features that produced tsunamis. These include the great earthquakes and tsunamis noted at the end of 16th century as well as tsunamigenic events in:

- 1530, MMI ≅ IX to X; 7.3-meter tsunami
- 1543, MMI ≅ IX
- 1766, MMI ≅ IX
- 1853, MMI ≅ X, Ms ≅ 6.7
- 1929, MMI ≅ IX, Ms ≅ 6.9

It may be possible to extrapolate to other events reported with the same high-intensity earthquakes and comparable features. Since tsunamis accompanied those events, the events with striking similarities may also have had tsunamis. For the same location in Venezuela, the following instances can be given:

- 1794, MMI ≅ IX, violent shock with its epicenter at Cumaná, Venezuela
- 1797, MMI ≅ IX, four-fifths of Cumaná destroyed by a vertical shock
- 1805, MMI ≅ VIII to IX

Chances are that a tsunami may have been produced by the violent earthquakes in these other instances with comparable circumstances. Examples from the Caribbean Tsunami History can similarly be suggested for Santiago de Cuba, Port-au-Prince, Haiti, and Port Royal, Jamaica.

It is the risks, the local hazard, and any special circumstances of past tsunamis in the historical record that can be garnered from this compendium. Learning from past events is critical to understanding the tsunami hazard in this vulnerable region.

Perhaps, an apt analogy for the value of using the 500-year Caribbean history of tsunamis may be in its comparison with a light bulb. The value of a light bulb derives from its power to radiate light. What is important about a light bulb is not the filament or glass, but the light rendered. What is important about the Caribbean Tsunami History is not how much scientific data is rendered, especially from the pre-instrumental era, but what may be learned about tsunamis.

5.4. Growing comprehension of the hazard

Tsunamis have only recently been recognized as a potential hazard, and still are not well understood by those directly affected, such as harbor masters and cruise ship captains, coastal businesses, resort managers, and the general populace. Education to alert the populace at risk is imperative (Lander and O'Loughlin, 1999c). More than one million tourists currently visit St. Thomas alone each year. Undoubtedly, many vacationers are unaware of the potential hazard, nor are they likely to know what to do or when to seek safety.

Early warning systems need to be in place to prevent populations at risk from the likelihood of suffering catastrophic harm and incalculable loss. Even a well-developed warning system could fail without a satisfactory response. People must be acquainted with what measures to take. They need to know when to evacuate, where to go for safety, and when it is safe to return.

Evidence of the growing awareness of the tsunami hazard in the Caribbean is exemplified by the recommendations from the 1997 Mayagüez, Puerto Rico, Tsunami Workshop, attended by 130 scientists and engineers from the United States, Russia, and throughout the Caribbean. These recommendations included the establishment of a warning system before the occurrence of the next tsunami disaster.

A comprehensive history of tsunami occurrences is essential to understand and assess the unique nature of the local hazard for each area, and for designing effective plans for early warning, evacuation, and emergency response systems. Table VI, after Curtis and Pelinovsky (1999), summarizes essential element for tsunami risk assessment. These must be considered in evaluating specific risks in an area and determining what action, if any, may be warranted. Since rescues at sea are often possible, mitigation plans should include search and rescue for people who are swept into the ocean from land or off piers. In the great 1867 event at St. Thomas, ships that survived the tsunami were able to rescue people

Table VI. Summary of key factors for tsunami risk assessment

To evaluate specific risks in an area and determine what action, if any, may be warranted. (After Curtis and Pelinovsky, 1999.)

Key Items for Tsunami Risk Assessment		
Event Probability		*Action*
Geological facts Historical information Recorded data	By source	Crude probability statistics developed from these data
Effective Warning Probability		*Options*
Local: < 0.4 hours		None: Education for own action
Regional: 0.4 > < 2 hours		Maybe: Quick response system
Distant: >2 hours		Good: Warning centers, evacuation systems, zones, education
Location Risk		*Action*
Coast unknown		Most areas
< 15 meters *re* mean sea level		Easily determined
In known inundation zones		In progress; all marine facilities

washed into the sea from boats that were swamped or sunk. In one instance, the U.S. squadron flagship, *Susquehanna,* was responsible for the rescue of several people from the still-surging waters in the harbor.

5.5. Awareness of tsunami indicators and risks

The most effective defense is awareness of tsunami indicators. This is especially true for near-field, or local, tsunamis. Locally generated seismic tsunamis are accompanied by a "natural" warning in the strong quaking of earthquake tremors. Although mitigation of destructive effects from local tsunamis would be difficult without an early warning system, the time before the arrival may be too brief for adequate warning, since it is limited by the short distances from the tsunamigenic origin. Only regional warning centers can provide the type of coverage necessary to be able to warn populations close to the source of a major tsunamigenic earthquake. It may never be possible to warn those who are within a travel time of about 10 minutes of the tsunamigenic source, but a regional warning center is best positioned to educate the populace in their region of what measures to take should they sense a very large earthquake and are located in a tsunami-vulnerable area (Blackford, 1996).

A tsunami cannot be outrun. People need to be able to recognize the "natural warning indicators" at once. The time for evacuation may be just a few minutes depending on the distance to the origin. Evacuation from coastal areas is vital and must be immediate after an earthquake. Probably the most effective measure is the education of the public to evacuate the shore or seek higher ground immediately if they encounter any of the natural warning indicators (Lander and O'Loughlin, 1999c). To save themselves from disastrous consequences, these are the three most helpful indications for people to be aware of are:

- Feel earthquake tremors while near the coast.
- Notice unusual movement or withdrawal of the sea from the shore.
- Hear a roaring of the sea.

For the general populace, the need for awareness is paramount. Tsunamis generated nearby will have relatively short periods, typically 5 to 15 minutes between waves. Regional tsunamis may have periods of 15 to 20 minutes or more between the waves, and teletsunamis may have intervals of approximately an hour. People sometimes return to their homes or boats after the first wave to view the damage, only to be subjected to the hazard of the oftentimes greater the second wave.

In some instances, when the sea had withdrawn a great distance, people went out to catch the fish trapped on the exposed area—a risk simply not worth taking. Subsequent waves may arrive minutes to an hour later. Amazed and intrigued by fish flapping all over the seabed left bare by the receding waters, countless numbers of people have been drowned during such folly.

Most tsunamis are small, as indicated by many marigram recordings with amplitudes of 60 to 70 centimeters. Large cresting waves, such as those pictured in the popular idea of a tsunami, are rare. Figure 21 shows an exaggerated representation of a giant cresting wave. Nevertheless, even relatively small waves can cause strong currents, and minor tsunamis may overflow the low-lying inhabited areas or damage small craft, docks, and so on, not so insignificant to owners if repair costs are not minor.

Tsunamis cause destruction in a number of ways. Small craft can be driven ashore. The force of the waves can destroy buildings, piers, bridges, cables, and other structures. In the harbors of San Francisco and Los Angeles, California, waves have caused millions of dollars in damage, principally due to yachts and fishing boats breaking free and colliding with each other and harbor structures. Damage from tsunamis can be caused through battering by water-carried debris, such as docks, logs, branches, cars, and furniture. The retreating waves can scour the supports for bridges, piers, and breakwaters, causing their failure. Communications, electricity, gas, and water services can be interrupted. Waves can cause chemical spills and ruptured storage tanks. Fires can be started in any number of ways from the onslaught, to say nothing of the devastating ecological impact that can occur to estuaries, bays, and beaches.

Another important, but often unrealized aspect of tsunamis is that the formidable backwash from a large influx can be nearly as severe as the incoming torrent of seawater. After surging ashore, its retreat is often filled with all manner of debris washed along with the tumult. The power of the retreating

Figure 21. This exaggerated rendering of the well-known Japanese depiction of a giant overcasting wave is a popular image of a tsunami, but in reality, such a wave is extremely uncommon. Great Wave Off Kanagawa, by Katsushika Hokusai (1760–1849), a color woodcut from the series "Thirty-six Views of Mount Fuji," late 1820s.

wave and its arsenal of wreckage can do incalculable damage to life and property.

The nature of tsunamis is a big factor in adding to losses. Knowing the potential size of a tsunami and how much time may be available between the occurrence of an earthquake and the arrival of a tsunami is indispensable for determining what types of mitigation plans and warning systems are needed.

Such plans necessarily should be based on a thorough comprehension of the characteristic tsunami hazard peculiar to each area within the vast Caribbean region, and should incorporate:

- Research into the history of prior events
- Marigraph records
- Amplitude estimates
- Periodicities of tsunami and earthquake occurrence
- Travel time and inundation maps
- Graphical runup modeling for each region
- Topography and near-shore bathymetry data
- Seismically active areas and fault locations
- GPS stations for crustal motion monitoring
- Loss estimation

Tsunamis and their damage recorded in historical documents are useful for studying past earthquakes and tsunamis for which no instrumental data are available.

Long term seismicity and the recurrence interval of earthquakes and tsunamis can be studied from these historical documents. Using the Caribbean Tsunami History for further study could help to:

- Provide data on historical runup and inundation
- Ascertain areas needing marigraph stations and warning centers
- Discern if correct waveheights have been reported historically
- Identify non-tectonic events, such as storm surges
- Indicate populations at risk
- Furnish information on the need to ameliorate the risks
- Provide a review for identifying areas in jeopardy
- Assist with assessment of the hazard
- Aid with evaluating tectonic and volcanic tsunamigenic potential

For instance, evaluating volcanic tsunamigenic potential would pertain especially to the increasing threat of eruption from the Kick-'em-Jenny Submarine Volcano, near Grenada, as it gradually, but inexorably, grows closer to the ocean surface.

Addressing the horrendous chances for loss of life from a Caribbean tsunami, George Maul (1999) remarked, "Do we need many more lives to be lost before taking preemptive action? Tsunami events in the Caribbean region, and indeed the North Atlantic basin, are not a matter of 'if,' but of 'when.'"

5.6. Concluding remarks

A comprehensive history of tsunami occurrences can be a valuable resource as an educational tool for local authorities and the general public. Historical and current conditions need to be evaluated for the sake of those potentially at risk. Compared to the U.S. West Coast, Hawai'i, and Alaska, considered tsunami-prone regions of the Pacific, the Caribbean has had almost five times as many fatalities from tsunamis in the past 150 years.

Caribbean Tsunamis—A 500-Year History from 1498–1998 currently contains 127 tsunami reports, including 53 tsunamis judged to be authentic occurrences. This chronicle reveals that the Caribbean has the potential for a sweeping catastrophe:

- A repeat of the 1755 Lisbon teletsunami could again reach the Caribbean Sea in 7 to 8 hours, with up to 7.6-meter waves in the Antilles.
- A landslide-generated tsunami of submarine origin, triggered by the 1867 cataclysm, could cause another awesome 19.8-meter, 5-kilometer-wide wave at Sainte-Rose, Guadeloupe, that reached the sea coast within minutes.
- St. Thomas could be deluged by a recurrence, as in 1867, of devastation from a wall of water 18.3 meters high penetrating the harbor and crashing ashore like an avalanche.
- St. Croix could face a recurrence, as in 1867, of a 9-meter tsunami that carried a warship over low buildings on the shoreline, dumped it on the first street, then the rushing backwash swept it back over the wharves, and cast it on the edge of the shore, where it remained stranded for six months.
- Puerto Rico could suffer yet again from terrible 6-meter swells along its northwestern coasts caused by a sudden collapse of the seabed.
- A magnitude 8.1 earthquake could result in tragic tsunami casualties, as in the Dominican Republic in 1946.

Without an early warning, little or nothing could be done to mitigate disaster from a repeat of any of these.

The state of tsunami preparedness in the Caribbean today is similar to that in the Pacific prior to the establishment of the Pacific Tsunami Warning System. Possibly, a tsunami warning system could be interconnected with the current hurricane warning system to conserve financial resources. The involvement of the UNESCO Intergovernmental Oceanographic Commission and its IOCARIBE (IOC Sub-Commission for the Caribbean and Adjacent Regions) has been very important for planning, coordination, and other assistance.

Knowing how much time may be available between an earthquake occurrence and the arrival-time of a tsunami is crucial for determining what types of detection and emergency response systems are needed. Emergency response plans should encompass search and rescue, since these can significantly impact loss of life. Warning and mitigation systems should be based on a thorough understanding of the unique tsunami hazard in each region, incorporating research into the history of prior events, marigraph records, travel time plots, inundation maps, graphical runup modeling for each region, topography, near-shore bathymetry data, tsunamigenic mechanisms, fault locations, seismic activity, amplitude estimates, periodicities, GPS stations for crustal motion monitoring, and loss estimations.

Marigraphic records can be used for modeling source regions and confirming major as well as the many minor tsunamis. Marigrams can give useful information on the travel times and the relative amplitudes to be expected. Information about the location of any archived records is needed. Improvements and procedures are needed for checking and saving marigrams for past and current earthquake and tsunami events.

Proper mitigation requires the awareness and education of officials and the public, coordinated among engineering, land use, and insurance programs as well as the numerous political divisions and countries in the area as well as school children, beach-goers, boaters, and those that live or vacation at the seashore. Updated fatality data amount to a death toll of 2,556 victims in the past 150 years. Dreadful as this is, it could be doubled or trebled because of increases in population and coastal development throughout the region. The tsunami risk for the Caribbean is now more significant than it has been in the past. While it often happens that the effort to mobilize a warning system comes after a catastrophe strikes, the Caribbean has the unique opportunity to be prepared in advance of the next deadly tsunami.

Education may be the most effective defense against the risk of locally generated tsunamis. Individuals need to be able to recognize tsunami warning signs at once. Earthquake tremors are a "natural" tsunami warning sign in coastal areas. People should evacuate or seek high ground immediately if they:

- Feel an earthquake tremor
- Hear a roaring sound
- Notice a recession or unusual movement of the sea.

People must be cautioned, so that they do not return to areas of risk before the danger is past. The first wave to threaten the coast may not necessarily the largest, and the backwash from a large influx can be nearly as ruthless as the incoming surge. The backwash may be filled with every kind of debris washed along with waters rushing back to the sea.

It would be wise if individual observers were encouraged to log any sightings of wave action, times, runup heights, and markings left by the seawater with as much detail as possible, and inform local authorities. These data could be added to the store of information about each area's unique form of the hazard.

Rising sea levels worldwide, the extensive development of coastal areas and intensive building near the seashore, the growth in tourism-related industries, and the vast increases in population, cruise ships, and tourists all leave the Caribbean in a state of incalculable vulnerability.

- Dire predictions of potential waves up to 10 meters from Kick-'em-Jenny Seamount eruptions, gradually growing nearer to the sub-surface limit of ocean-dampening restraint.
- Even more dire predictions for the flank of La Palma, the most volcanically active island in the Canaries, to collapse, potentially generating 50-meter mega-tsunami that would wreak unparalleled catastrophe in the Caribbean.

With the prospect of perilous tectonic activity in this seismically active region, the potential for a catastrophic tsunami is greater now than in the past. The vast area of the Caribbean is at severe risk for a terrible disaster. The relative lack release of built-up energy increases the impending threat of devastation from the hazard. The dearth of major earthquakes in recent years increases the chance that the eventual relief of strain may be sizeable. The potential danger from a major tsunami cannot be underestimated.

The implacable tick… tick… tick… of this natural hazard time bomb hastens ever onward. Only the inscrutable seismic forces active in the region will control when a devastating tsunami may overwhelm the shores and harbors of the Caribbean.

A catastrophic tsunami will strike the Caribbean.

It is only a matter of time.

津波

APPENDICES

Appendix A. Islands of the West Indies

Caribbean: *The Charibee*; French: *La Mer des Caraïbes;* Spanish: *Mare de las Antillas*
West Indies: French: *Les Antilles;* Spanish: *Las Antillas;* German: *Westindischen Inseln*
Greater Antilles: French: *Les Grandes Antilles;* Spanish: *Antillas Mayores;* German: *Die Großeren Antillen*
Lesser Antilles: French: *Les Petites Antilles;* Spanish: *Antillas Manores;* German: *Die Kleineren Antillen*
Leeward Islands: French: *Les Îles Sous-le-Vent;* Spanish: *Islas de Sotavento*
Windward Islands: French: *Les Îles Sur-le-Vent;* Spanish: *Islas de Barlovento*

The left column contains the islands and in brackets with some of their smaller islands. The right column lists the country or political association. Archaic or alternate names found in the sources are listed here in brackets. During the early history of the Caribbean, islands were sometimes called by the name of the principal settlement as with San Juan for Puerto Rico, Road Town for Tortola, Bassin (an early name for Christiansted) for Saint Croix, or St. John's for Antigua. This last may have precipitated Saint John in the U.S. Virgin Islands sometimes being mistaken for Antigua or vice versa.

A.1. Greater Antilles

Cuba [Isla de la Juventud (Isle of Pines), Archipiélago de Camagüey, Archipiélago de Sabana, Cayos de San Filipe, Archipiélago de los Canarreos, Archipiélago de los Colorodos, Jardines de la Reina]	República de Cuba
Jamaica [Pigeon Island, Morant Cays]	Jamaica [La Jamaïque, Jamaika, Jamaco]
Haiti (western Hispaniola) [Île de la Tortue, Île Pierre-Joseph, Île de la Gonâve, Grande Cayemite, Île à Vache]	République d'Haïti [Hayti, Santo Domingo, Saint-Domingue, Hispaniola, Española]
Dominican Republic (eastern Hispaniola) [Isla Beata, Isla Catalina, Isla Saona]	República Dominicana [Santo Domingo, Saint-Domingue, Haiti, Hispaniola, Española]
Puerto Rico [Isla de Vieques (Vieques; Crab Island): Caballo Rock, Corona Reef, Cucaracha Rock; Isla de Culebra: Isla de Culabrita, Cayos Norte, Luis Pena, Southwest, Northeast, Cross, Palada, The Washer, Pilot Rock; La Cordillera, Islas Hicacos, Palominos, Piñeros, Caja de Muertos, Magueyes, Mona, Monito, Desecheo]	Commonwealth of Puerto Rico (U.S. Protectorate) [Porto Rico, Portorico, San Juan]

A.2. Lesser Antilles

A.2.1. Leeward Islands (The Leewards)

British Virgin Islands:	[The Virgins, Les Îles de Vierges, Les Vierges]
Anegada	British West Indies, British Dependency
Virgin Gorda [Fallen Jerusalem, Great Dog, Mosquito, Prickly Pear]	British West Indies, British Dependency
Tortola [Norman, Salt, Peter, Cooper, Ginger, Beef, Scrub, Dog Islands, Great Camanoe, Guana, Great Thatch, Little Thatch, Frenchman's Cay]	British West Indies, British Dependency
Jost van Dyke [Little Jost van Dyke, Great Tobago, Little Tobago, Pelican Cay, Green Cay, Sandy Cay]	British West Indies, British Dependency

United States Virgin Islands:	[The Virgins, Les Îles de Vierges, Les Vierges, Danish West Indies]
Saint John [Leduck, Flanagan, Whistling Cay, Durloe Cays, Steven Cay]	United States Organized Unincorporated Territory [Saint Jan]
Saint Thomas [Water, Prince Rupert's Rock, Hassel, Little Saba, Turtledove Cay, Savana, Cockroach, Outer Brass, Inner Brass, Hans Lollik, Thatch Cay, Grass Cay, Mingo Cay, Lovango Cay, Congo Cay, Dog, Great Saint James, Little Saint James, Capellas, Buck, Frenchcap Cay]	United States Organized Unincorporated Territory [Saint Thomæ, San Tomas]
Saint Croix [Buck, Long Reef]	United States Organized Unincorporated Territory [Bassin, Sainte-Croix, Santa-Cruz]

Anguilla [Sombrero, Dog, Prickly Pear Cays, Scrub]	British West Indies, British Dependency (Formerly of Saint Kitts–Nevis–Anguilla)
Saint-Barthélemy [Île Fourchue, Île Frégate, Île Chevreux au Bonhomme, La Tortue, Île Coco]	Antilles Françaises, Guadeloupe Overseas Department de France (French West Indies, French Islands) [Saint Barts]

Saint-Martin (northern Saint Martin) [Île Tintamarre, Pine]	Antilles Françaises, Guadeloupe Overseas Department de France (French West Indies, French Islands)
Sint Maarten (southern Saint Martin)	Nederlandse Antillen (Netherlands Antilles)
Saba	Nederlandse Antillen (Netherlands Antilles)
Sint Eustatius	Nederlandse Antillen (Netherlands Antilles) [Statia]
Saint Kitts [Booby]	Saint Christopher–Nevis [Saint Christopher]
Nevis	Saint Christopher–Nevis [Nieves]
Barbuda	Antigua and Barbuda
Antigua [Long, Great Bird, Guana, Crump, Green, Middle Reef]	Antigua and Barbuda
Redonda	Antigua and Barbuda [Rockadunda, Redondo]
Montserrat	British West Indies, British Dependency [Mont Serrat]
Guadeloupe [Basse-Terre, Grand-Terre, Marie-Galante, Îles de la Petite Terre, La Désirade, Îlet-à-Fajou, Îlet-à-Kahouanne, Îlets-à-Goyaves, Îles des Saintes (Fond-du-Cure, Îlet-de-Cabrit, Terre-de-Haut, Terre-de-Basse) Grant Îlet]	Antilles Françaises, Guadeloupe, Overseas Department de France (French West Indies, French Islands) [La Guadeloupe, La Guadalupa]
Dominica	Dominica [La Dominique, Sainte-Dominique]

A.2.2. Windward Islands (The Windwards)

Martinique [Rocher de la Caravelle, Îlet Ramville, Îlet Cabrits, Rocher du Diamant]	La Martinique, Antilles Françaises, Overseas Department de France (French West Indies, French Islands) [Martinico]
Saint Lucia [Maria, Pigeon]	Saint Lucia [Sainte-Luche, Sainte-Lucie]
Barbados [Cobbler's Reef]	Barbados [Barbadoes, La Barbade]

Saint Vincent and The Grenadines **The Grenadines:** Bequia, Petit Nevis, Isle à Quatre, Pigeon, Battowia, Baliceaux, All-Awash, The Pillories, Mustique, Petit Mustique, Savan, Petit Canouan, Canouan, Catholic, Mayreau, Tobago Cays, Sail Rock, Union, Prune (Palm), Petit Saint Vincent	Saint Vincent and The Grenadines
Grenada **The Grenadines:** Petit Martinique, Carriacou, Saline, Frigate, Large, Bonaparte Rocks, Diamond, Les Tantes, Île de Ronde, The Sisters (near Kick-'em-Jenny Submarine Volcano), Caille, London Bridge, Bird	Grenada [Grenade, Île de la Grenade] Other Islands of Grenada: Green, Sugar Loaf, Sandy, Pearl's Rock (Conference), Marquis, Calivigny, Hog, Glover]

A.2.3. Southern Antilles

Tobago [Saint Gile's, Little Tobago]	Trinidad and Tobago [Tabaco]
Trinidad [Chacachacare, Huevos, Monos, Gaspar, Grande]	Trinidad and Tobago [La Trinidad, Île de la Trinité]
Bonaire [Klein Bonaire]	Nederlandse Antillen (Netherlands Antilles)
Curaçao [Klein Curaçao]	Nederlandse Antillen (Netherlands Antilles)
Aruba	The Netherlands
Dependencias Federales (Venezuelan Dependencies): Isla de Aves (Bird Island), Los Testigos, Nueva Esparta (Isla de Margarita, Isla Cubagua, Isla Coche), Isla La Tortuga, Los Frailes, La Sola, Los Hermanos, La Blanquilla, La Orchila, Los Roques, Islas de Aves, Los Monjes	República de Venezuela

A.3. Other Caribbean Islands in the Historical Record

Great Inagua [Little Inagua]	Commonwealth of The Bahamas
Grand Turk [Gibb's Cay, Long Cay, Turks Islands]	Turks and Caicos British West Indies, British Dependency
Caicos Islands [West Caicos, Providenciales, North Caicos, Middle Caicos, East Caicos, South Caicos, Ambergris Cays]	Turks and Caicos British West Indies, British Dependency
Navassa (U.S. Wildlife Refuge)	U.S. Protectorate
Cayman Islands [Grand Cayman, Little Cayman, Cayman Brac (Caiman-Brac)]	British West Indies, British Dependency [The Caymans]

A.4. Caribbean Continental Country Islands

Belize (Belize)
Ambergris Cays, Mangrove Cays, Turneffe Islands, Lighthouse Reef, Glover Reef, Ranguana Cay, North spot, Grass Cay

Colombia (Colombia)
San Andrés y Providencia (Isla de San Andrés, Cayos Albuquerque, Isla de la Providencia), Islas San Bernardo, Islas Rosario

Honduras (Honduras)
Islas Santanilla, Swan Islands (Islas del Cisne), Arrecife de la Media Luna, Cayos Cocorocuma, Cayos Vivorillo, Cayos Becerro, Cayos Cajones, Cayos Cochinos, Islas de la Bahía (Bay Islands): Isla de Utila, Isla de Roatán, Isla Barbareta, Isla de Guanaja

Mexico (Estados Unidos Mexicanos)
Isla Mujeres, Isla Cancun, Isla Cozumel, Banco Chinchorro

Nicaragua (Nicaragua)
Isla Venado, Islas del Maíz (Corn Islands), Cayos de Perlas, Cayos Miskitos, Cayos Tyras, Edinburgh Reef

Panamá (República de Panamá)
Archipiélago de San Blas, Sapodilla Islet, Archipiélago de Bocas del Toro, Islas Colón, Bocas del Toro, Bastimentos, Popa, Cayo Agua, Isla Escudo de Veraquas

Venezuela (República de Venezuela)
Isla San Carlos, Isla de Patos, Isla Antica, Isla Cotorra, Isla Venado, Isla Redonda, Isla Mariusa, Isla Iduburojo, Islas Caracas, Islas Chimanas, Islas Borrachas, Islas Píritu, Farallón Centinela

Appendix B. Interval Periods for Major Tsunamis

Significant tsunamis that have been used to tally the intervals between major tsunamis in the northern and eastern Caribbean subduction zones.

			MAJOR NORTHERN AND EASTERN CARIBBEAN TSUNAMIS			
Year	Interval (years)	Location of effects	Description and reported destruction Fatalities listed with an asterisk (*) include fatality data that does not distinguish between earthquakes and tsunamis	Max. height (m)	Max. sea recession (m)	Fatalities reported
1690		Nevis & St. Thomas	A terrible earthquake assigned an Ms ≅ 8+ because of the large mesoseismal area, one of the highest-magnitude earthquakes in the eastern Caribbean. Some persons were killed at Antigua, and Guadeloupe and the French Islands suffered as severely; also Redonda and St. Kitts. Recession of 201 m at Charlestown, Nevis, as a result of the eruption of Nevis Peak. Earliest tsunami reported at St. Thomas, which had a 16.5–18.3-m recession reported.		201	"some"at Antigua
1692	2	Jamaica	Earthquake and subsidence destroyed 90% of the buildings in Port Royal. Of the 3000 fatalities, 2000 people were killed by the earthquake and probably 1000 were due to the tsunami.	1.8	274	1000–2000*
1751	59	Dominican Republic	A violent earthquake dealt a fatal blow to Azua de Compostela [MMI ≅ VIII]. Azua was engulfed by a tsunami that overwhelmed the city destroying all the houses rebuilt since its entire destruction in 1691. Azua's situation at the head of Bahía de Ocoa increased the destructiveness of the wave. Azua was abandoned and rebuilt farther inland. Santa-Cruz del Seybó underwent the fate of Azua, and its inhabitants were left to find a new location.			
1755	4	Saba	Waves of 5–10 m reached the Caribbean from the great teletsunami generated by the massive earthquake near Lisbon, Portugal. Waves at Saba as high as 7.6 m were reported. Waves also at Antigua, St.-Barthélemy, St. Martin, Guadeloupe, Martinique, Barbados, and Dominican Republic.	7.6	1.6	
1770	15	Haiti	A terrible Ms ≅ 7.5 earthquake demolished Port-au-Prince [MMI ≅ X] where 200 killed. Waves inundated 7.2 km inland at Cul-de-Sac.			200*

1842	72	Haiti St. John	City of Port-de-Paix was covered by 5 m of water, killing 200–300. Extensive destruction was caused by the earthquake and 2-m tsunami on the north coast of Haiti killing 4,000–5,000. St. John experienced a 3.1-m tsunami.	5	60	200–300 4,000–5,000*
1867	25	Puerto Rico, St. Thomas, St. Croix & Guadeloupe	A 19.8-m wave, 5 km broad (the largest ever reported in the Caribbean) returned after the sea receded 100 m at Guadeloupe. Probably due to a submarine landslide triggered by the giant [Ms $\cong$ 7.5] shock in the Anegada trough. Waves of 9.1–18.3 penetrated 76 m inland at St. Thomas. Fatalities at St. Thomas and St. Croix totaled at least 23. Puerto Rico, Vieques, and Culebra suffered 6.1-m tsunamis. Widespread effects throughout the Antilles, occurred at Tortola (British Virgin Islands), Antigua, Saba, St.-Barthélemy, St. Martin, St. Kitts, Nevis, Martinique, Dominica, Bequia, and Grenada.	19.8	100	23
1902	35	Martinique & St. Vincent	Volcanic tsunamis produced by eruptions of Mont Pelée and The Soufrière. All save one of Saint-Pierre's 38,000 citizens were killed by the pyroclastic flow from Mont Pelée's violent eruption. The tsunami convulsed ships in Saint-Pierre Harbor. The tsunami from The Soufrière eruption wiped out the Carib Indians at St. Vincent; a few remained at St. Lucia and Dominica.			Carib Indians wiped out
1906	4	Venezuela	Tsunamis occurred at Cumaná, Carúpano, Río Caribe, and Nueva Esparta. 500 tsunami deaths. Tsunamis produced in both the Pacific and Atlantic.			500
1907	1	Jamaica	Damaging earthquake and tsunami at Port Royal Tsunami details and damage were reported at many bays on the northern coast of Jamaica.	2.5	73–93	
1918	11	Puerto Rico	A great earthquake [M $\cong$ 7.5] caused a wave of 2.4–3.3 m above sea level drowning 32 at Aguadilla, and 8 at Punta Agujereada.	6.1		91–100
1946	28	Dominican Republic & Puerto Rico	The town of Matancitas was severely damaged by an Ms $\cong$ 8.1 earthquake and 2.4-m tsunami. The ocean receded from the coast; people went to collect the stranded fish; the returning 2.4-m wave killed 1,790, washing inland 1 km. A 4.6-m tsunami pounded the northern coast of the Dominican Republic.	4.6		1,790
1946	0	as above, D.R. & P.R.	A huge Ms $\cong$ 7.9 aftershock and tsunami 4 days later killed 75 and left 20,000 homeless at the Dominican Republic. 60-cm waves at San Juan tide gauge.	0.6		75

** Includes data from both earthquakes and tsunamis.

Appendix C. Equivalent Linear Conversions

Metric equivalents listed by the non-metric unit of length. Equivalents are shown with archaic and foreign expressions in Italics. The foot-to-meter conversion factor is 0.3048.

Unit of Length	Meter (m)	Kilometer (km)	Other Equivalents plus Archaic and Foreign Terms
1 inch	0.0254		2.54 centimeters; 1/36 yard; 1 *pouce*
1 link	0.2012		7.92 inches; ¼ *vara*; ⅛ *brasse*
1 foot	0.3048		30.48 centimeters; 12 inches; 0.92 *pieds*
1 yard	0.9144		91.44 centimeters; 3 feet; ½ fathom
1 span	0.229		9 inches; 22.9 centimeters; 1 hand's spread
1 pace	1.524		5 feet; 2 steps, left and right, of a Roman legion
1 fathom	1.829		6 feet; 8 spans; 1 *estado*; 1 *faden*; 0.94 *toises*
1 rod	5.029		16½ feet; 1 pole; 1 perch; 5½ yards; ¼ chain; 25 links; 6 *varas*; 15 *pieds*
1 chain	20.12		4 rods; 66 feet; 100 links; ⅛ chain; 10 *toises*
1 furlong	201		⅛ mile; 1 *achtel meile*; 660 feet; 220 yards; 40 rods; 1 furrow plowed by an ox team; 10 chains
1 cable length	219		120 fathoms; 720 feet; 2 *cuadros*; 11 chains
1 cable length*	185		0.1 nautical mile; 608 feet *old British mariners'
1 statute mile	1,609	1.61	5,280 feet; 8 furlongs, 1760 yards; 1 *milla*
1 English mile	1,524	1.52	5,000 feet; 1000 paces; 0.95 miles; 303 rods
1 nautical mile (Admiralty)	1,852	1.85	6,076 feet; 1.2 miles; 1 *see meile*; 1 *marina* or *náutica* millas; 1 *mille*; 92 chains
1 league	4,828	4.83	3 miles; 2.6 *milles*; 1 *lieue*; 1 *legua*
1 nautical league	5,556	5.56	3 nautical miles; 3.5 miles; 1 nautical *lieue*; 1.2 *lieues*; 3 *marinas;* 300 cables' length
1 *pied*	0.333		⅓ meter; 1 *pie*; 13 inches
1 *vara*	0.846		33⅓ inches; 2 ¾ feet; ½ *brasse*; 4 links; 2½ *pies*
1 *brasse*	1.678		2 varas; 5 *pieds*; 5 *pies*; 5.5 feet; 1 *braza*
1 *toise*	1.949		6.4 feet; 1.1 fathoms; 10 links; 1/10 chain
1 *cuadro*	110		360 feet; 5.5 chains; 22 rods; ½ cable's length

METRIC EQUIVALENTS

Unit	Metric Equivalent	Other Equivalent Non-Metric Units
0.394 inch	1 cm	1/30 foot
7.9 inches	20 cm	1 link
2 feet	0.61 m	60.96 centimeters; 3 links
3.3 feet	1 m	100 centimeters; 1.1 yards; 39 inches
4 feet	1.22 m	¾ fathom; ¼ rod; ¼ pole; ¼ perch
8 feet	2.45 m	1⅓ fathoms; ½ perch; 1½ *brasses*; 1¼ *toises*
12 feet	3.67 m	4 yards; 2 *estados*; 2 fathoms; 1.88 *toises*
66 feet	20.12 m	4 rods; 100 links; 1 chain
96 feet	29.24 m	15 *toises;* 16 fathoms
0.621 miles	1 km	5 furlongs; 200 rods; 515 *toises (toisen)*
5 miles	8 km	1.67 leagues; 4.34 *milles;* 4.34 *náutica millas*
30 miles	48 km	10 leagues; 8.64 nautical leagues; 26 *milles*

Appendix D. Earthquake Scale Definitions

Although modern technology has superseded the intensity scales, the method is often used to "fill in the blanks" of the seismic historical record.

D.1. Modified Mercalli Earthquake Intensity Scale: MMI

Modified Mercalli Earthquake Intensity Scale (1931) values are given to the left; corresponding Rossi-Forell (R–F) equivalent values are the right. Used by Fiedler as MCS. Abridged. (After Richter, 1958, p. 137; Blocksma, 1989, p. 58.)

	Modified Mercalli Earthquake Intensity Scale	Equivalent R–F
I	Generally not felt; detectable only by seismographs	I
II	Feeble; noticed by few people; objects may swing if suspended	I–II
III	Slight; felt by few people, mostly indoors; similar to vibrations of a passing truck	III
IV	Moderate; felt by many people indoors but few outdoors; windows, dishes, and doors rattle; parked cars rock	IV–V
V	Rather strong; felt by nearly everyone; sleepers awakened; small unstable objects may fall and break; doors move	V–VI
VI	Strong; felt by everyone; trees sway and visibly shake; people walk unsteadily; some heavy furniture moves; some damage; windows and dishes break; books fall from shelf	VI–VII
VII	Very strong; general alarm; walls crack; difficult to stand; moderate to heavy damage to poorly constructed buildings; plaster, loose bricks, tiles, and stones fall; small landslides along slopes; water becomes turbid	VIII
VIII	Destructive; weak structures damaged; walls fall; difficult to steer cars; damage to good unbraced masonry; chimneys, monuments, towers, and elevated tanks fall; tree branches break; steep slopes crack	VIII+
IX	Ruinous; some houses collapse as ground cracks; extensive building damage; good masonry damaged seriously; foundations crack; underground pipes break serious damage to reservoirs;	IX+
X	Disastrous; most masonry, frame structures, and foundations destroyed; numerous large landslides; water thrown on banks of rivers and lakes; railroad tracks bend slightly	X
XI	Very disastrous; few buildings survive; landslides; railroad tracks bend severely; many bridges destroyed; underground pipelines completely inoperative	X
XII	Catastrophic; nearly total destruction; ground forms waves; large rock masses displaced; objects thrown in the air	X

D.2. Rossi-Forell Earthquake Intensity Scale: R–F

Abridged. (Grases, 1971. After Richer, 1958, p. 650; Reid and Taber, 1919b, p. 98)

ROSSI-FORELL EARTHQUAKE INTENSITY SCALE	
I	Microseismic shock recorded by seismographs
II	Extremely feeble shock felt by very few persons at rest
III	Very feeble shock felt by many persons at rest
IV	Feeble shock; felt by persons walking; movement or rattling of doors, windows, shutters
V	Shock of moderate intensity; felt by nearly everyone; furniture moved
VI	Fairly strong shock; ringing of bells, stopping of pendulum clocks, general awakening of those asleep; bushes and trees visibly agitated; small cracks in a mamposteria buildings of poor construction and weak mortar; overthrow of a few relatively unstable objects
VII	Strong shock; overthrow of moveable objects, such as vases, bottles; fall of plaster in many buildings; some walls of mamposteria badly cracked; cracks in mortar of some buildings of brick, stone, or concrete blocks; general panic
VIII	Very strong shock; Fall of chimneys; mamposteria walls badly shattered; some walls partly or wholly thrown down; small cracks in some concrete buildings
IX	Extremely strong shock; Partial or total destruction of mamposteria buildings; cracks passed through bricks and concrete blocks where mortar is strong; cracks in many concrete buildings, and some concrete walls thrown down
X	Shock of extreme intensity; great disaster; ruins; disturbance of the strata, fissures in the ground, rockfalls from mountains; general destruction of all buildings except those especially adapted to resist earthquake shocks; some trees broken and uprooted

D.3. Other Earthquake Intensity Scales

D.3.1. Hall Earthquake Intensity Scale

(Hall, 1907, p.10)

HALL EARTHQUAKE INTENSITY SCALE	
I	Light shock
II	Well marked shock
III	Shock sufficient to make houses rock
IV	Shock sufficient to crack walls of houses
V	Shock sufficient to throw down a few houses
VI	Shock sufficient to throw down almost all houses

D.3.2. Milne Earthquake Intensity Scale

(Grases, 1971. After Milne, 1908, p.19) Used by Montadon.

MILNE EARTHQUAKE INTENSITY SCALE	
CLASS I	Intensity sufficient to crack walls or break chimneys; destructivity usually confined to a single town or village
CLASS II	So far in excess of class I that a few buildings have fallen; the area may be represented by a radius of 20 miles
CLASS III	Earthquakes which have destroyed towns or caused wide-spread disasters; usually severe 50 to 60 miles around their origin and have produced effects similar to class I at distances of 200 to 300 miles

D.3.3. Ramírez Earthquake Intensity Scale

(Grases, 1971. After Ramírez, 1957 and 1969, p. 191)

RAMÍREZ EARTHQUAKE INTENSITY SCALE	MODIFIED MERCALLI EQUIVALENT	ROSSI-FORELL EQUIVALENT
I	III, IV, V	I, II, III, IV
II	VI, VII, VIII	V, VI, VII
III	IX, X, XI, XII	VIII, IX, X

D.3.4. Robson Earthquake Intensity Scale

(Grases, 1971. After Robson, 1964, p.785)

ROBSON EARTHQUAKE INTENSITY SCALE	
V	All earthquakes qualified by smart, severe, strong, sharp, violent
IV	All earthquakes without any qualifying word
III	All earthquakes qualified by words such as slight, mild, or feeble

Buildings erected before 1930 would fall within in the Richter Masonry C or Masonry D categories given in Grases (1971). (After Richter, 1958, p. 137)

Masonry C	Ordinary workmanship and mortar; neither reinforced or designed against horizontal forces
Masonry D	Weak materials, such as adobe; poor mortar; low standards of workmanship; weak horizontally

D.3.5. Rockwood Earthquake Intensity Scale

(Grases, 1971. After Rockwood, 1886, p. 7)

Rockwood Earthquake Intensity Scale		Rossi-Forell Equivalent
I	Very light	II–III
II	Light	IV
III	Moderate	V–VI
IV	Strong	VI–VII
V	Severe	VIII
VI	Destructive	IX–X

D.3.6. Rudolph Earthquake Intensity Scale

Abridged; in German. (Grases, 1971. After Rudolph, 1895, p.609)

Rudolph Earthquake Intensity Scale		Equivalent R–F
I	Ganz schwaches Erzittern....	III
II	Schwaches Erzittern....	IV
III	Erzittern in ganze Schiff....	IV
IV	Mässig starke Erschütterung....	IV
V	Ziemlich starke Erschütterung....	IV
VI	Starke Erschütterung....	V–VI
VII	Recht starke Erschütterung durch Stösse; so dass das Gebälk kracht und es unmöglich ist, sich auf dem Verdeck aufrect stehend zu erhalten.	VII
VIII	Sehr starke Erschütterung durch Stösse; Masten, Raaen, und Takelwerk, sowie alle Gegenstände auf deck werden durch die stossende Beweng erschüttert, der Compass in Gehäuse abgeworforn....	VIII
IX	Ausserordentlich starke Erschütterung durch Stösse; das Schiff wird auf die Seite gestossendund geräth in Schwanken, es wird gehaben, verlight im Gange....	IX
X	Zerstörende Wirkung; Leute werden an deck niedergeworfen, schwere Gegenstände in die Höhe geschleudert; Kanonen von den Lafetten die Fugen des Decks springen auf, das Schiff wird leck.	X

Appendix E. Richter Earthquake Magnitude Scale

Each level indicates a tenfold increase in ground motion from the previous level in the *Richter Earthquake Magnitude Scale.* This scale of 1 to 9 represents a ground motion increase of 1 to 100 million. An approximate Richter–Modified Mercalli comparison is shown beneath the Richter Earthquake Magnitude Scale.

(After Blocksma, 1989; p.59)

RICHTER EARTHQUAKE MAGNITUDE SCALE

Richter Value	Increase in Ground Motion
9	100,000,000
8	10,000,000
7	1,000,000
6	100,000
5	10,000
4	1,000
3	100
2	10
1	1

RICHTER–MODIFIED MERCALLI COMPARISON

MMI Equivalent	MMI Characteristics	Approximate Richter Equivalent
I	Detectable	Less than 3.5
II	Feeble	3.5
III	Slight	4.3
IV	Moderate	4.5
V	Rather strong	4.8
VI	Strong	5.4
VII	Very strong	6.1
VIII	Destructive	6.5
IX	Ruinous	6.9
X	Disastrous	7.3
XI	Very disastrous	8.1
XII	Catastrophic	Greater than 8.1

Appendix F. U.S. Caribbean Territories Affected by Tsunamis

Data and Area	Tsunami Effects in Present U.S. Territories in the Caribbean in the Past 500 Years
Apr. 6, 1690 St. Thomas, U.S. Virgin Islands	First tsunami ever recorded at any present U.S. territory. A violent earthquake devastated Antigua, the "French Islands," and St. Kitts. The sea receded 16.5–18.5 m at St. Thomas. A landslide on the Nevis Peak Volcano at Nevis caused a 1-furlong [201-m] recession at Charlestown [MMI ≅ IX].
May 7, 1842 St. John Puerto Rico U.S. coasts	Devastating earthquake [Ms ≅ 8.1] and tsunami at Port-de-Paix, Haiti, with thousands of fatalities. A 3.1-m tsunami occurred at St. John, U.S. Virgin Islands. Effects at Puerto Rico, and St. Thomas. A *Raz-de-marée* was observed along the U.S. Atlantic seaboard.
Feb. 8, 1843 St. Thomas, U.S. Virgin Islands	Strongest intensity [MMI ≅ IX] earthquake [Mw ≅ 8.3] in the eastern Caribbean. At Pointe-à-Pitre, Guadeloupe, 1,500 people perished. Effects at St. Thomas, U.S. Virgin Islands, and the Antilles, from Haiti to Trinidad. Effects reported at South Carolina on the U.S. East Coast. Notably similar to the April 1690 event.
Nov. 18, 1867 St. Croix St. Thomas, U.S. Virgin Islands Puerto Rico	Greatest tsunami experienced at any U.S. Caribbean territory. Puerto Rico and the (then) Danish West Indies were affected with waveheights of 6.1–7.3 m at Puerto Rico and waveheights of 9.1–18.3 m at St. Thomas and 7.6–9.1 m at St. Croix, U.S. Virgin Islands. At least 23 were drowned at St. Thomas and St. Croix. These enormous tsunamis were generated by a massive Ms ≅ 7.5 earthquake that occurred at a point centered in a triangle formed by St. Thomas, St. Croix, and the islands of Vieques and Culebra. Closest to the tsunamigenic origin, Puerto Rico suffered as severely as St. Thomas and St. Croix, but details are scant.
Mar. 17, 1868 St. Thomas, U.S. Virgin Islands Puerto Rico	An earthquake, considered nearly as severe as the one on Nov. 18, affected Puerto Rico and Isla de Vieques [R–F ≅ IX]. A small flux and reflux was observed with inundation along the western and southern coasts at Arroyo and Naguabo. At Charlotte Amalie, St. Thomas, U.S. Virgin Islands, the recession caused the sea to drop 1.4 m, and waves of 60 cm and flooding were experienced.
Aug. 13, 1868 Puerto Rico	Tsunamis were experienced at Juan Griego, Isla de Margarita, and Río Caribe, Venezuela. The San Juan tide gauge recorded 60-cm waves. A huge earthquake in Peru generated Pacific-wide tsunami with 21-m waves at Chile, which carried the *USS Watertree* 4.8 km up the coast from Arica and 805 m inland, where it still lies today.
Mar. 11, 1874 St. Thomas, U.S. Virgin Islands	A submarine shock southeast of St. Thomas, U.S. Virgin Islands, shook the island, especially distressing ships in the suddenly turbid harbor, which had been calm just prior to the sudden onset. A little later, strong ripples from the south, lasting some time, agitated the water surface. Heavy rollers in the harbor at Dominica disrupted communication with the shore. The waves may not have been associated with the earthquake local to St. Thomas, although a submarine event may have been triggered nearer to Dominica.

Aug. 27, 1883 St. Thomas, U.S. Virgin Islands	A tidal wave was reported at St. Thomas, U.S. Virgin Islands. The sea receded from shore 3 times. Tsunamis generated by the immense volcanic eruption of Krakatoa, in the Sunda Straits near Java, propagated worldwide. At the English Channel and at Table Bay, Cape of Good Hope, Africa, 36-cm waves were recorded.
Oct. 11, 1918 Puerto Rico St. Thomas, U.S. Virgin Islands	The northwest coast was devastated by one of the most severe earthquakes [Ms ≅ 7.5] to hit Puerto Rico. Tsunami fatalities were estimated from 91–108. Punta Agujereada had a tsunami of 5.5–6.1 m; Aguadilla had waves of 2.4–3.7 m; Punta Higüero had 5.5 m waves; Punta Borinquen had 4.6 m waves; and others occurred throughout Puerto Rico. Krum Bay, St. Thomas, experienced a 1.2-m tsunami, and at Charlotte Amalie, 0.45-m waves occurred.
Oct. 24, 1918 Puerto Rico	An aftershock [R–F ≅ VIII–IX] occurred nearly as great as the Oct. 11 event. The tsunami was recorded on the tide gauge at Galveston, Texas. A steamship rolled heavily southwest of Isla Mona, Puerto Rico. Submarine cables again were cut as on the 11th.
Aug. 4, 1946 Puerto Rico	A violent earthquake [Ms ≅ 8.1] occurred at the Dominican Republic. Tsunamis of 4–5 m were experienced along the northeast coast. A 2.4-m tsunami at Matancitas drowned 1,790 people. A small tsunami was observed 36 minutes later along the western and northern coasts of Puerto Rico. The San Juan tide gauge recorded a series of 66-cm waves.
Aug. 8, 1946 Puerto Rico U.S. coasts	Aftershock [Ms ≅ 7.9] nearly as strong as the earthquake on Aug. 4, occurred at the Dominican Republic. At Santo Domingo 75 drowned. The sea withdrew 76 m at Mayagüez and a small tsunami hit 35 minutes later. Aguadilla experienced a 24-m recession and small tsunami. The San Juan tide gauge recorded a 60-cm wave. Probably at least a small wave was experience in the U.S. Virgin Islands, since waves were also recorded at tide gauges at Daytona Beach, Florida, and Atlantic City, New Jersey.
Nov. 1, 1989 Puerto Rico	Minor earthquake [Ms ≅ 5.2] occurred in the Mona Passage that generated a small tsunami. The Puerto Rico Civil Defense reported a notable augmentation of the sea level in the area of Cabo Rojo at Bahía de Boquerón, Puerto Rico.
Apr. 22, 1991 Panamá Puerto Rico St. Croix	An earthquake and a 2–3 m tsunami occurred along the Atlantic Seaboard of Central America from Panamá to Costa Rica. Most places had sizeable inundations along the low-lying terrain. Disturbance of the water lasted 4–5 hours. At Canal de El Matina, near Moín, 2 people drowned. At Isla de Magueyes, Puerto Rico, and at Limetree, St. Croix, U.S. Virgin Islands, a train of 7-cm waves were recorded 3½ hours after the shock.

THE CARIBBEAN TSUNAMI HISTORY

ORIGIN DATA	TSUNAMI DATA		
DATE & TIME LAT. & LON. SOURCE AREA	LOCATION OF DISTURBANCE	MAX HT. (m)	DESCRIPTION OF TSUNAMI AND SEISMIC EFFECTS VALIDITY AS V TSUNAMI MAGNITUDE SCALE AS I–I EARTHQUAKE INTENSITY SCALES AS MMI, R–F, HALL, MILNE, ETC.

1498 08 2/3 10.2N 61.2W Venezuela	Venezuela: Boca de la Sierpe, Pedernales, Golfo de Cariaco, Cumaná & Península de Araya		Singer reported a wave that affected Boca de la Sierpe, Pedernales, and added the notation: *"maremoto?"* [tsunami]. Humboldt reported, "Rough waves smashed the natural dike between the coasts of Cumaná and Península de Araya, opening a connection to the sea. The Gulf owes its existence to a rent of the continent attended by an irruption of the ocean." Water entered 14 leagues [68 km] into the gap, forming Golfo de Cariaco. Many dwellings collapsed, and many native Indians perished. This earliest tsunami report for the Caribbean pre-dates many other chronicles, which usually begin at 1530. Humboldt placed the rupture that opened up the Gulf of Cariaco, otherwise attributed to the 1530 event, as earlier due to the remembrance preserved by the inhabitants of Isla de Margarita and along the coasts of the Cumaná region, referred to as "very recent at the time of Christopher Columbus' third voyage," which took place in 1498. Humboldt & Bonpland, 1818; Singer *et al.*, 1983. V4.
1530 09 01 [10:00 LT (Local Time)] 10.7N 64.2W Venezuela	Venezuela: Península de Paria, Cumaná, Isla Cubagua, Isla de Margarita, Nueva Cádiz, Península de Araya, Umbral de Araya, Golfo de Cariaco, Puerto Cabello & Cariaco	7.3 7.3 7.3	Along the coasts of Paria, Cumaná, and Isla Cubagua, "On a serene and tranquil day, in an instant, at 10 a.m., the sea rose up 4 Fathom in height [7.3 m]," according to Oviedo. (At times earlier texts Fathom was used as a plural form.) Las Casas said the first shock lasted an eighth of an hour [7½ minutes] and that the sea surpassed its ordinary limits by 4 *estados* [Old Spanish fathom; 1 *estado* = 1.83 m, thus 7.3 m]: "*Grande y caudal* [vast and voluminous], it covered the whole plain, encircling half-a-league [4.8 km] around the slope of the mountains." Perrey (1843) noted the sea climbing 4 *brasses* [Old French fathom; 1 *brasse* = 1.68 m, thus 6.7 m]. Silgado said, "The sea rose like a miraculous thing to see, fell forcefully, and entered the land with a roaring that continued along with the temblors from 9 a.m. to noon." Hoff said, "People experienced the Day of Atonement in Cariaco by the land being torn to pieces and penetrated by the oceans that rose up." He noted, "the entire region, and mainly the seaboard between Puerto Cabello and Cariaco," was often subjected to the recurrence of earthquakes. At Cubagua, immersion of the coasts reduced the extent of the land mass, according to Fiedler, and the terrible movement caused the ruin of Nueva Cádiz, and devastated Cumaná [MMI ≅ IX–X], Paria, [MMI ≅ IX], and the neighboring vicinity. The Montañas de Cariaco were cleft and an immense breach was left. Humboldt referred to this

DATA	LOCATION	(m)	DESCRIPTION OF TSUNAMI AND SEISMIC EFFECTS
1530 09 01 *[continued]*			cataclysm and tsunami as earlier than 1530. Herrera said that a huge volume of "black fœtid salt water welled up out of the micaceous schist, mixed with asphaltum and accompanied by a sulphurous *[sic]* odor." The massive earthquake was experienced along the entire northern coast of South America. Ramírez placed the epicenter off the north coast of Venezuela. Singer reported subsidence at Umbral de Araya where coastlines remained altered. The sea inundated the land and entirely destroyed the new fort built near Boca del Río Manzanares by Capitán Jácome Castellón at Nueva Toledo (Cumaná), according to Benzoni. Castellanos (1959) included a canto about this event in his 1589 poems, *Elegies de Verones Illustres des Indias.* Shepherd said earthquake effects at Trinidad [MMI ≅ VII] could be inferred from comparisons with similar later events and possible effects could be inferred for the Lesser Antilles. Tegg reported an earthquake at Cuba in 1530. Grases queried whether it had been confused with this event. Rappaport, citing Pérez, reported a hurricane the day before, on Aug. 31, at Puerto Rico, with an "uncounted number of deaths by drowning." The earthquake at Cuba may have been triggered by the hurricane due to the stresses from the violence of a hurricane. Brooks stated, "A 10-foot [3-m] rise of water caused by a hurricane adds about a million tons of weight to each square mile of Earth's crust. This load effect upon an unstable sea bottom often provides the necessary trigger-action to set off an earthquake and the tidal wave that follows." The drownings at Puerto Rico could have been from a storm surge or a tsunami. It is possible that the hurricane served to trigger the tectonic movement that, in turn, may have generated the earthquake at Península de Paria, attended by the tsunami along the entire northern coast of Venezuela. Another possibility is that the hurricane trigger-action produced a local earthquake unrelated to the events at Venezuela. Grases reported Sep. 11 as the corrected Gregorian date. Benzoni, 1565; Brooks, 1919; Fiedler, 1961; Grases, 1971; Heck, 1947; Herrera y Tordesillas, 1740; Humboldt & Bonpland, 1818; Las Casas, 1876; Mallet, 1853; Milne, 1912; Oviedo y Valdes, 1547; Pérez, 1971; Perrey, 1843 & 1858; Poëy, 1857; Ramírez, 1969; Rappaport & Fernández-Partagas, 1995; Robson, 1964; Schubert, 1994; Shepherd & Lynch, 1992; Silgado, 1985; Singer *et al.*, 1983; Tegg, 1854; von Hoff, 1840. V4.

1539 11 24 [23:00 LT] 16.2N 88.5W Belize	Honduras: Cabo de Higueras & Golfo de Honduras		Molina reported a local tsunami due to an earthquake at Belize, with effects along Golfo de Honduras, generated in the North American-Caribbean Plates Boundary tectonic region. "A violent seaquake" felt by sailors aboard three ships, 40 leagues [160 km] off Cabo de Higueras, who were waiting for violent *El Norte* winds to cease, feared damage because "the sea shook" and felt as if they had crashed against rocks. Feldman, citing Pérez [1851], said, "In fear that the sea was against them," the flotilla turned and headed for port. Movement that began at the "edge of the rivers" advanced, slowly along 84 meters from N–S, wiping out massive amounts of land. Arce *et al.,* 1998; Feldman, 1993; Fernández *et al.,* 1993; Grases, 1990; Molina, 1997; Montessus de Ballore, 1888; Rojas *et al.,* 1993; Sutch Osieki, 1981. V4.
1541 12 25 10.7N 63.8W Venezuela	Nueva Cádiz (Cumaná), Isla Cubagua, Venezuela		Singer reported a possible tsunami at Isla Cubagua, with an uncertain link to an earthquake. It possibly destroyed Nueva Cádiz (although it may have occurred in 1543). An extant legend about Spanish explorers who discovered the opulent pearl beds of Cubagua in 1499, recounted their exploitation of the native divers for some 40 years—then nature took a hand, leveling the island and destroying the pearl beds by earthquake and tidal wave in 1541—possibly lending credence to 1541 for a tsunami. Humboldt & Bonpland, 1818; Singer *et al.,* 1983. V2.
1543 09 01 [23:00 LT] 10.7N 64.1W Venezuela	Venezuela: Cumaná (Nueva Cádiz) Península de Araya & Isla Cubagua		At Cumaná, a tsunami was reported, and "The sea was much higher than the land." Fiedler said the same epicenter as in 1530 was probable. Nueva Cádiz was possibly destroyed at this time. Shepherd characterized an earthquake as "doubtful," but cited Fiedler's MCS ≅ IX for Cubagua and Cumaná and "The city of Cumaná was destroyed." Singer reported subsidence and coastlines altered at Península de Araya and Cumaná, though he queried whether the effects occurred in 1541. Berninghausen, 1968; Centeno-Graü, 1969; Fiedler, 1961; Grases, 1971; Heck, 1947; Robson, 1964; Shepherd & Lynch, 1992; Singer *et al.,* 1983. V3.
End of the 16th century 10.7N 64.1W Venezuela	Cumaná, Venezuela		Frequent strong earthquakes and giant waves were noted for the latter part of the 16th century. Oral traditions in the Cumaná area said that the sea "very frequently" rose 15 to 20 fathoms [27–37 m] inundating the shores. Herrera reported waves rearing up 90 to 100 *pieds* [Old French unit; 1 *pied* = 33 cm, a foot, thus 30–33 m] that overwhelmed the shores. Hoff said, "Frequent inundations by the sea over the low shore, even climbing as high as 15 to 20 *toises*

DATA	LOCATION	(m)	DESCRIPTION OF TSUNAMI AND SEISMIC EFFECTS
End of the 16th century *[continued]*			[Old French unit; 1 *toise* = 1.95 m, or 1.1 fathoms, thus 29–39 m]. These inundations induced the inhabitants to build the part of town that stands on the declivity of the mountain." The "frequent inundations" were not specified individually, so they are given as one tsunami. Humboldt recounted a bizarre reason to account for the lack of data: "As no record exists at Cumaná, and its archives—on account of the continual devastations of the termites, or white ants—contain no document that goes back farther than 150 years, we are unacquainted with the precise dates of ancient earthquakes." His study of "equinoctial regions," conducted during travels from 1799–1804, means no archives prior to the 1650's. Centeno-Graü, 1969; Grases, 1971; Herrera y Tordesillas, 1726; Humboldt & Bonpland, 1818; von Hoff, 1840. V4.
1687 02 19 [08:00 LT] 17.8N 76.7W Jamaica	Jamaica: Spanish Town & Port Royal		Sloane reported a shock at Spanish Town (Santiago de la Véga) [MMI ≅ VI]. It was "felt all over the island at the same time, or near it. Houses were near ruin with few escaping injury." Ships in the harbor at Port Royal [MMI ≅ VI] felt it. The ground was observed "to rise like the sea in a wave as the earthquake passed along, and then it went northwards," according to the Colonial Privy Council. Undulation may have taken place on the seafloor analogous to the severe land undulations. At Port Royal, many tsunamigenic events were similar. Taber said, "The Spaniards had learned that precautionary measures were desirable in the construction of buildings. The English, who took Jamaica in 1655, did not seem to have learned this lesson." Sloane said, "The Spaniards who inhabited this island, and those neighboring, built their houses very low, consisting only of ground-rooms, their walls being made of posts, which were as much buried underground as they stood above, purposely to avoid the danger from earthquakes that attended another manner of building." *Acts of the Privy Council,* 1690; Grases, 1971; Shepherd & Lynch, 1992; Sloane, 1694; Taber, 1920. V1.
1688 03 01 17.6N 76.5W Jamaica	Port Royal, Jamaica		Sloane said, "In the span of a minute three shocks were felt through the whole island at the same time [Milne ≅ II] accompanied by a thunder-like subterranean noise and were felt at St. Christopher and the Lesser Antilles. Rudolph said, "The land appeared like *die Wellen des Meeres* [wave undulation of the sea]." Mallet said ships lying in the roadstead of Port Royal were "much injured," and a ship at sea east of Jamaica was greatly damaged by a hurricane

1688 03 01 *[continued]*			(in Mar. highly unusual). Terrain undulation and damage to the ships suggest the possibility of a wave, though not whether a seaquake, seiche, storm wave, or tsunami. Grases used Feb. 19 for both years. Hoff said Feb. 19 was an alternate date. Shepherd said Sloane "was in Jamaica both years and reported a minor earthquake on Feb. 19, 1688, but did not mention this one [Mar. 1]," concluding it was spurious. The Feb. 19 and Mar. 1 confusion may be a Julian day issue. Specifics vary somewhat between the 1687 and 1688 reports, but they may be for the same event. Berninghausen, 1968; Grases, 1971; Hall, 1907; Humboldt, 1842; Mallet, 1853; Montadon, 1962; Perrey, 1843; Pinkerton, 1812; Poëy, 1857; Robson, 1964; Rudolph, 1895; Salmons (no date); Sloane, 1694; Shepherd & Lynch, 1992; von Hoff, 1840. V1.
1690 04 06 [16:00 LT] 17.2N 62.5W Antigua	Charlestown, Nevis, & Charlotte Amalie, St. Thomas, Danish West Indies (U.S. Virgin Islands)		A terrible earthquake "laid some buildings in rubbish and some persons were killed" at Antigua [MMI ≅ IX], and at Guadeloupe [MMI ≅ IX], "The French Islands suffered as severely," according to Codrington. Dorel assigned a magnitude Ms ≅ 8+ to this violent earthquake because of "the large mesoseismal area," making it one of the highest-magnitude earthquakes in the eastern Caribbean. Oldmixon said, "The earth opened 9 feet [2.7 m] in many places on St. Kitts [MMI ≅ IX], and buried solid timber, sugar mills...." At "a rocky islet called Rockadunda (Redonda), a great part of it split and turned into the sea." The 1843 event that also extended throughout the Antilles was notably similar, with a nearly identical description of Redonda: A mighty chunk of the rocky mass of Redonda broke off and fell into the sea, which resulted in a surging rise of the ocean. Smith reported landslides on Nevis Peak that caused the sea to recede a furlong [201 m] from Charlestown [MMI ≅ IX], returning after 2 minutes. Substantial receding of the sea resulted at many locations. "The earth in the lower part of Charlestown was fissured and water discharged." Taylor said that, at Charlotte Amalie, St. Thomas, subsequent to the terrible shock at 4 p.m., "lasting a long time and cracking the walls of the company's warehouse, the sea receded a while after, to a distance of 9 or 10 fathoms [16.5–18.5 m]." Taylor's report is the earliest record of a tsunami affecting any present U.S. territory. Two letters confirming the report have been translated by Danish Archivist Poul Olsen. In *Breve og dokumenter fra Vestindien 1674–1698* (98), one dated

DATA	LOCATION	(m)	DESCRIPTION OF TSUNAMI AND SEISMIC EFFECTS
1690 04 06 *[continued]*			"Christiansfort of 8 April, 1690," from the governor of St. Thomas to the Danish West Indian and Guinea Company Board stated, "Last Sunday, which was April 6, this island has been hit by a terrible earthquake lasting one-half of a quarter-of-an-hour, so that people feared their houses would come down. Some hours after that the sea receded so much that one could walk about 9 to 10 fathoms into the sea and pick out fishes on dry land." Schubert said "retreated vertically" 16.5–18.5 m. Martinique, Barbados, and St. Lucia were less strongly affected. Montadon translated *Barbade* as Barbuda. This event has striking parallels with the great event in 1843. Both were similar in magnitude and encompassed a comparably vast area. The Gregorian date for this event would be Apr. 16. Details of this event and two earthquakes in The Leewards, dated elsewhere Feb. 26 and Apr. 10, were lumped together by Shepherd, dated Apr. 5. For Feb. 26, Mallet reported Antigua, Montserrat, Barbadoes, and St. Christopher's. "Ditto," for Apr. 10, "and at Martinique and St. Lucia." Robson reported Barbados, Martinique, and St. Lucia [all MMI ≅ IV] for Apr. 10. Codrington, 1901; Dorel, 1981; Grases, 1971; Lander & Lockridge, 1989; Mallet, 1853; Milne, 1912; Montadon, 1962; Oldmixon, 1741; Olsen, 1988; Perrey, 1843; Poëy, 1857; Robson, 1964; Robson & Tomblin, 1966; Schubert, 1994; Shepherd & Lynch, 1992; Smith, 1745; Taylor, 1888. V4.
1692 06 07 [11:43 LT] 17.9N 76.9W Jamaica	Jamaica: Port Royal, Liguanea, St. Ann's Bay & Runaway Bay Cuba	1.8	Leslie reported, "The most dreadful calamity [Ms ≅ 7.5] shook down and drowned nine-tenths of Port Royal [MMI ≅ X] in 2 minutes' time; all the wharfs *[sic]* sunk at once. There were soon several Fathom Water [5.5–7.3 m] where the streets stood." The street that suffered least was "so overflowed that the water swelled as high as the upper rooms of the houses." Mallet said, "The island rose in waves like the sea, and sank a little permanently. Louis Gelday, an inhabitant, was caught in one of the fissures of the earth, and thrown out again uninjured by a second shock." Many were swallowed up or caught by the middle and pressed to death in yawning crevasses that cleft the streets, opening and closing with incredible violence before the entire waterfront slid into the sea. Leslie noted, "About 2,000 Whites and Blacks perished in this Town;" Port Royal, "the finest Town in the West Indies, was almost intirely *[sic]* destroyed: so that in the place where rich merchants lived and fine warehouses stood, now is

1692 06 07 *[continued]*			20 Fathom Water [37 m]." A 1720 British Admiralty chart said simply: "Port Royall Sunk." It remains the best-preserved record of an early settlement in the Western Hemisphere. A watch recovered from under water was stopped at 11:43, possibly the exact time of the cataclysm. Infamous Port Royal, "wallowing in riches and abandoned to wickedness," was in an ideal location to protect its notorious commerce stemming from the days of officially sanctioned privateering. Enclosed by the Palisadoes Sandspit, more than 20 vessels were moored in the harbor when a 6-foot [1.8-m] wave rose out of the sea. Myles said it "came not from the open ocean, but from within the bay itself." What yet remained above water was engulfed. Leslie said, "The number of people who perished on this fatal day was upwards of 3,000;" more than 2,000 drowned. Capsizing ships as it traversed the bay, it caused "near total destruction of ye merchantmen and pinnaces." Sloane cited, "One Captain had one of his boys roul'd *[sic]* out of a port-hole of his ship in its motion, by which he was drowned. Afterwards seamen noted that they rode in fewer Fathom water." Leslie reported, "Almost all the ships and sloops were overset and lost in the Harbour; nor had even the Men-of-War a better fate. Myles noted, "The violent motion of the sea drove the *Swan* frigate, over the tops of many houses, and beat in her round-house; she did not overset." The strength of the tsunami is manifest by this remarkable "voyage of the *Swan,*" which had been "hove down on the beach for caulking and scraping." Myles said, "Even though she was heeled over when the wave struck, some seamen must have managed to get aboard, perhaps through inadvertent entanglement in her rigging," which "may account for the near-miraculous survival of the *Swan.*" He said the account was "well-authenticated" and noted that many of the structures were 2 to 4 stories in height. If a story was about 3 m, to rise over the houses would need a noteworthy 6- to 8-m wave. Myles added, "Mansions of wealthy colonists had even higher ceilings to allow the heat of the tropics to rise." Taking into account the great wreckage, loss of life, and breadth of coastline affected, the corresponding tsunami magnitude was I–I ≅ 2 [4–6 m] or higher. Preceded by a 1.6-km withdrawal, the tsunami reached St. Ann's Bay almost immediately. Waves arrived on the north coast sooner and with greater heights than elsewhere. Reflections between Jamaica

DATA	LOCATION	(m)	DESCRIPTION OF TSUNAMI AND SEISMIC EFFECTS
1692 06 07 *[continued]*			and Cuba accounted for the numerous waves there. A French privateer, raiding St. Ann's at the time, told of the impressive tsunami hitting the north coast of Jamaica. Leslie said that at Runaway Bay, "the filibusters on the beach were thrown to the ground." ("Filibusters" were pirates.) He noted, "For all the mishap and violence at Port Royal, the shake was stronger in the country than in the town; the mountains split, and all plantations and every house destroyed; from the openings whole rivers spouted out to a vast height in the air, accompanied with stenches and offensive smells. In the north-side, above 100 acres of land sunk, and 13 people with it. All the houses were thrown down over the island, and the surviving inhabitants were forced to dwell in huts. The two mountains at Bog Walk fell and met and so stopped up a river that it was dry for a whole day; by which means vast quantities of fish were taken up, to the great relief of the distressed." He said that at *Liganiea* [Liguanea] (Kingston), "The sea retired so from the land, that the bottom appeared dry for up to 274 m. In a minute or two, it returned and overflowed a great part of the adjacent shore." Water was ejected from wells up to heights of 40 feet [12 m]. At *Yellows* [Yallahs], a great mountain split and fell into the level land, covered several settlements, and destroyed 19 White people." A plantation was removed half-a-mile [805 m] from the place where it formerly stood. At Port Morant, a high mountain was swallowed up and in its place was left a great lake of 4 to 5 leagues across [19–24 km]. Perrey said, "The highest mountain on the isle was tumbled into the sea. One could be persuaded that the entire isle of *Jamaïque* was lowered a little." Shepherd said, "The high epicentral intensity over a relatively small area suggested a moderate-magnitude earthquake at shallow depth beneath the eastern part of Jamaica." (Perhaps moderate is a bit understated for $M_s \cong 7.5$.) Leslie concluded, "Although Jamaica suffered most, yet this terrible shock was not confined to this island alone; and felt near the same time in most parts of the habitable world." Berninghausen, 1968; Chuy and Pino, 1982; Grases, 1971; Hall, 1922; Leslie, 1740; Mallet, 1853; McDonald, 1999; Milne, 1912; Murty, 1977b; Myles, 1985; Perrey, 1843; Poëy, 1857; Rubio, 1982; Shepherd & Lynch, 1992; Sloane, 1694; Taber, 1920 & 1922b; Texas A&M nautical archives website, 2000; Tomblin & Robson, 1977; von Hoff, 1840. V4.

1701 11 09 18.7N 72.6W Hispaniola	Haiti: Île de la Gonâve & Péninsule de Tiburon		An earthquake struck Léogâne [MMI ≅ VII] and Plaine du Cul-de-Sac. Taber placed the epicenter in the Swan Island–Jamaica–South Haiti Trough, possibly in the strait between Île de la Gonâve and Péninsule de Tiburon, suggesting a submarine landslide. Since this low-lying area was frequently inundated from similar seismic occurrences, further investigation is needed to rule out the possibility of a tsunami. The road leading along the seashore from Léogâne to Petit Goâve, "sank into the sea." Grases, 1971; Moreau de Jonnès, 1843; Moreau de Saint-Méry, 1797; Scherer, 1912; Shepherd & Lynch, 1992; Taber, 1922b. V1.
1702 09 14.2N 61.1W Martinique	Martinique		Violent shocks threw down houses at Martinique [MMI ≅ VIII], were felt at sea off the coast, and felt at Guadeloupe and Barbados. A possible tsunami cannot be discounted. Charlevoix, 1730a; Labat le Pers, 1722; Mallet, 1853; Perrey, 1843; Poëy, 1857; Robson, 1964. V1.
1718 03 06 St. Vincent Martinique	St. Vincent Martinique		Very violent shocks at St. Vincent and Martinique occurred the night of Mar. 6–7, according to Mallet (1853). Near Martinique a piece of land rose from the sea with a terrible noise and sank again. At St. Vincent, the earthquake was accompanied by a "furious hurricane" and the eruption of Morne-Garou Volcano. The conjunction of earthquake, volcanic eruption, and hurricane almost certainly produced waves, but of unknown type. (A hurricane in Mar. is very unusual.) The noisy emergence of land may well have been accompanied by a tsunami. As Brooks suggested, hurricane-triggering of earthquakes may have also played a part. Mallet (1851) regarded the occurrence of earthquakes with volcanoes and vice versa as common, and laments, the lack of detail available. Brooks, 1919; Eyriès, 1824; Mallet 1851 & 1853. V2.
1726 10.6N 64.3W Venezuela	Venezuela: Salina de Araya & Península de Araya		One of only two very large "sea-waves" in Venezuela, according to Schubert, the other a 10-m tsunami in 1900. By extrapolation, 10 m could reasonably be inferred for this similar wave. The *irrupción del mar* [bursting or upsurge of the sea] destroyed "la Salina" (the saltworks), and partially ruined a Spanish fort at Península de Araya. Since no date was specified, Singer was unsure of an earthquake association. Tannehill reported only four historical hurricanes passing near Venezuela prior to 1950 [1802, 1877, 1892 & 1933]; Singer added 1813 & 1847. Since none of them were reported for 1726, it was probably not storm surge. Berninghausen, 1968; Neumann *et al.*, 1988; Schubert, 1994; Singer *et al.*, 1983; Tannehill, 1950. V3.

DATA	LOCATION	(m)	DESCRIPTION OF TSUNAMI AND SEISMIC EFFECTS
1750 10.5N 64.3W	Cumaná, Venezuela		Singer reported a wave at Cumaná, which he associated with an earthquake within the city of Cumaná itself. Despite the lack of corroborating reports (see End of the 16th century), this is likely to have been a tsunami. Schubert, 1994; Singer *et al.*, 1983. V3.
1751 09 15 [22:00 LT] 18.6N 72.8W Hispaniola Martinique Antilles			Heavy shocks were felt at *Saint-Domingue* (Haiti), Martinique [MMI ≅ IV], and the Antilles leading up to the heaviest on Nov. 21. Difficulty telling effects apart led to uncertainty about how many major shocks occurred. Taber distinguished at least two major earthquakes for the period. Different epicentral intensities were estimated for Sep., Oct., and Nov. Scherer said, "the earth trembled violently from one extremity to the other at Hispaniola." Reports of subsidence "off the coast" for this date appear to belong in Nov. Grases, 1971; Mairan, 1752; Perrey, 1843; Scherer, 1912; Shepherd & Lynch, 1992; Taber, 1922b; von Hoff, 1840. V0.
1751 10 18 [15:30 LT] 18.3N 70.7W Hispaniola	Dominican Republic: Azua, Bahía de Ocoa & Santa-Cruz del Seybó		A violent earthquake dealt a fatal blow to Azua de Compostela [MMI ≅ VIII]. Azua was engulfed by a *raz-de-marée* [tsunami] that overwhelmed the city, destroying all the houses rebuilt since its entire destruction in 1691. Tippenhauer said, "Azua was annihilated." Taber concluded that Azua's situation at the head of Bahía de Ocoa increased the destructiveness of the wave. Azua was abandoned and rebuilt farther inland. Santa-Cruz del Seybó [MMI ≅ VIII] "underwent the fate of Azua and its inhabitants were left to find a new location." Santo Domingo and other towns near the south coast were severely damaged. Earlier, two violent shocks were reported at 2 p.m., at Croix-des-Bouquets, Le Cap (Cap-Haïtien), and Port-au-Prince. Poëy reported shocks at 2, 3:30, and 5 p.m. The conditions that increased the destructive power of the tsunami at the bay at Azua could pertain to the total ruin of Azua in 1691, and perhaps to the earlier devastations at Azua reported by Scherer for 1673 and 1684. Grases, 1971; Moreau de Saint-Méry, 1844; Perrey, 1843; Poëy, 1857; Rubio, 1982; Scherer, 1912; Shepherd & Lynch, 1992; Taber, 1922b; Tippenhauer, 1893. V4.
1751 11 21 [07:50 LT] 18.4N 72.8W	Haiti		The heaviest shocks wreaked havoc on this date. The Ms ≅ 8.0 mainshock were felt as far as the Antilles. Tremors lasted 1 minute and then a violent 2-minute shock ensued. Violent earthquakes since Sep. 15 had stricken Hispaniola (Langrené used *l'île de Sainte-Domingue*). Hoff said, Santo Domingo, the Gravel Sea region, Mirebalais, and Port-au- Prince were hit

1751 11 21 *[continued]* Hispaniola Antilles			hardest. Scherer said, "Houses and factories were thrown down at St.-Marc, Léogâne, and Plaine du Cul-de-Sac. Crevices formed and abundant springs of nauseous water broke forth. Great landslips occurred and the beds of the rivers changed direction." Even the strongest buildings crumbled; "one alone, which remained standing, fell the next day, Nov. 22, from a new shock. A noise like that of a canon underneath the ground was heard. The population of Port-au-Prince [MMI ≅ X] lived in tents from Nov. 22 to Dec. 8." Mallet said that at Santo Domingo, "during some of the violent shocks [Sep. 15 & 29; Oct. 8, 12, 18, 19 & 29; Nov. 19, 21, 22 & 24; Dec. 5 & 12], the earth opened and threw out hot, fœtid water; the weather was generally fine and calm, but each shock was preceded by a rushing noise like a sudden gust of wind." Taber reported, "On Nov. 21 and 22, severe earthquakes destroyed Port-au-Prince and overthrew buildings on the neighboring plains. Lyell refers to a "submergence of part of the coast at this time so as to form a bay of the sea, but it has not been possible to verify the statement. Hoff said, "At the same time, a part of the coast 20 leagues [97 km] in length, sank into the sea, which ever since then formed a bay." The collapse of a 60-mile section of coastline, thence creating a bay, could very well have produced a tsunami. Grases, 1971; Heck, 1947; Langrené, 1752; Lyell, 1830; Mairan, 1752; Mallet, 1853; Perrey, 1843; Scherer, 1912; Shepherd & Lynch, 1992; Taber, 1922b; Tippenhauer, 1893; von Hoff, 1840. V1.
1755 11 01 [09:38 LT] 39.0N 11.0W Lisbon, Portugal	West Indies Antigua Barbados Saba St. Martin St.-Barthélemy Martinique	5.0– 10.0 3.7 3.7– 6.0 6.4– 7.6 4.6	A teletsunami generated by the catastrophic Lisbon, earthquake [Ms ≅ 8.75 to 9.0] propagated across the North Atlantic Ocean. Its epicenter was placed near the Gorringe Bank off the Iberian Coast, near the Azores–Gibraltar Ridge. Myles gave the source of this colossal tectonic release as the subduction by the Eurasian Plate of the African Plate from Syria as far as the Mid-Atlantic Ridge. Davison reported that the tremendous earthquake lasted 5 minutes. Schubert estimated a travel time for the teletsunami of 9.3 hours to Antigua, at 600 kilometers per hour. Reaching the Caribbean with runup heights of 5 to 10 m, according to Davison, "The seawaves swept across the Atlantic and were observed in several of the West Indian Islands, where the usual rise of the tide is little more than 2 feet [60 cm]. At Antigua (3,540 miles) [5,700 km], they were first noticed at 7.38 p.m., Lisbon, time. The sea rose 12 feet several times, and

DATA	LOCATION	(m)	DESCRIPTION OF TSUNAMI AND SEISMIC EFFECTS
1755 11 01 *[continued]*	Guadeloupe French West Indies Dominican Republic: Samaná Bay Santiago de Cuba France Holland Friesland England Ireland Cape Bonavista, Newfoundland Atlantic Coast of North America	 3.7 2.8–3.7 1.7	every five minutes afterwards until 10 p.m., when it rose 5 feet without any violent disturbance on the surface of the water. At St. Martin, the sea retired so far that a sloop, riding at anchor in 15 feet [4.6 m] of water, was laid dry on her broadside." According to Bevis, "At Barbadoes *[sic]*, more than 1200 leagues distant [5,800 km] from Portugal, at 5:14 p.m. Lisbon, time, the sea flowed over the wharves and streets, and continued ebbing and flowing for 8 hours." At Saba, lying 5,700 km from the Iberian Coast, according to Myles was the report that "dispatches sent from the extinct volcano, Saba, spoke of waves 21 to 25 feet [6.4–7.6 m]." He said habitations on the extinct volcano were not affected because the capital of Saba, aptly named The Bottom, was built in the floor of the crater 244 m above sea level. The Captain of the Man-of-War, *Advice,* spoke of the *Agitation of the Sea at Antigua* in a letter to a friend in Portsmouth, England, one of the key sources of information about waves in the West Indies cited in many later accounts. Captain Affleck stated, "On the 1st of November last, I find you had a remarkably sudden flux and reflux of the sea and other parts of the coast which was agitated in like manner on the coast of America and all these islands. The tide rose here 12 feet [3.7 m] perpendicular several times, and returned almost immediately: the same at Barbadoes. At Martinique and most of the French Islands, it overflowed the low land, and returned quickly to its former boundaries. The people at Barbadoes were never more astonished; the rising water in Carlisle Bay appearing as black as ink, instead of clear sea-green." Humboldt noted, "At Barbadoes, where the tides rise only from 24 to 28 inches [60–70 cm], the water rose 20 feet [6.1 m] at Carlisle Bay," and it "became at the same time 'as black as ink' because, without doubt, it was mixed with petroleum or asphaltum, which abounds at the bottom of the sea, as well as on the coast of Cariaco, as near the island of Trinidad. In the West Indies, this extraordinary motion of the waters was observed 6 hours after the first shock was felt at Lisbon." Affleck went on to say, "Since I wrote this, I have taken a more particular account of the flux and reflux above-mentioned, from an observing man of this island; who remarked, that here it began at 3:30 p.m., and flowed every 5 minutes, 5 feet [1.5 m] perpendicular, till as much after 6 p.m., without any violent disturbance on the surface of the

1755 11 01 *[continued]*	Portugal: Lisbon, ,Cabo de São Vicente, Alvor & Sagres Cádiz, Spain Tangiers, Morocco Gibraltar Arquipélago da Madeira Azores	 15.2 9.1 9.1 9.1 18.3 15.2 2.7 15.2 14.6	water." The reference is ambiguous. It may refer to Antigua, the original "here" in his letter, or to Barbados," mentioned in the preceding sentence. Subsequent accounts used one or the other location. Affleck's postscript said: "In Martinique, in that remarkable flux and reflux of the sea, it was in some places dry for an [English] mile [1.5 km]; and, in others, flowed into the upper rooms of houses, and destroyed much coffee: At the island of Sabia *[sic]*, it flowed 21 feet [6.4 m]." At Barbadoes," it rose 5 or 6 feet [1.5–1.8 m], and ebbed and flowed every 5 minutes for 3 hours. The movement recurred 64 times from 2 to 10 p.m." The 5-minute periods reported for Antigua or Barbados" may have been due to a seiche set up by the teletsunami or other phenomena, since the long wavelength of a transoceanic teletsunami would have had a greater period between waves. Waves were 12 feet [3.7 m] high at Samaná Bay, Dominican Republic. At 3 p.m., the sea almost completely inundated Santiago de Cuba, according to Salterrain. Rubio considered the tsunami to be associated with "the great earthquake that battered Portugal." On the Iberian Coast of the Atlantic at Portugal, Spain, and Morocco, and at Madeira, tsunamis of 50 feet [15.2 m] were reported. At *Lisboa* [Lisbon, 40 to 60 minutes after the terrible convulsion, Myles noted that the estuary at Oeirás Bay, "suddenly and with fantastic speed emptied itself to a distance 1.6 km, causing craft of various tonnages to bottom and heel over, stranding hundreds of wriggling fish and exposing rocks never before seen." A vast throng of people, fled over the rubble to the waterfront to escape from the enveloping heat of the spreading fires caused by the earthquake. When the wave loomed up at sea, the multitude panicked. Myles stated, "Its black waters rose hellishly as the rapidly shoaling bottom piled its awful crest higher and higher." Calculated at more than 50 feet [15.2 m], the raging surge swept half-a-mile [805 m] inland, "its thunderous mass carried all before it, hurling great ships against shattered structures, leveling them, drowning and crushing thousands. Minutes later, many who had somehow survived the initial onslaught were caught in the all-embracing sweep of the backwash and drowned. The toll of lost lives was enormous [60,000 lost]. Two more waves struck about an hour apart, finding little left to ravage." Waveheights reached 30 feet [9.1 m] at Alvor, Sagres, and Cabo de São Vicente.

DATA	LOCATION	(m)	DESCRIPTION OF TSUNAMI AND SEISMIC EFFECTS
1755 11 01 *[continued]*			Ships were overturned and entangled; three times water rose and fell 16 feet [4.9 m]. Mallet said, at Gibraltar, the sea rose 7 feet [2.1 m] higher than usual, and 15 minutes after, fell extraordinarily low; the ebb and flow, "constantly becoming weaker, lasted until the following morning." For Cádiz, Spain, Humboldt cited, "A mountain of water 60 feet high [18.3 m], was seen 8 miles [12.9 km] distant at sea that threw itself impetuously on the coasts." According to Myles, the giant wave at Cádiz was said to originate from a second focus off the coast of North Africa. He said the tsunami struck the Moroccan Atlantic litoral. At *Tanger* [Tangiers], Morocco, the first immense wave of 50 feet [15.2 m] flowed 1½ miles [2.4 km] inland "along its extensive shoreline followed by 17 more waves, decreasing in height" at a cost of 9,000–11,000 lives. Waves inundated *Saffé* [Safi], Ceuta, and Oran along the North African coast. At the *Açores* [Azores], at a place with an elevation of 15 m, a 48-foot [14.6-m] wave entered inland 300 m. Myles reported the north coast of Madeira heavily damaged by 15.2-m waves. Davison reported that at Funchal, on the south coast, "At 11:45 a.m., the sea rose with a great swell and, suddenly advancing, overflowed the shore and entered the city. The height of the wave was estimated to be 18.5 feet [5.6 m] above the water at the time. After fluctuating 4 or 5 times, it gradually subsided. On the north side the swell was much greater and lasted longer than at Funchal. At first, the sea retired more than 100 paces [152 m], and then suddenly overflowed the shore, breaking down walls of the store-house and carrying away their contents." According to Bevis, "Great quantities of fishes were left on the shore and in the streets of the village of Machico." Tsunami effects were detailed from France; the "Low Countries," where the violent sea at Holland and Friesland dashed vessels together and broke buoys from their chains; England, where the force of 2.8- to 3.7-m seriously damaged or sank small craft, and sent waves to heights of 1.8 to 2.4 m to many ports, mostly on the north and south coasts of Cornwall; Ireland, where waves rose to 1.7 m, and "a large mass of water suddenly poured into Kinsale Harbor with such force that it broke cables of two vessels, each moored with two anchors, and other locations. Bevis allowed the possibility "that the seawaves were also observed at three places near the east coast of

1755 11 01 *[continued]*			England, but without reported arrival times, he was doubtful whether the water movements were those of seiches or tsunamis. Waves also reached the coasts of North America. Smallwood reported that at Cape Bonavista, Newfoundland, was subjected to unusually high seas accompanied by a phenomenon that drained the basin of Bonavista Harbor. In 10 minutes the sea returned, overflowing parts of the community. This became the subject of a popular Newfoundland folk song entitled, "A Great Big Wave Hove in Long Beach." Humboldt said "The shock was felt at almost the same moment on the coast of Sweden, Lake Ontario [between the U.S. and Canada], and *Martinico* [Martinique]." The transatlantic velocities of the seawaves, according to Davison, were calculated by D. Milne (1841) along nine paths, taking the epicenter to be in Latitude 39° N, Longitude 11° W. To Antigua, he calculated 528 feet per second [161 meters/second] and 642 feet per second [196 m/s] to Barbados. The seven other paths were calculated at 148 m/s to Lisbon, Portugal; 134 m/s to Ayamonte, Spain; 99 m/s to Funchal, Madeira; 97 m/s to Cádiz, Spain; 73 m/s to Kinsale, Ireland, and Mount's Bay, Cornwall; and 56 m/s to Plymouth, Cornwall. Berninghausen appears to have substituted feet for meters in some instances (50 feet [15.2 m] cited as 15 feet; 60 feet [18.3 m] cited as 18 feet). Acosta, 1792; Affleck, 1757; Agostinho, 1935; Bevis, 1757; Bolt *et al.*, 1975; Borlase, 1756; Belcher, 1756; Berninghausen, 1964; Bewick, 1809; Brigham, 1871; Coffman *et al.*, 1982; Chuy & Pino, 1982; Davison, 1936; Davy, 1756; d'Ulloa, 1756 & 1809; Fowke, 1755; *Gentleman's Magazine*, 1755; Heberden, 1809; Heck, 1947; Humboldt & Bonpland, 1818; Kant, 1756; Kendrick, 1956; Lyell, 1875; Mader, 1997; Mallet, 1853; Milne, 1841; Mitchell, 1761; Moreira, 1993; Myles, 1985; Pereira, 1888; Pereira de Sousa, 1911, 1913, 1914 & 1919; Perrey, 1843; Pezuela, 1866; Poëy, 1857; Reid, 1914; Richter, 1958; Robson, 1964; Rubio, 1982; Salterrain y Viñes, 1883; Schubert, 1994; Shepherd & Lynch, 1992; Smallwood, 1980; Southey, 1827; Taber, 1922a; Wolfall, 1809. V4.
1755 11 18 [04:15 LT] 42.6N 70.0W	St. Martin		Two violent shocks at 4:15 a.m. were felt in New England [MMI ≅ VIII], especially in the provinces of Massachusetts and New Hampshire (prior to statehood), according to Mallet, citing half-a-dozen sources. "A remarkable ebbing and flowing of the sea at St. Martin's

DATA	LOCATION	(m)	DESCRIPTION OF TSUNAMI AND SEISMIC EFFECTS
1755 11 18 *[continued]*			Harbour in the West Indies was supposed to be connected with this disturbance." He also reported, "The water was much agitated in the harbours of North America, and quantities of dead fish were observed." Mallet also said a ship in the Atlantic Ocean, 70 leagues [389 km] from Cape Anne, Massachusetts, experienced this earthquake. He reported that "Its total extent seems to have been about 800 miles [1,287 km] from northeast to southwest, by 550 miles [885 km] from east to west, the centre being in 43° North Latitude; it was felt in New York, Philadelphia, Chesapeake Bay in Maryland, at Annapolis Royal, and in New Scotland, at Halifax, Canada, and at Lake St. George to the west." The earthquake had been "preceded by about a minute by explosions like distant thunder." In Boston, Massachusetts, 100 chimneys were thrown down and windows and furniture much agitated. Mallet noted that an eruption was reported at Scituate, 20 or 30 miles [32–48 km] south of Boston, and that violent shocks were also felt in Portugal, Italy, and Morocco throughout that day. *Gazette de France,* 1755; *Gentleman's Magazine*, 1756; *Le Collection Académique*, 1755; *Le Journal Historique,* 1755; Mallet, 1853; Rothman, 1968. V2.
1761 03 31 [12:05 LT] 36.0N 11.0W Lisbon, Portugal	Barbados	1.2	A teletsunami traversed the North Atlantic reaching Barbados, 1,200 leagues [5,800 km] from Portugal, about 8½ hours after the great earthquake at the Galicia Bank near the epicenter of 1755. Davison reported that the shock was of great estimated duration [5 minutes], as in that Lisbon, earthquake. Effects, manifest far beyond the disturbed area, were much less widely noticed. Davison estimated the felt area as "half-a-million to a million square miles." Ships at anchor, though riding in "some" fathoms [1 fathom = 1.83 m, perhaps 7 m], were left dry at intervals at *Lisboa* [Lisbon]. Mallet noted that the sea fell and rose 8 feet [2.4 m] vertically, 1¼ hours after the terrible shock, continuing to ebb and flow to this extent at 6-minute intervals until evening. Tsunamis occurred widely in The Canaries, France, Holland, and the British Isles [a 2.7-m tsunami at Cornwall, a 1.2-m wave at the Scilly Isles, and waves of 0.6–1.2 m at Ireland]. Berninghausen said, in the Azores at Ilha de Fayal and Ilha Terceira, "The sea rose to great heights and fell so low the quays were left dry. All the boats that were hauled onto the beach at Portorico, were swept into the bay and destroyed." *Portorico* is an

1761 03 31 *[continued]*			unclear reference. It may refer to a port (?) in the Azores or, since it is an archaic spelling for Puerto Rico, *Portorico* could refer to the West Indies. This would be an additional report for the Caribbean besides Barbados. Berninghausen also reported that waves were noted "throughout the British Isles and the Low Countries as well as on Barbadoes." According to Davison, at Barbados [MMI ≅ IV], "The sea, then near low water, suddenly retired from the shore and in 3 minutes returned rising to the height of 4 feet [1.2 m]. This flux and reflux continued until 6 a.m. the next morning." Perrey reported "an earthquake together with a remarkable ebbing and flowing of the tide," at Barbados. Mallet said, "At Madeira and Terceira violent agitation was observed, and at Barbadoes (no land shock) from 4:30 p.m. to 6 the next morning. $4^h\ 30^m$ Barbadoes time equals $8^h\ 30^m$ Lisbon time; hence the agitation of the waves at Barbadoes" occurred about 8½ hours after the shock at Lisbon." Schubert listed a 12-m tsunami (most likely a decimal point error). Berninghausen, 1964; Borlase, 1761; Davison, 1936; Heck, 1947; Heberden, 1761; Mallet, 1854; Mason, 1762; Perrey, 1843; Poëy, 1857; Robson, 1964; Schubert, 1994. V4.
1766 06 11 [23:45 LT] 20.0N 75.5W Cuba Jamaica	Santiago de Cuba Port Royal, Jamaica		At Santiago de Cuba [MMI ≅ IX], the "most violent" 7-minute earthquake killed 120 people, injuring 600, and "threw down a multitude of edifices." It was felt at El Cobre [MMI ≅ VIII], Bayamo and Holguín [both MMI ≅ VII], Sancti Spiritus, and La Habana; 30 aftershocks followed. Especially severe at Port Royal [MMI ≅ VI], a 1½-minute shock at midnight caused undulation north to south throughout the island. Houses all through Jamaica were greatly shaken. Rudolph said a wave affected a ship 3 *see meilen* [5.6 km] from the coast of Jamaica. "It rolled so strongly that its gunwales and decks were immersed." Perhaps it was in shoaling waters, since the long wavelength and low waveheight of a tsunami propagating at sea would not typically affect a ship in this way. Several shallow banks dot the area. Mallet reported ships at sea "a league-and-a-half [8.3 km] from Jamaica" that felt the shock. Bacardí, 1925; Chuy, 1982; Chuy & Pino, 1982; *Diario de la Habana*, 1853; Grases, 1971; Hall, 1907; Mallet, 1854; Pezuela, 1866; Poëy, 1857; Rudolph, 1887; Salterrain y Viñes, 1883; Shepherd & Lynch, 1992; Taber, 1922b. V1.

DATA	LOCATION	(m)	DESCRIPTION OF TSUNAMI AND SEISMIC EFFECTS
1766 10 21 [03:24 LT] 06.5N 67.4W Venezuela Colombia	Venezuela: Punta Delgado & Río Orinoco		An earthquake uplifted a huge area at Cumaná [MMI ≅ IX] and left it in ruins. Hoff reported that in the sea by Cariaco, the ground was seen to elevate; consequently, Punta Delgado was enlarged, transforming the coastline. A reef jutted out in Río Guarapiche. An *islote* [islet] near Aravacoto Rock in Río Orinoco sank and disappeared beneath the waters. Disturbances of the surface were produced. Isla de Margarita, the epicentral area, San Juan Bautista, and Trinidad all had MMI ≅ VIII intensities. According to Centeno-Graü, undulations continued 1 hour, and "coincided with temblors in the Antilles." Herrera said, "The ground opened and hurled out *agua hirviendo* [boiling water]." The inhabitants camped in the streets for 15 months due to the constant tremors. Many persons perished. It was felt at Cayenne (French Guiana) [MMI ≅ VI] at 5:15 a.m., and at the same time in Trinidad, according to Perrey (1858). In Surinam [MMI ≅ VI] at 5:30 a.m., a great earthquake with violent movement occurred, undulating from east to west. Subterranean noise was heard with the first and third of three strong shocks. Venezuela and eastern Colombia were heavily affected, and it was felt with lessening intensities at Caracas [MMI ≅ VII], La Asunción (Isla de Margarita), Grenada, Barbados, Martinique, Guadeloupe, and other islands. Many seismic effects took place, including eruptions of sulfurous water near Casanay, and rockfalls at Cerro Paurarí, Península de Paria. According to Shepherd, "This was the biggest known historical earthquake in the region, remarkable for its high epicentral intensities, very large total felt area, both of which imply a high magnitude [Ms ≅ 7.5–8.0] and an intermediate focal depth." An earthquake felt along the coasts from Bogotá to French Guiana, with terrain elevation and subsidence, may have produced a wave, especially in areas prone to tsunamis. Baralt & Díaz, 1841; Centeno-Graü, 1969; Grases, 1971; Herrera y Tordesillas, 1726; Humboldt & Bonpland, 1818; Lugo & Giraldo, 1988; Mallet, 1854; Perrey, 1843 & 1858; Ramírez, 1957 & 1969; Robson, 1964; Shepherd & Lynch, 1992; Silgado, 1985; Singer *et al.*, 1983; von Hoff, 1840. V1.
1767 04 24 [06:30 LT] 06.1N 50.1W	Martinique Barbados" Surinam		Mallet said that at Martinique and Barbadoes, "The sea ebbed and flowed in an unusual way." Perrey (1858) said, "it seemed the turbulent sea would hurl itself on the shore, due to violent shocks at Surinam. At the same time, at Martinique, the waters of the sea were very agitated.

1767 04 24 *[continued]* Surinam			It was felt strongly near 7 a.m. "in the Cayéconchiennes that separate the waters of the Oyapac [also Oiapoque River] from those of the Maroni [also Marowijne River]," the mountains of French Guiana. *Le Journal Historique* gave the date as Apr. 14, probably the Julian date. *Gazette de France,* 1767; Grases, 1971; Heck, 1947; *Le Journal Historique*, 1767; Mallet, 1854; Perrey, 1843 & 1858; Robson, 1964. V2.
1769 06 03 Haiti	Port-au-Prince, Haiti		"The sea inundated 1.5 leagues [7.2 km] from the shore," at Port-au-Prince. Schubert's report appears to duplicate almost exactly reports for Jun. 3, 1770, most likely a spurious report, although a tsunami cannot be entirely ruled out. Berninghausen, 1968; Schubert, 1994. V1.
1770 06 03 [19:15 LT] 18.6N 72.8W Hispaniola	Haiti: Arcahaie, Golfe de la Gonâve & Plaine du Cul-de-Sac		An earthquake [Ms ≅ 7.5], preceded by a dull noise, like rumbling, demolished all the buildings at Port-au-Prince [MMI ≅ X], and devastated many other places. Scherer cited, "The first two terrible shocks lasted 4 minutes, and made the circle of the compass. The earth boiled and became liquid; its movement resembled waves of the ocean. Stifling dust obscured the air and made it hard to breathe. The earth was open in a thousand places, "hot springs appeared and "earth fissures emitted noxious vapors," which Vivenzio said resulted in "a volcanic eruption." Perrey noted "the opening of a volcano in the Rapion." Waves along the Golfe de la Gonâve and Arcahaie prompted Taber to place in the gulf, "the sudden vertical displacement of the sea floor." The sea inundated Plaine du Cul-de-Sac 1.5 leagues [7.2 km] from the shore. The base of La Selle Massif was partly submerged at Grand Goâve. Scherer said, "The misfortunes in the country were no less. It was believed that the whole island had reached its hour of destruction. In Léogâne [MMI ≅ X], well-built public buildings that had survived the 1751 earthquake fell to the ground. The rest presented only ruins; 50 persons died in the catastrophe. Petit Goâve did not escape unharmed. A single house remained standing. All houses from Lake Miragoâne to Petit Goâve [MMI ≅ IX] were razed." At Les Cayes [MMI ≅ VII], part of the shore sank 6 inches [15 cm], and emitted sulfurous water. Rivière Des Carpes rose 9 feet [2.7 m] and overflowed. Landslips and rockfalls made roads impassable. Severe shocks were felt at Croix-des-Bouquets, Môle St.-Nicolas, Port-de-Paix, Cap-Haïtien, and Île Pierre-Joseph. Santo Domingo and Azua felt it strongly, but with little

DATA	LOCATION	(m)	DESCRIPTION OF TSUNAMI AND SEISMIC EFFECTS
1770 06 03 *[continued]*			damage. Long said a few dwellings at Jamaica tumbled down. "They were felt at *l'île du Vent, dans l'île entière* ["the Isle of Wind, in the entire isle"]," according to Perrey. Which is Isle of the Wind? The Windwards are *Les Îles du Vent* (plural). Possibly Île de la Tortue? Navassa Island, off Anse d'Hainault, or even Jamaica lying at one end of *Paso de los Vientos* [Windward Passage] may be *l'île du Vent*, or perhaps it is Great Inagua capping the other end of the passage. Moreau de Saint-Méry reported that authorities forbade any construction except with wood, "in fear that the lesson of the frightful earthquake would be forgotten, as was that of 1751." Thus, "No foundation of masonry could be more than 2 feet [0.6 m] above the ground. The enclosures of premises must be of live hedges, unless sufficiently isolated that in their fall they could not touch any structure, or else open spaces were to be left." Currejolles reported that at Cap-Français, a water barometer was observed to sink 2½ inches [6.4 cm] immediately before the earthquake. Currejolles, 1770; Grases, 1971; Heck, 1947; Humboldt & Bonpland, 1818; Long, 1774; Mallet, 1852 & 1854; Moreau de Saint-Méry, 1797; Neumann, 1953; Perrey, 1843; Poëy, 1857; Rubio, 1982; Scherer, 1912; Shepherd & Lynch, 1992; Taber, 1922b; Tomblin & Robson, 1977; Vivenzio, 1788; von Hoff, 1840. V4.
1771 09 03 [08:00 LT] 17.6N 76.5W Jamaica	Jamaica: Port Royal & Kingston		A violent 30-second shock at *La Jamaïque* [Jamaica], caused houses to collapse and did considerable damage at Port Royal and Kingston [Hall ≅ V]. It was felt on board vessels in port. Since features of known Port Royal tsunamis were similar and ships in the harbor were affected, the possibility of a tsunami cannot be ruled out. *Gazette de France,* 1771; Grases, 1971; Hall, 1907; Long, 1774; Mallet, 1854; Perrey, 1843; Poëy, 1857; Taber, 1920. V1.
1775 02 11 Hispaniola Cuba	Haiti Cuba		Southey reported, "Three shocks of an earthquake were felt at *Española* [Hispaniola], several storehouses were thrown down, and great damage done by the sea." Taber and others noted the absence of a precise location in the Southey report. Shepherd cited this quote from Southey for Feb. 11, at Haiti [MMI ≅ VI], and also listed Santiago de Cuba [MMI ≅ VII], where shocks "destroyed houses and damaged churches from there to Santa Catalina." Berninghausen, 1968; Chuy & Pino, 1982; Heck, 1947; Neumann, 1953; Pezuela, 1866; Poëy, 1857; Rubio, 1982; Shepherd & Lynch, 1992; Southey, 1827; Taber, 1922b. V4.

1775 03 Hispaniola Martinique	Dominican Republic		Poëy reported three shocks for Mar. 1775 [no day], at *Saint-Domingue,* citing Southey, and also mentioned Martinique. Rubio reported a tsunami for Mar. 1775 at the Dominican Republic and noted an earthquake "felt over the whole island, causing damage for most." He listed this event separately from an event at "Haiti" [no month or day] where "damage was produced by waves following three earthquakes." For both reports, he noted that the locale affected was not given. Rubio's Haiti event was dated 1755 [no month or day], but listed between 1770 and 1775 events (probably a typo since a previous chronologically correct 1755 entry was given). Heck reported damage caused by a tsunami, subsequent to three earthquakes at Haiti for 1775 [no month or day]. Locations could be confused transliterations of "Santo Domingo," since the city of Santo Domingo, the Dominican Republic, Haiti, and all of Hispaniola were all at one time or another referred to as Santo Domingo. Grases defined *Española* as "the island of Haiti." Montadon aptly queried about this confusion for "St. Domingo" in 1771 asking: "*S'agit–il de la ville, du pays ou de l'île*?" ["Is it about the town, the country or the island?"] Grases, 1971; Montadon, 1962; Poëy, 1857; Rubio, 1982; Southey, 1827; Taber, 1922b. V1.
1775 12 18 Hispaniola Cuba	Haiti Cuba		Great damage from waves following three shocks at Haiti and Cuba was reported for Dec. 18. Fiedler reported an MCS ≅ VIII+ earthquake in Dec., due south of Hispaniola [09.3N 70.4W], local to Trujillo, Venezuela; Centeno-Graü said Trujillo was ruined. All event reports for 1775 had scant, similar details, muddled together with a hodgepodge of locations and dates [year only, month only, Feb. 11, and Dec. 18]. Most were apparently based on Southey's report. In unraveling these perplexing reports, validity assessments were made as follows: Assuming only one tsunami occurred, Southey's initial report, cited by Shepherd for Feb. 11, was assigned the highest validity [V4]. If two tsunamis occurred, Dec. 18 is the most likely date, thus it was assigned the validity for a possible occurrence [V2]. A third event cannot be ruled out, perhaps occurring in Mar. [V1], though three events seem doubtful. Centeno-Graü, 1969; Chuy & Pino, 1982; Fiedler, 1961; Grases, 1971; Heck, 1947; Pezuela, 1866; Poëy, 1857; Rubio, 1982; Shepherd & Lynch, 1992; Southey, 1827; Taber, 1922b. V2.

DATA	LOCATION	(m)	DESCRIPTION OF TSUNAMI AND SEISMIC EFFECTS
1780 10 03 [22:00 LT] 18.1N 78.1W Jamaica	Savanna-la-Mar, Jamaica	3.2	An earthquake "ruined Savanna-la-Mar, and tumbled down Fort-Saint-Georges, and greatly agitated the sea," according to Poëy. A 3.2-m wave swept away the town of Savanna-la-Mar, following several shocks. "The inhabitants of Savanna-la-Mar were gazing with astonishment at the sea swelling as it never had before; on a sudden, bursting through all bounds, and surmounting all obstacles, it overwhelmed the town, and swept everything away so completely on its retreat, as not to leave the smallest vestige of man, beast, or habitation behind. About 300 persons perished in this dreadful irruption. The sea flowed half-a-mile [805 m] beyond its usual limits; and so sudden and unavoidable was the destruction, although it took place at noon-day, that of the inhabitants of one gentleman's house, consisting of 10 whites and 40 Negroes, not a soul escaped." This was reported elsewhere as 10 deaths. In that era, regrettably, slaves, natives, and indentured servants were often not included in the counts. Ludlum reported 200 in Savanna-la-Mar and "several white people and some hundreds of Negroes killed in the whole parish." Seon reported upwards of 1,000 deaths in Jamaica; Evans and Millas indicated 300. Rappaport put the death toll at 1,115 for both Cuba and Jamaica. Southey said, "Where the sea did not reach, the destruction was nearly as effectual by the succeeding earthquake and hurricane. The gale began from the southeast at 1 p.m. (after the wave at noonday); at 4 p.m. it veered to the south and became a perfect tempest, which lasted at full force until near 8 p.m.: it then abated. At 10 p.m., a smart shock followed and the waters subsided. All vessels in the bay were dashed to pieces or driven ashore." Rubio said a great wave devastated the west coast of Jamaica at 10 p.m., following the hurricane, but "This is a little unsatisfactory; the wave was probably produced by effects from the tempest." Murty noted the possibility of a storm surge. Shepherd reported it as spurious. Other reports indicated an uncertain wave source. A severe hurricane took place, from Oct. 2–5. Beckford, writing about the hurricane that wrecked his house, concluded, "The sudden swelling of the sea we may reasonably attribute to the heavings of the earthquake to which likewise the general ruin of our houses may be in some measure attributed." Lyle attributed the wave to a submarine earthquake. Taber said the wave might have been seismic in origin. Since a shock

1780 10 03 *[continued]*			was distinctly reported before the huge wave, it could have been generated seismically even if a hurricane was present. As Brooks stated, "The stress from the violence of a storm causes a load effect upon an unstable sea bottom that often provides the necessary trigger-action to set off an earthquake and the resulting tidal wave." Southey's eyewitness report appears to support this (the authority given is the *Annual Register*): an earthquake at noon, ensuing wave, 7-hour hurricane, and then an aftershock and subsidence of the waters. Bochet noted earthquakes and hurricanes often occurring together in the Caribbean. Although not unheard of, this hurricane moved from east to west. *Annual Register,* 1781; Beckford, 1790; Bochet, 1843; Brooks, 1919; Cotte, 1807; Evans, 1848; Grases, 1971; Henderson, 1852; Ludlum, 1963; Lyell, 1832; Millas, 1968; Montadon, 1962; Murty, 1977b; Musson, 1999; Örsted, 1867; Piddington, 1852; Poëy, 1857; Rappaport & Fernández-Partagas, 1995; Rubio, 1982; Shepherd & Lynch, 1992; Southey, 1827; Taber, 1920; von Hoff, 1840. V3.
1781 10 02 Jamaica			Mallet reported that several severe shocks occurred and, "The sea rose to the height of 10 feet [3 m] at half-a-mile [805 m] from its ordinary beach, and swept away a number of houses." He said that it was accompanied by a tremendous hurricane. This is likely the same event as Oct. 1780, the preceding year. Mallet also noted that "von Hoff, on the authority of Cotte, gives the date as Oct. 2, 1780." Cotte, 1807; Mallet, 1854; von Hoff, 1840. V0.
1785 07 11 [03:00 LT] 17.2N 61.5W Tortola	Tortola Antigua		A violent shock at Tortola [MMI ≅ VIII], caused huge clefts in the rocks, completely separating part of the island to form a new island. Although no tsunami reports are known, such a massive failure almost certainly would have produced a tsunami. The most violent shock known at Antigua [MMI ≅ V] up until then, it was also felt at St. Thomas, St. Kitts, and "on board the ships in the neighborhood of these isles." Moreau de Saint-Méry reported it as strong at Trou, *Saint-Domingue*. Grases, 1971; Mallet, 1855; Milne, 1912; Moreau de Saint-Méry, 1797; Poëy, 1857; Reid & Taber, 1919a; Robson, 1964; Tomblin & Robson, 1977. V2.
1787 10 27 [14:20 LT] 18.3N 77.5W	Montego Bay, Jamaica		A wave was reported and vessels shaken about in the harbor. This may have been seaquake effects; a doubtful tsunami. Rubio said, "the earth undulated slightly and then shook for some time." Berninghausen, 1968; Mallet, 1855; Rubio, 1982; Tomblin & Robson, 1977. V1.

DATA	LOCATION	(m)	DESCRIPTION OF TSUNAMI AND SEISMIC EFFECTS
1798 02 22 [15:00 LT] 10.2N 82.9W Costa Rica	Barra de Matina, Costa Rica	0.5	After an earthquake, the sea became agitated at Barra de Matina [MMI ≅ VI+]. A magnitude I–I ≅ −1.0 tsunami was generated off the Caribbean coast of Costa Rica. González reported that between 7 and 8 p.m., "The sea was in a state of roaring that left all the inhabitants completely astonished." Then in 6 to 8 minutes, the ground began a trembling that lasted for 15 minutes, continuing with slight oscillations until 8 a.m., "causing people to expect death at any moment. At once the peak of the sea, rising high up, looked as if it would come upon us, making us realize the tendency for this place to be inundated by the sea, and consequently considered an appalling place by all the retreating inhabitants seeking high ground." Arce *et al.*, 1998; Boschini & Montero, 1994; Camacho & Víquez, 1993; González Víquez, 1910; Molina, 1997; Peraldo and Montero, 1994; Rojas *et al.*, 1993; Tabor, 1967. V4.
1802 03 19 17.2N 62.4W	Antigua St. Kitts		An earthquake at Antigua [MMI ≅ IV], St. Kitts, and the other Leewards, was the strongest of several in Feb. and Mar. Mallet said it was "accompanied by great agitation of the sea." A tsunami may have possibly occurred. Berninghausen, 1968; Heck, 1947; *Le Moniteur Belge*, 1802; Mallet, 1855; Robson, 1964. V1.
1802 08 15 10.3N 64.5W	Río Orinoco, Venezuela		At Cumaná [Milne ≅ II], 3 shocks were reported. The waters of Río Orinoco rose so high as to leave a large part of riverbed dry. A boat had its rudder ripped off. The ground undulated like waves of the sea. A section of ground 12 m × 130 m subsided leaving a pool in its stead. Grases, 1971; Humboldt, 1842; Mallet, 1855; Milne, 1912; Perrey, 1843 & 1858. V1.
1812 03 26 [16:07 LT] 11.5N 66.9W Venezuela Colombia	La Guaira, Venezuela		A *mar de leva excepcionale* [exceptional rise or threat of the sea] occurred at La Guaira. "Gigantic waves" broke stretches of the sea wall protecting the coast and ships in port felt the effects "as if on rocks." Nine-tenths of Caracas [MMI ≅ X–XI] was destroyed by the terrible earthquake that extended over a mammoth area. The epicenter was off La Guaira, which felt it strongly, as did Puerto Cabello, Tocuyo, Maiquetía, and other towns along the coast. Fiedler said that it was probably the worst earthquake in Caracas' history with 10,000 killed there, 4,000–5,000 at Barquisimeto, 5,000 at Mérida, 3,000 at La Guaira, and 3,000 at San Filipe. Lago de Maracaibo's level was lowered and the coastline altered. Masses of rock were hurled from the mountains and widespread seismic phenomena resulted, including the *deslizamientos*

1812 03 26 *[continued]*			[slipping] of an entire town. Humboldt reported cathedrals 50 meters high with 4-meter diameter columns destroyed at Altagracia and Trinidad. Volcanic eruptions began 35 days later at St. Vincent. Buros (no date); Centeno-Graü, 1969; Delpech (no date); Eyriès, 1825; Fiedler, 1961; Grases, 1971; Humboldt, 1819; Humboldt & Bonpland, 1818; Perrey, 1843 & 1858; Poëy, 1857; Ramírez, 1969; Rudolph, 1895; Silgado, 1985; Singer *et al.*, 1983. V4.
1812 11 11 [05:50 LT] 18.0N 76.5W Jamaica	Jamaica: Annotto Bay & Kingston		The sea was much agitated following 3 rapidly succeeding shocks at Kingston [MMI ≅ VII], and all of Jamaica. "A shock of 40 seconds' duration; almost all the houses collapsed," according to Mallet, and the Nov. 12 report that "the sea was much agitated," was probably the same account. A submarine landslide is suggested since anchorage ground sank at Annotto Bay. The anchor and 90 fathoms [165 m] of cable from the ship, *Experiment,* suddenly ran out with such force that, had the bulkhead not given way, the ship would have foundered. Hall, 1913; *Le Moniteur Belge,* 1813; Mallet, 1855; Milne, 1912; Moreau de Jonnès, 1819. V2.
1821 04 10 10.2N 82.9W	Barra de Matina, Costa Rica		A tsunami at Barra de Matina was reported. Cartago and San José sustained damage. The similarity in detail may mean this report is a duplicate for 1822; however, the month and day are different. Corroborating data could confirm this tsunami. Bermúdez León *et al.*, 2000. V3.
1822 05 07 [23:00 LT] 09.5N 83.0W Costa Rica Nicaragua Panamá	Costa Rica: Barra de Matina Nicaragua: Punta Chica & Mosquito Coast Panamá: Bocas del Toro & Istmo de Darién	0.5	A magnitude I–I ≅ −1.0 tsunami [0.5 m] at Matina, Costa Rica, was generated locally in the North Panama Deformed Belt tectonic environment. A violent earthquake [Ms ≅ 7.6] caused liquefaction with damage at Cartago and San José. Salty water and black sand erupted from deep fissures. Tremors lasted 24 hours. Rivers and bays grew and flooded; people left for Cartago, which was ruined by the earthquake. At Punta Chica (Monkey Point) on the coast of Nicaragua, lagoons dried; floating canoes were left dry. Most huts were violently cracked and twisted. The ground was rent in places. Land deformations occurred (new dunes and sand furrows and ridges), transforming the Mosquito Coast. Damage was caused at Santiago de los Caballeros, Panamá. Istmo de Darién and Bocas del Toro were both affected by the tsunami. A 30-second shock was felt at Cuba. Arce *et al.*, 1998; Bermúdez León *et al.*, 2000; Boschini & Montero, 1994; Camacho & Víquez, 1993; González Víquez, 1910; Güendel, 1986; Mallet, 1855; Molina, 1997; Montero, 1986a; Roberts, 1829; Tomblin & Robson, 1977. V4.

DATA	LOCATION	(m)	DESCRIPTION OF TSUNAMI AND SEISMIC EFFECTS
1823 11 30 [3:10 LT] 14.2N 61.1W	Saint-Pierre, Martinique		A strong undulation occurred at Martinique [MMI ≅ V] at 3:10 a.m., followed by a "tidal wave." Mallet said, "The sea rose after the shock, and occasioned some damage in the harbours." He noted that it was "preceded by a very intense noise. The heat had been suffocating all day. Abundant rain followed the earthquake and lasted ten days." *Annales de Chémie et de Physique,* 1823; Heck, 1947; Mallet, 1855; Perrey, 1847b; Robson, 1964. V4.
1824 09 13 16.5N 62.0W Guadeloupe	Plymouth, Montserrat		Mallet reported several shocks at Guadeloupe and a hurricane on the night between Sep. 7th and 8th, and more shocks at 10 p.m. on the 9th at Basse-Terre [MMI ≅ V]. Mallet also noted, "On the 13th a remarkable and irregular rise and fall of the tide at Plymouth [most likely Montserrat]." On Sep. 14–15, a hurricane caused 100 deaths in Georgia on the U.S. East Coast. If the immense pressure on the seafloor from the hurricane provided the necessary trigger-action to set off an earthquake, a tsunami may have caused this rise and fall, or the rise may have been tidal, an exceptional moon-related gravitational effect, or the rise may have been a storm wave. *Archive des Découvrirs,* 1824; Brooks, 1919; *Le Constitutionnel,* 1824; Ludlum, 1963; Mallet, 1855; Rappaport & Fernández-Partagas, 1995; Robson, 1964. V1.
1824 11 30 [03:30 LT] 14.2N 61.1W Martinique	Martinique		A severe shock occurred at Saint-Pierre, Martinique [MMI ≅ V]. Mallet (1852) noted, "A very high tide threw many ships upon the strand. It was accompanied by subterranean noise that appeared at first to proceed from the atmosphere. The earthquake was preceded by great heat, which ceased after the shock, and heavy rain began, which lasted for 10 days." Mallet said he "did not consider it superfluous to quote remarkable atmospheric conditions existing during earthquakes," noting several instances at Martinique as in this one: "The temperature of the air (which before had been very high) fell very considerably." Heck cited this account, but dated it the year before. The Nov. 30, 1823 event may be the same event. Mallet reported similar noise and weather occurrences for both events, but cited a different *tome* [volume] of *Annales de Chémie et de Physique* among several differing source reports. Subsequent accounts use one or the other phrasing or dates. A tsunami may or may not have occurred. Yet another event was reported for Martinique on Nov. 30, three years later in 1827. *Annales de Chémie et de Physique,* 1824; Férussac, 1824; Heck, 1947; Mallet, 1852 & 1855. V2.

1825 02 [19:00 LT] 12.0N 75.0W Honduras Belize	Honduras: Golfo de Honduras & Isla de Roatán		An earthquake [Ms ≅ 5.5] near Isla de Roatán [no day in Feb. specified] was reported in Molina's tsunami catalog, with the North American–Caribbean Plates Boundary Zone as the tectonic environment. Passengers aboard the ship, *Recovery,* near Roatán in the bay were "alarmed by a rumbling noise as if the vessel had been running over a reef of rocks." The shock was felt as if it had struck on a bank between 7 and 8 p.m. aboard the ship and at the same time at Belize and on the Yucatán. The description may be seaquake effects. Arce *et al.,* 1998; Grases, 1971; Mallet, 1855; Mallet & Mallet, 1858; Molina, 1997; Montessus de Ballore, 1888; Ramírez, 1969; Roberts, 1829; Rockstroch, 1902; Rojas *et al.,* 1993; Sieberg, 1932; Sutch Osieki, 1981. V1.
1825 09 20 [21:45 LT] 10.4N 61.3W Venezuela Trinidad Guyana	Demerara, Guyana		Centeno-Graü reported temblors on the coast of Güiria, Venezuela, and the ruin of Trinidad [MMI ≅ VIII]. Robson said that "scarcely a house escaped damage," at Port-of-Spain, *Île de la Trinité* [Trinidad]. The violent shocks were also felt at Tobago, St. Vincent, Barbados, and "in all the Antillean archipelago." Mallet said the shock at 10 p.m. was the most severe felt for many years in Demerara, and that it was felt at Barbadoes and Trinidad at the same time. Perrey (1858) said that in the Demerara region of Guyana, "A phænomenon lasting 3 to 4 minutes began with an oscillatory motion accompanied by a muffled sound. It ended with a movement resembling waves of the sea." Mallet said, "The sea was agitated by an oscillatory motion analogous to that felt on shore." If the adjacent sea bottom was also agitated by an oscillatory motion "as that felt on shore," there may have been a tsunami. In 1851, Mallet discussed whether "the originating impulse of an earthquake arises under the sea or on dry land gives rise to some difference in the nature and succession of the phænomena constituting the whole phase of one complete shock." He described the successive earth, sound, and sea waves "if the origin be inland"—inferring not far inland, since no great sea-wave had "been observed to take place in any earthquake whose center of impulse was inland, however violent." He called this sea wave a "forced sea-wave," and defined it as "the small wave produced on the beach at the moment the earth-wave either plunges under the sea, or *vice versâ,* emerges from it when the origin is under the ocean." Mallet then commented, "Thus

DATA	LOCATION	(m)	DESCRIPTION OF TSUNAMI AND SEISMIC EFFECTS
1825 09 20 *[continued]*			Darwin says, 'In almost every severe earthquake the neighboring waters of the sea are said to have been greatly agitated; the disturbance seems generally to have been of two kinds: first, at the instant of the shock the water swells high up on the beach with a gentle motion and then as quickly retires; secondly, some time afterwards the whole body of the sea retires from the coast and then returns in waves of overwhelming force;' Darwin well says, 'surely the waters close to land, even a steep coast, would partake of the motion of the land.'" Mallet noted that records of many of "these peculiar wave phænomena" are mentioned. *Allemange Konst en Letterbode,* 1825; *Barbadian,* 1825; Centeno-Graü, 1969; Grases, 1971; *Guyana Chronicle,* 1825; Mallet, 1852 & 1855; Mallet & Mallet, 1858; Perrey, 1843 & 1858; Poëy, 1857; *Port-of-Spain Gazette*, 1825; Robson, 1964; Shepherd & Lynch, 1992; Sieberg, 1932. V2.
1827 11 30 [03:00 LT] 14.4N 64.0W Martinique Guadeloupe	Martinique		A powerful earthquake [Ms ≅ 6.5] was felt at Martinique, Guadeloupe, Marie-Galante, Antigua, and "Santo Domingo," according to Mallet. "A violent shock of 30 or 40 seconds' duration. In Martinique, it was the most severe shock there remembered. It was also said to have affected *Terra Firma* (Colombia and Venezuela). This is said to be the tenth earthquake in the West Indies within six months." Perrey said the shocks were preceded in some places, as at Guadeloupe, by a strong *bourrasque* [violent squall of wind (Mallet's translation)]. Rudolph reported that it was felt 100 nautical leagues [564 km] west of Martinique, aboard the ship, *Le Martiniquais,* "in a place specified as shoals on the charts." Seaquake effects or a tsunami may have disturbed the ships there. Mallet said, "All the vessels near Pointe-à-Pitre and in the roadsteads of Saint-Pierre and Port Royal (probably Fort Royal [Fort-de-France] Martinique) also experienced the shock." Shepherd reported MMI ≅ VI for Fort-de-France and Saint-Pierre; MMI ≅ V for both Dominica and Barbados"; MMI ≅ IV for Guadeloupe, and MMI ≅ III for Antigua, and said "the effects of this earthquake seem to be less than previously supposed." Poëy's mention of *Saint-Domingue,* the French colony of *Haïti* thence translated as Santo Domingo may have been mixed up with Dominica, which lies between Martinique and Guadeloupe. Shepherd said Poëy translated Dominica as *La Dominique*, which Mallet then rendered back into English as Santo Domingo, and which was subsequently taken to

1827 11 30 *[continued]*			mean the Dominican Republic, "increasing the area of perceptibility by a very large factor." *Saint-Domingue* translated as Santo Domingo was common usage at the time, though usually place names, such as Cul-de-Sac or Trou (Haiti), confirmed the location. Reports are lacking from islands between Hispaniola and Antigua. If Haiti was affected, then the area of impact would be considerably broadened. The area was otherwise reported outward from Martinique to Antigua and down through The Windwards to Guyana. The breadth of area affected remains unclear. Reports of the date and time were muddled as well. Robson reported the same 3 a.m. time, but dated Dec. 1. Mallet reported a separate shock at Martinique at 10 a.m. on Dec. 1. Perrey noted, "Nov. 29, in the night." This may mean after midnight on Nov 30. Robson may have taken note of 3 a.m., Nov. 30 and deduced the wee hours of Dec. 1. The seeming discrepancy would not exist. Shepherd gave the date as Dec. 21, although he cited from the *Barbadian* of Dec. 12: "Great alarm was caused in Fort Royal. The walls of several houses were cracked." Corresponding details indicate the same event was under discussion. Shepherd also said that two hurricanes affected the area in Aug., but that "December was well outside the range of the hurricane season," apparently referencing Dec. 21. *Annales de Chémie et de Physique,* 1828; *Barbadian,* 1827; Dorel, 1981; Feuillard, 1982; Grases, 1971; Mallet, 1851 & 1855; Milne, 1912; Perrey, 1843; Poëy, 1857; Robson, 1964; Rudolph, 1887; Shepherd & Lynch, 1992; Tomblin & Robson, 1977. V1.
1829 10 26	Martinique		In a report for Sep. 14 for Martinique of "two shocks, of which the second was violent," Mallet noted, "On the 26th and 27th of Oct., a violent *raz-de-marée* [tsunami] on the coast of Martinique." This may be report for a tsunami, or even more than one, but more likely indicates waves due to a storm lasting over both days. No separate chronological earthquake report(s) were given. *Le Constitutionnel*, 1829; *Le Moniteur Belge*, 1829; Mallet, 1855. V1.
1830 04 14 [18:30 LT] 15.1N 61.2W Hispaniola	Hispaniola, Santo Domingo		"Two other shocks, more violent than those of Mar. 29 and 30," were reported by Mallet, "accompanied by a noise like distant thunder when it loses itself in the echoes of mountain ravines." Houses of brick and stone suffered severely. "The most violent shock yet felt at *Saint-Domingue* (Haiti) [MMI ≅ VII], surpassing in violence any earthquake for 20 years,"

DATA	LOCATION	(m)	DESCRIPTION OF TSUNAMI AND SEISMIC EFFECTS
1830 04 14 *[continued]*			according to Moreau de Jonnès. It lasted 4 to 5 seconds and "was accompanied by a noise like the crack of lightning, which echoed and disappeared in the mountains." Perrey said ships felt the shocks *en rade et même au large* [in the roadstead and the same at sea]. In port (apparently Santo Domingo or, if Haiti, perhaps Cap-Haïtien), waves may have affected the vessels, but they may have been a seaquake. Perrey said the effects at *Saint-Domingue* were akin to those in the widespread event of 1827. If a tsunami could be ascertained for one of the events, it might help determine probability for the other; area size for the two events could also be worked out (translation issues notwithstanding). Tippenhauer specified *Haïti,* but did not ports mentioned for 1827. Grases, 1971; Mallet, 1855; Moreau de Jonnès, 1843; Perrey, 1843; Poëy, 1857; Rudolph, 1887; Shepherd & Lynch, 1992; Tippenhauer, 1893. V1.
1831 12 03 [19:40 LT] 12.4N 61.5W Grenada	St. Kitts Trinidad		Heavy shocks [Ms ≅ 7.0] were reported at Grenada, at 7:40 p.m., and at St. Kitts, Trinidad, St. Vincent, and Guyana, along with a smart shock at Tobago. According to Mallet, at Trinidad and St. Kitts (note the large distance between reporting areas), "The sea was reported to be in a state of violent agitation and on board ship the shocks were felt as well as on land." Rudolph said, "The sea got itself into a state of fierce *Erregung* [excitement]." Mallet said that when the second shock occurred, "which was much more terrible than the first, the earth seemed to rise and fall like waves of the sea, and the strongest as well as the slightest quivered to the ground." He said a violent 3-second shock, 10 minutes later at Trinidad [MMI ≅ VII], "was followed by an oscillation perceptible for 4 to 6 seconds. After the noise which succeeded this, the second shock occurred which was even more terrible." The possibility of a tsunami cannot be ruled out. Shocks of lesser violence were felt at 10 p.m. and 2 a.m. An account in the *Grenada Free Press* reported, "St. George's [MMI ≅ VII] church bell rang and chandeliers danced." Mallet again reported, as for Montserrat, Sep. 13, 1824, unbearable heat earlier in the evening and "some heavy showers of rain followed." Robson reported the events at 7:40 p.m. and 7:50 p.m., but dated Dec. 4. Grases, 1971; *Grenada Free Press,* 1831; Kelleher *et al.,* 1973; Leonhard & Bronn, 1833; Mallet, 1855; Perrey, 1843; Poëy, 1857; *Port-of-Spain Gazette,* 1831; Robson, 1964; Rudolph, 1895; Shepherd & Lynch, 1992. V1.

1837 07 26 [12:51 UT] 14.2N 61.1W Martinique	Martinique		Mallet reported several shocks accompanied by a "terrible *raz-de-marée* [tsunami] during a dreadful hurricane." The H.M.S. *Spay* ship's log said, "in the midst of the hurricane, shocks of earthquakes were felt," and noted St. Thomas, where it "concentrated all its power, force, and fury." The *raz-de-marée* may have been a tsunami or a storm surge. The intense pressures on the sea floor can trigger earthquakes, and consequently a true tsunami. A hurricane Jul. 26–Aug. 3 caused 141 deaths at St. Thomas and Puerto Rico, and 57 at Martinique. Reid reported two hurricanes in 1837 near Santo Domingo, drowning 3 people, plus "three Haytian vessels on the coast, and only one man saved." Brooks, 1919; Mallet, 1855; Pérez, 1971; Piddington, 1852; Poëy, 1857; Rappaport & Fernández-Partagas, 1995; Reid, 1841. V2.
1839 01 11 [06:00 LT] 14.4N 61.1W Martinique	Martinique		A strong earthquake at Martinique [MMI $\cong$ IX], left Fort Royal] (Fort-de-France) in ruins, sending up vast clouds of dust and leaving 1 m of debris in the streets and 400 dead. At Castries, St. Lucia [MMI $\cong$ VIII], most buildings collapsed. Ships on the high sea at a distance of 112 km felt the shock. The shock was felt at Barbados" [MMI $\cong$ VII], Guadeloupe, St. Kitts, Surinam at 6:15 a.m., and Guyana, where bells were set ringing in Georgetown. According to a report from the French corvette, *La Recherche,* "In the harbor of Martinique, all the ships in this place were violently shaken about during the shock that lasted almost 40 seconds. The sails were whipped about like bamboo canes." This may have been seaquake effects, although the possibility cannot be ruled out that a tsunami added to the tumult. "Then we beheld on shore a kind of smoke cloud rising up. An optical illusion was created that it rose up out of clefts in the earth, although it was from the collapse of the buildings. Several hours elapsed before the mountains could be seen again." Brière de l'Île, 1839; Grases, 1971; L'herminier, 1843; Mallet, 1855; Moreau de Jonnès, 1839; Pacine, 1839; Perrey, 1847b & 1858; Rudolph, 1895; Robson, 1964; *Surinaamsche Courant,* 1839. V1.
1842 05 07 [17:00 LT] 19.5N 72.1W	Haiti: Port-de-Paix Île de la Tortue,	 4.6 2.0	Cap-Haïtien was violently shaken by an Ms $\cong$ 8.1 earthquake at 5 p.m., which occurred in the Northern Depression of Hispaniola. It extended from Samaná to Môle St.-Nicolas, and crossing the Windward Passage, was felt along the Sierra Maestra in Cuba, and throughout the Antilles. The shocks annihilated the northwest Cape, and killed 5,000 people (about half the

DATA	LOCATION	(m)	DESCRIPTION OF TSUNAMI AND SEISMIC EFFECTS
1842 05 07 *[continued]* Hispaniola Cuba	Môle St.-Nicolas, Cap-Haïtien Fort-Liberté Dominican Republic: Santo Domingo Danish West Indies: (U.S. Virgin Islands) St. John & St. Thomas	2.0 2.0 2.0 1.6 3.1	population) at Cap-Haïtien alone. Scherer (1912) reported, "It is at Cap-Haïtien that destruction came with the most frightful shock. The sea, also shaken to its depths, dashed its waves on the buildings along the quay, but fortunately did not advance farther. The cloud of dust settled and the shocks diminished after 4 minutes. Of all earthquakes felt till then in the Antilles, the most terrible; so sudden that several thousand who were inside their houses, found their graves there; those in the streets were enveloped in such a whirlwind of dust that it grew dark and they were nearly suffocated. In less than one minute the towns of Cap-Haïtien, Santiago de los Caballeros, Port-de-Paix, Môle St.-Nicolas, and Fort-Liberté [all MMI ≅ IX] were reduced to heaps of ruins." The *Le Journal de la Belgique* reported, "The destruction of Cap-Haïtien is complete. Not two houses remain standing." Concepción de la Véga and Santiago de los Caballeros were punished by "shocks were not less violent than at Cap-Haïtien," according to Scherer. He also said, "The ocean first drew back 200 feet [61 m] from the shore; the *raz-de-marée* [tsunami] threw itself upon Port-de-Paix, covering it more than 15 feet [4.6 m] in depth. Nothing remained standing, not even the church." Heck reported Port-de-Paix covered with 5 m of water. Of Port-de-Paix's 3,000 inhabitants, 200–300 perished. The northern coast of Haiti was struck by a disastrous 2-m tsunami. According to Scherer, "A large part of the misfortunes were due to a great wave of translation which piled up in the channel of Île de la Tortue, and poured over on the beach." It caused catastrophic damage at Fort-Liberté and Môle St.-Nicolas where, "of the city only a *plan par terre* [patch of ground] remained and nothing more. Of the hamlets, the villages destroyed, nothing is said: Little matters, alas! have no history." He reported consulting many "older persons whose memories were still full of recollections," to recover traces of information about the catastrophe. "For them everything dated from the *évèntment* [event]: "I was born so many years before the *évèntment;* I took my first communion before the *évèntment.*" St. John experienced a 3.1-m tsunami. At Santo Domingo [MMI ≅ VI], 1.6-m waves were observed. Scherer reported that "The *raz-de-marée* was felt all along the length of the Atlantic Coast." Taber (1922b) said, "The distribution of intensities and the phenomena of the accompanying sea- wave indicate

1842 05 07 *[continued]*			that the earthquake resulted from a vertical displacement along the fault passing between Île de la Tortue and Péninsule de St.-Nicolas. The displacement may have continued eastward for it is said that the bed of the Yaqui River was suddenly heaved up, driving its waters in a torrent both up and downstream." This shock, "a physiographic manifestation of the Cayman Islands–Sierra Maestra–North Haiti Fault Zone, was the most destructive up to that time, though the intensity may have been no higher than during the earliest reported earthquake at Haiti in 1564." Shepherd said, "The epicentral intensity and total felt area are entirely comparable with those of 1946, Aug. 2. The instrumental magnitude in 1946 was $Ms \cong 8.1$, so it can be reasonably concluded that the May 1842 earthquake was of similar magnitude. At Cuba, this was the strongest shock felt in 80 years, with intensity MMI ≅ IX at Baracoa and MMI ≅ VII at Santiago de Cuba and along the length of Sierra Maestra." Schubert reported waves of 0.9–18.3 m at Hispaniola. (Such a wide range was perhaps a decimal error, intending rather 0.9–1.83). At Spanish Town, Jamaica [MMI ≅ V], "it took place a few minutes to 5 p.m. without regrettable results," according to *Le Journal de la Belgique*. *Le Moniteur Belge* reported, "The effects extended to Jamaica, *Porto Rico,* and almost all the Antillean isles, and as far as the *Montagues Roqueses* [the Rocky Mountains], in the proximity of the Pacific Ocean." Destructive effects were reported at Grand Turk at 4 p.m., and St. Thomas [both MMI ≅ V]. Mallet said they were felt on board ships in the roadsteads, and had been preceded by very sultry weather. Many houses were thrown down. Of the two principal shocks, the second lasted 3 minutes, with the first not as long. He also said another account reported the shocks lasting 85 seconds. They were felt at Santo Domingo, especially at *Cape Haytien* [Cap-Haïtien], and they extended to Jamaica, *Porto Rico,* and almost all the West Indies. *Annual Register,* 1842; Ardouin, 1853; Berninghausen, 1968; Grases, 1971; Heck, 1947; *Le Journal de la Belgique,* 1842; *Le Moniteur Belge,* 1842; Mallet, 1855; Milne, 1912; Neumann, 1953; Perrey, 1843; Poëy, 1857; Reid & Taber, 1919b; Rubio, 1982; Rudolph, 1895; Scherer, 1912 & 1913; Shepherd & Lynch, 1992; Taber, 1920a&b; Tegg, 1854; Tippenhauer, 1893; Tomblin & Robson, 1977. V4.

DATA	LOCATION	(m)	DESCRIPTION OF TSUNAMI AND SEISMIC EFFECTS
1843 02 08 [10:35 LT] 16.1N 62.2W Guadeloupe	Antigua Guadeloupe	1.2	A catastrophic Mw ≅ 8.3 earthquake devastated Pointe-à-Pitre, Guadeloupe [MMI ≅ IX], leaving it in total ruins. One-third of the inhabitants of Pointe-à-Pitre perished, variously estimated as 4,000–6,000. Montadon said Pointe-à-Pitre suffered losses of 110,000,000 francs and 1,500 deaths. The earthquake also ravaged Marie-Galante, Dominica, Montserrat, "and above all, Antigua." The violent 90-second shock at 10:35 a.m. was felt "in all the Antilles." The *Barbadian* reported that columns of water 100 feet high [30.5 m] and several feet thick [1–1.2 m] were observed to be ejected from fissures in the ground at Guadeloupe. This description is remarkably similar to that of a submarine volcano eruption between Guadeloupe and Marie-Galante, on Feb. 17. Rudolph said the disastrous earthquake was "very heavy east and south of Guadeloupe at St. Dessada *(La Désirade?)*, Marie-Galante, and Dominica, just as at Martinique, and lesser at St. Lucia and St. Vincent." Poëy noted the mean direction as west to east at Trou, *Saint-Domingue* (Haiti). This vast area encompasses both Dominica and Haiti, noteworthy since confusion exists about the extent of several events, as in 1827. The *Antigua Weekly Register* reported that St. John's, Antigua [MMI ≅ IX], suffered greatly, as well as the whole island, including the loss of the sugar crop; 12 to 40 fatalities were estimated. After the earthquake, the sea rose 1.2 m, remaining calm throughout, and sank again immediately. At English Harbour [MMI ≅ IX–X], the wharf built on "made-ground sank and afterwards had an undulating appearance like waves on the sea." The *Saint Vincent Observer* reported buildings and sugar mills severely damaged at Montserrat, Nevis, and St. Kitts [all MMI ≅ IX]. At Sint Eustatius [MMI ≅ VIII], an 805-m × 25-cm crack appeared. Severe MMI ≅ VIII shocks were felt at Barbuda, St.-Barthélemy, St. Martin, Saba, and Dominica. Martinique and St. Lucia both experienced MMI ≅ VIIs. Tortola [MMI ≅ V]; St. Thomas, St. Vincent, and Grenada, Barbados [MMI ≅ IV], Trinidad, and Guyana all felt the effects, and it was felt as far away as Bermuda and South Carolina on the U.S. East Coast. Violent trembling near 11:15 a.m., lasting 7 seconds at Paramaribo, Surinam, was reported in the *Surinaamsche Courant*. Rudolph cited a report from the French warship, *Juvencelle,* in sight of the *Felsinsel Redondo* [Redonda rock island] that said, "A mighty chunk of the rocky mass broke off and fell into

1843 02 08 *[continued]*			the sea, which resulted in a surging rise of the ocean." Many of the circumstances and effects have striking parallels with the Apr. 1690 event, especially the report of a great mass of rock plunging into the sea, giving rise to a wave. Both events were comparable in magnitude and their vast area. Capadose said the Mount Misery Volcano at St. Kitts, emitted a spiral column of white smoke and unusual activity from the crater fumarole. Robson said the volcanic activity was an eruption in the central crater with either solfataric activity or a phreatic explosion. Kelleher concluded that this tectonic event was almost certainly shallow since very high intensities were reported. *Antigua Herald,* 1843; *Antigua Weekly Register,* 1843; *Barbadian,* 1843; Blainville, 1843; Brière de l'Île, 1843; Capadose, 1845; Carnegie, 1843; Chocque, 1843; Duchassaing, 1843; Dutrouleau & Gonenet, 1844; Fayers, 1843; Grases, 1971; Itier, 1843; Jameson, 1843; Kelleher *et al.,* 1973; Landre, 1843; Mallet, 1855; Montadon, 1962; Perrey, 1843 & 1858; Poëy, 1857; Reboul, 1843; Reid & Taber, 1919a; Robson, 1964; Robson & Tomblin, 1966; Rudolph, 1887; Sainte-Claire Deville, 1843b & 1848b; *Saint Vincent Observer,* 1843; Scherer, 1912 & 1913; Shepherd & Lynch, 1992; *Surinaamsche Courant,* 1843. V4.
1843 02 17 16.1N 61.2W Guadeloupe	Marie-Galante Guadeloupe		A submarine eruption "half-way between Guadeloupe and Marie-Galante" was described by Blainville as "a large column of water, of blackish color, rose in a succession of jets, whilst steam was emitted around the column at the sea surface. The eruption continued for about half-an-hour." Successive jets spouting up from below the surface might have produced a tsunami, depending on the explosiveness of the eruption. The depth of the shallowest point in the area was 7 m below sea level, the depth of the surrounding sea floor 550 m deep; and it then falls away abruptly to great depth. Robson said the location referred to by Blainville may be near shallows called the Vaisseaux Bank on British Admiralty chart 885 (1882). The description is similar to the one for Feb. 8, in the *Barbadian,* of columns of water spouting 100 feet high [30.5 m] and several feet thick [1–1.2 m] observed as being ejected from fissures in the ground. Robson gave the date as Feb. 17. If this date was from sources using the Julian calendar, then it could be the identical phenomenon described for Guadeloupe of

DATA	LOCATION	(m)	DESCRIPTION OF TSUNAMI AND SEISMIC EFFECTS
1843 02 17 *[continued]*			100-foot columns of water being ejected from fissures in the ground, or *vice versa*—that the water columns could be this same submarine eruption, therefore placing it on the same day. Poëy gave the date as Mar. 17; Perrey said aftershocks from Feb. 8 lasted until Mar. 17. *Barbadian,* 1943; Blainville, 1843; Perrey, 1843; Poëy, 1857; Robson & Tomblin, 1966. V1.
1852 07 17 [07:20 LT] 19.6N 75.5W Cuba	Cuba: Santiago de Cuba & Bahía de Santiago de Cuba		At Santiago de Cuba [MMI ≅ V], a wave engulfed the bay, affecting vessels supplying the port, piers, and other port structures. The shock was felt throughout Cuba and at Kingston, Falmouth, and Montego Bay, Jamaica. Rubio said the Jul. 17 wave was probably produced by a shock that affected the American frigate, *USS Tropic,* 70 nautical miles [130 km] off Jamaica. Poëy (1857) gave 30 nautical miles [56 km] off Jamaica, on the 7th (possibly the Julian date). Perrey gave 210 nautical miles [389 km], on the 19th, and that the *Tropic* felt a "terrible seaquake," 2 minutes long, when the sea was otherwise "completely calm." Rudolph (1887) pointed out that the 3 reports from the *Tropic* for different days [Jul. 7, 17 & 19] at different distances [70, 30 & 210 km] from Jamaica were all reported for 7:20 a.m. Thus he concluded that these were the same event. Grases, 1971; Perrey, 1853; Pichardo, 1854; Poëy, 1855a&b & 1857; Rubio, 1982; Rudolph, 1887 & 1895; Shepherd & Lynch, 1992. V4.
1853 Trinidad	Trinidad		Singer reported terrain uplift south of Trinidad causing the appearance of an *isla de lodo efímera* [short-lived mud island]. Waves may have occurred as in the similar Nov. 1911 event off Trinidad, accompanied by "extraordinary waves" after the explosion of a mud volcano. Similar small islands appeared in 1874, 1928, and 1934. Singer *et al.,* 1983. V1.
1853 07 15 [14:15 LT] 12.1N 63.6W Venezuela	Venezuela: Cumaná, Sabana de Salgado, Puerto Sucre, & Sabana de Caiguire		A violent earthquake [Ms ≅ 6.7] occurred near Cumaná [MMI ≅ X+], followed by tsunamis at Sabana de Salgado, Sabana de Caiguire, Puerto Sucre, and Cumaná, which Singer called a *mar de leva excepcionale* [exceptional rise or threat of the sea]. Dwellings were destroyed at Puerto Sucre. Cumaná was devastated and 113 perished. Montadon reported more than 1,000 fatalities in the region. The total death tolls were estimated variously as 600, 800, 1,000, and 4,000. An active fault ruptured at Caiguire and Sabana del Peñón with displacement east to west along the length of the fault. Crevices opened in the terrain, which poured out salty water. El Dique, Cumaná, collapsed and sank 15 m. Centeno-Graü said that the earthquake,

1853 07 15 *[continued]*			followed by *maremotos* [tsunamis], had coincided with temblors occurring in the Antilles. Berninghausen, 1968; Centeno-Graü, 1969; Fiedler, 1961; Grases, 1971; Heck, 1947; Milne, 1912; Montadon, 1962; Perrey, 1854b & 1858; Robson, 1964; Singer *et al.,* 1983. V4.
1854 03 17 [01:00 LT] 19.6N 75.5W Cuba	Santiago de Cuba		Heavy shocks caused damage at Santiago de Cuba, especially to the vessels anchored in the roadstead. Tremors were not very intense, but had duration of 10 minutes. Waves may have abetted the harm done to the ships, but they were probably not gravity waves, rather those of a seaquake. Grases, 1971; Perrey, 1855c; Poëy, 1855a&b & 1857; Rudolph, 1895. V1.
1855 09 25 [10:45 LT] 16.2N 88.5W Honduras	Honduras: Trujillo Bay & Golfo de Honduras		A very heavy 15-second shock [Ms ≅ 6.0–6.5], ESE to WNW, occurred at Trujillo, which was heavily damaged. Molina noted that during the first shock, the *Simporonius,* in Trujillo Bay, at anchor in 7 to 13 *brasses* [Old French unit; 1 *brasse* = 1.68 m, roughly a fathom, thus 11–21 m], was unexpectedly lifted up and brusquely dropped "as if it were a mass of lead, creating a wave." Nine shocks were counted during the day. Molina said that it (a tsunami?) was locally generated in the North American–Caribbean Plates Boundary Zone. This may have been a seaquake. Arce *et al.,* 1998; Feldman, 1993; Kluge, 1863; Molina, 1997; Montessus de Ballore, 1888; Perrey, 1856b; Rojas *et al.,* 1993; Sutch Osieki, 1981. V1.
1856 08 09 16.2N 88.5W Honduras	Honduras: Puerto Omoa, Puerto Cortés, Atlantida, Trujillo & Criba Lagoon	5.0	On Aug. 9, the largest [Ms ≅ 7.0–8.0] of numerous earthquakes from Aug. 4 to 14, completely destroyed Omoa and settlements on the coast from Río Tinto to Río Ulúa, Livingston, Santo Tomás, Isla de Roatán, Trujillo, Belize, and Guatemala. A tsunami of magnitude I–I ≅ 2.0 was generated, which Cruz cited as "one of the largest tsunamis to have affected the Atlantic Coast of Central America," including Puerto Cortés, Atlantida, Trujillo, and Criba Lagoon (La Ceiba?), and "bringing about the complete ruin of Puerto Omoa, destruction of entire villages, and rivers changing directions." Solov'ev said, "At Omoa, the sea fell and rose 5 m in still weather; it reached the foot of the Spanish citadel, and added to the destruction begun by the earthquake." Citing Anthony [1856:167–171], Feldman quoted: "Then came a low rumbling sound, increasing in force as it came sweeping over the land from the south, so unlike thunder that we could not mistake it. The earth commenced trembling violently. Suddenly this changed, and a positive undulatory motion succeeded that threw us to the ground. The earth

DATA	LOCATION	(m)	DESCRIPTION OF TSUNAMI AND SEISMIC EFFECTS
1856 08 09 *[continued]*			rocked like waves of the sea." The water receded from the 8-km-broad Criba Lagoon, leaving the bottom dry. "Away it went far toward the sea. The little island rose in the centre like a conical hill in some vast plain; its rough sides in dark relief against the retreating water." Upon returning, "from every direction, the waters rushed to a common centre, where, meeting, they rose in the air in an immense column, filling the heavens with an ocean of spray. Then falling, they came toward the land with the force of an avalanche. On came the flood, carrying rocks from the bottom, now all foam as far as the eye could reach, and now an unbroken mass as solid as a rock. The destroyer tore off masses of earth in its passage, boiling up over the hill in a perfect whirlpool, in which whole trees, branches, and even stones were tossed about like feathers. I had been there about two hours—clinging to a tree, though I recollect when the water slowly turned toward its proper bed with but little less violence than when rushing upon the land. We were then knee-deep in the angrily moving flood. Natives (Sambo Indians) informed us that the water had swept back into the interior about 15 miles [24 km]. Very little of their property was destroyed, as most of the natives build their towns on rising ground. A respectable old Spaniard reported having seen a similar occurrence on the coast some years before. Then it was witnessed by many people, several of whom were carried out to sea by the retreating current." Heck mistakenly identified Omoa as being on the Pacific coast. Arce *et al.,* 1998; Boscowitz, 1885; Carr & Stoiber, 1977; Cruz & Wyss, 1983; Feldman, 1993; Fernández *et al.,* 1993; Grases, 1990; Heck, 1947; Iida *et al.,* 1967; Milne, 1911 & 1912; Molina, 1997; Montadon, 1962; Montessus de Ballore, 1888; Perrey, 1859 & 1872; Ponyavin, 1965; Rojas *et al.,* 1993; Sieberg, 1932; Solov‘ev & Go, 1984a; Sutch Osieki, 1981; Taber, 1922b. V4.
1860 04 08 [04:20 LT] 18.6N 73.3W Hispaniola	Haiti: Golfe de la Gonâve & Anse-à-Veau		A severe earthquake [Ms ≅ 7.5] at 4:20 a.m., with a second strong shock at 6 a.m., originated a little farther westward in the same fault zone as the shock of 1770. The epicenter was near Anse-à-Veau, where the sea also withdrew, and then broke with a crash on the shore, according to Taber. The shock was due to vertical faulting beneath the Golfe de la Gonâve. The accompanying tsunami, along the coast of the Golfe de la Gonâve, after first receding,

1860 04 08 *[continued]*			reversed itself, and with a powerful force, returned and surged over the land. At Anse-à-Veau, crevasses sliced across the streets and 124 houses were demolished; at Miragoâne, the bridge sank; at Petit Goâve, all the houses were abandoned; it was felt at Léogâne, Port Salut, St.-Marc, and Jacmel. Ships in the harbor of Les Cayes distinctly felt the shock, as did ships at sea. From Acquin and Les Cayes, the tremors crossed l'Asile valley from SSW to NNE. At Port-au-Prince, after the great shock at 10:30 p.m. that night, the terrified population passed the night outdoors. The preceding night, the inhabitants of Torbeck had abandoned their houses due to a shock after midnight, although at Les Cayes, 3 *lieues* [leagues] [14.5 km] away, nothing distressing was felt. Rudolph reported the sea appearing calm on the southern side of the island. Perrey said, "The tremors were not without some relationship to the volcanic eruption near Santa Fé de Bogotá," and he said it was rather curious that Venezuela and Curaçao would not also be afflicted. Berninghausen, 1968; Grases, 1971; Heck, 1947; Neumann, 1953; Perrey, 1862c & 1864b; Rudolph, 1887; Scherer, 1912; Taber, 1922b. V4.
1861 04 27 [15:15 LT] 18.3N 77.6W Jamaica	Kingston, Jamaica		Perrey described a heavy earthquake at Montego Bay [MMI ≅ VIII] and Cornduailles County as the most violent for many years. Rudolph said a captain was flung out of bed on a ship lying in the roadstead in Kingston *Hafen* [haven]. The effects may have been caused by a seaquake. Shepherd said Perrey might have confused this event with one at Cornwall, England. Grases, 1971; Milne, 1912; Montadon, 1962; Perrey, 1864a; Rudolph, 1895; Shepherd & Lynch, 1992. V1.
1864 05 19 18.1N 72.3W Hispaniola	Jacmel, Haiti		An earthquake was felt at the port of Jacmel [Milne ≅ II]. Many houses were destroyed along the coast of Haiti. The shocks, felt heavily in the harbor by the vessel, *George Prescot,* may have produced a wave. Grases, 1971; Milne, 1912; Perrey, 1867; Rudolph, 1895. V1.
1867 07 29 16.5N 62.2W Guadeloupe	Îles des Saintes, Guadeloupe St. Martin		At Îles des Saintes, a very strong *raz-de-marée* [tsunami] was reported, considered the result of a violent *coup de vent* [blast or gust of wind]. The *raz-de-marée* also ravaged St. Martin, 60 *lieues* [leagues] [290 km] north of Guadeloupe. The uncertain wave source may have been a storm surge. On the U.S. East Coast, two deaths were reported due to a hurricane on Aug. 3 and "all perished except the captain" on a brig off the coast. The hurricane possibly passed

DATA	LOCATION	(m)	DESCRIPTION OF TSUNAMI AND SEISMIC EFFECTS
1867 07 29 *[continued]*			Guadeloupe several days earlier on the 29th. Chapman, 1867; Rappaport & Fernández-Partagas, 1995; Romer, 1932. V1.
1867 09 or 10 or ? Venezuela	Carúpano & Isla de Margarita, Venezuela		Waves at Carúpano and Isla de Margarita were reported for "Sep. or Oct. or ?" with an uncertain link to an earthquake. That these waves may have been associated with the Nov. 18 event is an interesting possibility, since waves may well have traveled to the Southern Antilles and Venezuela. Reid said waves "must have reached the north shore of South America, but we have no record of them from there." Another alternative might be a storm wave emanating from the violent hurricane of Oct. 29. Reid & Taber, 1920; Singer *et al.*, 1983. V1.
1867 11 18 [14:50 LT] 18.1N 65.1W St. Thomas St. Croix Puerto Rico Dominica	**Danish West Indies** (U.S. Virgin Islands)		A violent earthquake [Ms ≅ 7.5], along the north scarp of the Anegada Trough, occurred 15 to 20 km southwest of St. Thomas [MMI ≅ X]. The two tremendous shocks were separated by an interval of 10 to 15 minutes. The earthquakes were teletsunami reported as occurring variously from 2:25 p.m. to 3:18 p.m. (A consequence of solar time at the different islands?) Characterized by initial vertical oscillations, probably due to down-drop of the seafloor, both shocks generated tremendous tsunamis. Since the tsunamis propagated in all directions with the trough in front of the crest, the leading trough caused the ocean to recede from shores on all sides of the convulsion. St. Croix, St. Thomas, and Isla de Vieques more-or-less formed a triangle around the epicenter. Near the middle, it is 2,500 fathoms [4.6 km] deep, the deepest part of the channel that passes from the Caribbean Sea to the Atlantic Ocean, more than 3 times as deep as the Grand Canyon, according to Reid. These islands were deluged by massive tsunamis of tremendous force. At Charlotte Amalie, 8-foot [2.4-m] waves plunged over the piers, which today might be lined with cruise ships and tourists. Demolishing dozens of ships and small craft, this momentous tsunami was the most destructive ever to strike the U.S. Virgin Islands. At least 23 people perished at St. Thomas and St. Croix. The actual death toll from the tsunami may have been much higher. The *Sanct Thomæ Tidende* said that, at Charlotte Amalie, "Loss of life does not exceed 50. Other islands in the Virgin group felt the shocks as severely. St. John suffered enormously." According to Watlington, accounts indicated that, within a 12-hour period, 89 earthquakes jolted the entire Virgin group as well

1867 11 18 *[continued]*	**St. Thomas:** Charlotte Amalie, St. Thomas Harbor, West Gregerie Channel Little Saba, Water Island, Hassel Island & Prince Rupert's Ledge **St. Croix:** Christiansted & Frederiksted **Puerto Rico:** Vieques Culebra, Arroyo, Salinas, Yabucoa & Fajardo	 2.4–4.3 9.1–18.3 9.0 12.0 7.0 4.9 7.0 7.6–9.1 6.1 6.1 6.1 6.1 6.1 6.1	as Puerto Rico and the Antilles. Fuchs reported an earthquake at St. Thomas and "in the sea such a large wave; simultaneously a shock being felt at St. Croix and Porto Rico, and extending itself as far out as the French Islands; this same shock was felt at Cuba and all the Antilles, and even Havana; likewise *St. Jean, Hauptstadt von Porto Rico* [San Juan, capital of Puerto Rico] was similarly damaged; also seawaves at Jamaika." The *Saint Croix Avis* noted, "Again it is reported, that while Guadeloupe, Dominica, Antigua, and some parts of the Spanish Main have escaped with but slight shocks, Jamaica has suffered to a fearful extent. True or not, we cannot vouchsafe for the correctness of this report. Some parties say, that news, by telegram, via America, has reached us, stating that the whole of Europe has been visited by the same calamity; and that Mexico has been dealt with severely." Reid said that effects could not have been observed west of Haiti. Sainte-Claire Deville reported, "The earthquake was extremely violent in *les îles Sous-le-Vent* [Leeward Islands], starting from *les îles Vierges* [Virgin Islands], and in all *les grandes Antilles* [Greater Antilles]." In translation, the word *grandes* was ignored, thus translated, "and in all the Antilles"—seeming to imply "on down the rest of the Lesser Antilles." If the area depicted meant was, "The Leewards from the Virgins downward and all the Greater Antilles," then the breadth of the area described would be considerably larger, and include Puerto Rico, Hispaniola, Jamaica, and Cuba. According to Perrey, "The same seismic wave may well have propagated in one direction as far as the coast of Mexico, and in the other up to about the Azores." The calamities began just days after "Thanksgiving Day," celebrating the termination of the hurricane season (traditionally Oct. 25). On Oct. 29, a brutal hurricane laid waste to the islands, killing 500 at St. Thomas, and sinking more than 60 ships in the harbor. About the hurricane, the *Avis* reported, "Portorico, St. John, and Crab-Island (Vieques), have all suffered to a very extensive degree." Sir Arthur Rumbold, President of the British Virgin Islands, remarked, "Anegada has escaped with little or no damage and forms an exception to this tale of woe." Anegada appears to be the only island untouched by this killer storm. Backshall commented, "To give an idea of the strength of the winds, the diving bell belonging to the

DATA	LOCATION	(m)	DESCRIPTION OF TSUNAMI AND SEISMIC EFFECTS
1867 11 18 *[continued]*	**British Virgin Islands:**		dredging apparatus—a bulk of about three tons—was lifted, carried over, and thrown onto a spar pit, a distance of at least a quarter of a mile [402 m]." Paiewonsky said, "The tidal wave that struck the Danish West India Islands 20 days after the violent hurricane was no mere coincidence." Commenting on Brooks' report about the immense pressure changes affecting the ocean floor from a hurricane (that can trigger an earthquake and resultant tsunami), Paiewonsky said, "Evidently the hurricane that passed over St. Thomas did just that."
	Virgin Gorda	1.5	
	Peter Island	1.4	
	& Road Town, Tortola	1.4	
	Leeward Islands:	1.4	**St. Thomas:** According to the *Tidende,* "Scarcely had we completed our painful account of the lamentable hurricane of the 29th October, than it becomes our most distressing duty again to record an amount of calamity that has no parallel in the annals of St. Thomas. On Monday, Nov. 18th, just one year that day, after the visitation of Cholera, at about 10 minutes after 3 o'clock (we judge of the hour, as we found the clock stopped at that time), this Island was visited by the severest Earthquake than it has ever at any time been its misfortune to experience. The loud rumbling of the earth and the severity of the shocks (there were many in a very brief space) soon emptied every store and building of its occupants." The panic in Main Street was described: people rushed to the comparative safety of the waterfront until the shocks appeared to have subsided, "but, again, worse was to come, this time from the sea," according to Backshall. The *Tidende* continued, "No sooner had the first terrific shocks ceased than a spectacle presented itself that will never be obliterated from the memory of those whose misfortune it was (or privilege if you will) to see it. An ocean of water (the height no one will undertake to compute) of snowy whiteness, and forming one wave was seen at the south west beyond our harbour, majestically, yet steadily rolling onward towards us. No army in high day review ever moved with more precision; or in battle array, ever advanced with more self-possession upon the adversary, than this ocean of water moved upon us. Had it been, that it was only our privilege to *see* it, no one would have declined the opportunity. Imagine as bright a sun as ever shewn *[sic]* out upon our planet, with no breath of air from out the heavens, our harbour as dull as the Dead Sea, and an advancing ocean or wave just before us, foretelling in its every movement, that it must engulf or ride down all
	St. Martin	3.1	
	St.-Barthélemy		
	Saba	1.4	
	Barbuda		
	St. Kitts		
	Sint Eustatius		
	Montserrat		
	Antigua: St. John's	2.4–3	
	Guadeloupe:		
	Sainte-Rose,	19.8	
	Deshaies,	19.8	
	Basse-Terre,	1.0	
	Pointe-à-Pitre,		
	Fond-du-Cure,	1.0	
	Îles des Saintes		

1867 11 18 *[continued]*	**Windward Islands:** Dominica: Prince Rupert's Bay Martinique St. Lucia: Layon, Anse-la-Raye & Rivière Dorée St. Vincent: Cumberland Bay & in The Grenadines: Admiralty Bay, Bequia Canouan & Carriacou	 3.0 1.2 1.2 0.6 1.8 1.8	things that would encounter it. It came on; in length, it must have been the span of our Island. It reached our fort points, that happily received and broke its shock; otherwise, an inevitable inundation of the whole town must have been our fate. That part of the wave that entered our harbor leaves a melancholy reminiscence for many, for it swept clear up into our principal or main street, and no store from King's Wharf downward but has had from 4 to 5 feet [1.2–1.5 m] of water within it," spoiling the goods. The torrent was seen "hauling sloops, schooners, buoys and what-not else of a marine kind" down the main street, thence to be seen high and dry in the streets. "Our merchants are fearful to go into their stores whilst the damage must keep increasing from want of ventilation. Fear still occupies the minds of all our inhabitants, and numbers, without regard to position, are encamped on the hills under tents." The *Tidende* continued, "The sea entered not only at the fort points, but passed into The Gregories *[sic]* from the west and east ends of Water Island, and then again into the harbour at the haulovers, so that the sea shore and the streets at the western end of the town, have suffered equally with those exposed to the wave that came in at the points main entrance to the harbor. The receding of the waters, after they had spent themselves on the shores, were almost as threatening as their coming, and for a time, looked as if the harbour and the Gregories, would have remained dried up." Large fish were seen "lying on the strand as the waters went down." As a final threat to the well-being of the town, it was reported that "The junction of the receding water with that on the outside of the fort points, for a time, really looked as if a reef had formed across the mouth of the harbor, which would have been the inevitable ruin of the place." Raupach reported "a fine blue West India sky and no hint of the great convulsions to come, when suddenly about 2:45 p.m., an underground rumbling noise was heard, immediately followed by a terrific earthquake. It seemed to come from S-by-W and pass on to N-by-E. The earth seemed as if composed of small waves rising and sinking under your feet, so that if you made a step forward, your foot seemed to meet higher ground, and if you put your foot backward, it also met higher ground. To stand in one spot was impossible. The underground sound, while the shock was going on, for about 1½ minutes was most dreadful."

DATA	LOCATION	(m)	DESCRIPTION OF TSUNAMI AND SEISMIC EFFECTS
1867 11 18 *[continued]*	**Grenada:** St. George's, Gouyave, Clarke's Court Inlet, Westerhall Bay & La Baye	 2.4–3.0 2.4–3.0 0.6 0.6 0.3	After the shocks, people on the hill above "beheld clouds of dust rising from the town, proving that many houses had fallen, and crowds of people running towards the hill, followed by horses and cows neighing and bellowing fearfully. The sea had receded, and half the bay was dry, with multitudes of large fishes jumping about on the bottom." Then, about 10 to 15 minutes after the first tremendous jolt, 4 to 5 km out to sea, an imposing wall of water was seen advancing toward the harbor. Raupach reported that the foaming wave "stood up like a straight white wall, about 15 to 20 feet [4.6–7 m], and advanced very fast into the harbor, sweeping or upsetting small vessels before it, and raising large Men-of-War and steamers to its top. The appearance of this wave was like a white masonry wall, erect and straight, as if built with a rule; it had not the appearance of ordinary waves. The rising of the waves was repeated a second time after an interval of 10 minutes; the second appeared to be even larger than the first and went farther inland. The shocks of the first day continued every few minutes as if hung together in one chain." *Harper's Weekly* reported that "The tidal wave was described by those who saw it as a monstrous breaker or rather a sea-wall, variously estimated at from 30 to 60 feet high [9.1–18.3 m], and flowing at a rate of at least 50 mph [83 kph]." Little Saba was reported to be in active eruption from an old volcano. "Froth spumed 100 m into the air," according to a letter cited by Sainte-Claire Deville (Maskell noted 100 feet). Diminished after engulfing Little Saba, the turbulent tsunami traveled onwards. Cresting at 12 m as it surged into The Gregeries, its fury broke against the wedge of Water Island. As a 9-m swell, it penetrated the harbor, crashing like an avalanche, and driving ships and everything in its path before it. Discrepancies in the various estimated waveheights may have had much to do with which area of the harbor was being viewed and the observer's location. Striking the other entrances to the harbor obliquely, the massive wall of water "not less than 23 feet [7 m], broke and retarded on Prince Rupert's Ledge, and caused a great rise in the waters of the harbor," according to Commodore Boggs of the American corvette, *USS De Soto*. "The majestic rollers hurled themselves upon the beaches and burst 8 feet [2.4 m], over the wharves sweeping up into the principal streets." Added to the wharf height, the wave

1867 11 18 *[continued]*			would have had to be greater than 14 feet [4.3 m] high. At the Hassel Island coaling wharf, a 4.9-m runup was reported. At Altona, "good-sized" cottages had been carried far inland by the surge, which inundated 76.2 m over the land. After the wave had spent itself inland, the rushing backwash, full of debris and charging back to the bay, was almost as menacing as the onrush. The *De Soto* was deposited on the iron pilings of the demolished American screwpile jetty by the wave that "swamped our boats at the booms; then receded as violently," according to Boggs. "At the first swing of the current we had 6 Fathom [11 m], under our stern, the ebb was so rapid that in a few minutes we took the bottom, having only 1¼ Fathom [2.3 m] under us," a drop of 29 feet [8.7 m]. "The second wave, 10 minutes later, came in much greater in volume and inundated even further onto the shores. The next roller swept her off again into deep water; this third wave was not as large as the first two, and the fourth rose not more than 12 to 15 inches [30–38 cm]. The current, full of whirlpools, swung the warship against the wharf, breaching 3 holes in her hull. The *De Soto* made at least 20 full swings around the compass before being able to check her movement around the harbor." Canvas was patched over the holes, and although not critically damaged, leaks were not controlled until after steam to run the pumps was "brought up in a record 1 hour and 5 minutes. Waves came in at irregular intervals until long after dark," when the cruiser was finally able to anchor safely. Rear Admiral Palmer, Commodore of the American frigate flagship, *USS Susquehanna,* reported sighting "the heavy wall of sea advancing towards the harbor with terrible power, with a skirmish line of tumultuous rollers in front. The sea was rushing in and receding with great violence." The *Susquehanna* met the tsunami "stem on" and was not damaged; "when the life boats could be manned," they were sent to rescue those drowning from the numerous capsized vessels; "several were picked up, one already dead." The *Tidende* said that the *De Soto,* "immediately proceeded to its mouth in the hope of saving the lives of those aboard the capsized vessels." Since the *Susquehanna* reported assisting with rescues and the *De Soto* was otherwise occupied rushing about with the currents, the article may have confused the names of the two U.S. Men-of-War. The *Derwent,* sunk during the great hurricane of Oct. 29,

DATA	LOCATION	(m)	DESCRIPTION OF TSUNAMI AND SEISMIC EFFECTS
1867 11 18 *[continued]*			was raised again by the tsunami. Palmer said the British Royal Mail Steamer, *La Plata*, lying at anchor in the West Gregerie off Water Island, "was as nearly being lost as possible. Her passengers spoke of having felt the shock as we did, and looking behind them saw a small islet in their rear was cleft in twain, flames and smoke issuing from the fissure—shortly afterwards the sea arose and came well-nigh carrying them down; their cables parted, but they were thrown most fortunately under a lee which placed them in comparative safety. The passengers, all but three (females) who were lost in the surf, reached the shore." A sketch in *Harper's Weekly* illustrated the Royal Mail Packet, rising upon a mighty roller with a coaling scow alongside. Paiewonsky vividly described the excitement surrounding the *La Plata* Royal Mail Packet and her passengers, quoting Maskell, a passenger aboard the steamer: "We were leaning over the rail at 2:15 p.m., watching the antics of the coal workers, 30 or 40 Negroes briskly carrying the coal on their heads and making the surroundings ring with their songs. (Had we been at the coaling dock instead of at anchor coaling from 3 scows alongside, women with baskets on their heads would have done the coaling, as at Jamaica; here only two women were working with the men.) Five minutes later, without warning, came the shock, the heaviest any of us had ever experienced. It was of no less than 2½ minute's duration. It was not one continuous shock. For some 4 or 5 seconds in the middle, there was an appreciable lull. Then it came again, if anything, with redoubled force. Our Captain, who ran on deck from his cabin at the initial shock, said the first things that met his eye were stones shot from the cliffs at Water Island like balls from a cannon! The first glance thrown seaward showed, if not the cause, at least one principal accessory to the shock. Little Saba lying 7.2 km southwest of our position was in active eruption! A cloud of dark smoke hung over it. People on shore told us afterwards that they saw a column of bright flame in our direction (in line with Little Saba) and thought the *La Plata* was on fire. Meanwhile, a strong smell of sulfur pervaded the atmosphere. A quarter of an hour passed, and at 2:35 p.m., a second cloud of smoke, this time white like steam, appeared over Little Saba. Then it was we saw, looming behind that island, like a gigantic wall, a tremendous tidal wave rushing towards us! A small

1867 11 18 *[continued]*			steamer passing outside the harbor turned her head to sea and rode over the wave; probably because she was in deep water. A moment more and the great wave reached Little Saba, and in an instant that rock was a mass of foam 100 feet [30.5 m] high. We had 5 minutes to prepare for the shock of the terrible mountain of water that rolled swiftly upon us. The confusion on board was indescribable. Some passengers climbed into the rigging, so as, if possible, to float clear of the ship; some went forward to the bows or on the paddle boxes with the same intention. Half wild with terror before, about 20 of the Negro workmen, blind with fear, rushed headlong into one of the life-boats. Their action induced 2 or 3 lady passengers and children to follow them. One instant before the great wave struck us, the forward tackle of the lifeboat broke, and it fell nearly perpendicular. Down on our ship came the roaring sea, literally piled up like a wall, rolling up on us at the rate of 80 kph, with a perpendicular face of fully 40 feet [12 m]. Fortunately, half-a-mile [805 m] outside our position, the point of Water Island, exposed to the full force of the wave, met it like a wedge. It broke and dropped, so that when the wave struck our ship, it was perhaps not more than 10 feet [3 m] higher than the taffrail [3 m above a rail of about 1.5-m on the stern, roughly 4.5 m above the waterline, adds up to a 9-m wave]. The water, receding from the shore like a mill race to meet the wave, turned our ship around, so that we took the onslaught on our stern. Three waves in one it came, with the roar of 100 Niagaras; and contrary to all expectations, the *La Plata* rose over it like a duck and was saved! The main force of the sea smashed the starboard quarter-deck to ribbons, twisting davits like pinwire and flooding all the saloons and after cabins. The coaling scows had been cast loose; two floated away and the third sank like a stone. It is a mystery how the lady passengers and children hung on in safety to the lifeboat that had fallen just before the wave struck us. Almost all the Negro workers were thrown out instantaneously. Only 3 to 4 of them were recovered afterwards. Probably 10 or 11 of them were drowned, amongst them one woman worker. Had it not been for the point of Water Island, which broke the full force of the wave, it is more than likely that the *La Plata* must have gone—and the chances of saving life in that sea almost none. In a short time, the wave beat in an awful fury

DATA	LOCATION	(m)	DESCRIPTION OF TSUNAMI AND SEISMIC EFFECTS
1867 11 18 *[continued]*			on the shore, its sound mingling with its echoes from the hills. After the great wave passed, one of our boats was lowered immediately to pick up the Negro survivors. In view of possible consequences of another convulsion, the Captain considered it advisable to send the passengers ashore at St. Thomas. About 5:30 p.m., we landed ashore; before us was every imaginable kind of wreck and debris, flooded, overturned, or smashed by the great wave. We had a mile and a half [2.4 km] of mud and mire to trudge across before gaining the town; the usual roads and paths were entirely washed away. The main street, running parallel to the beach, presented an extraordinary sight. Though fully 30 feet [9 m] above the water's edge, and 200 to 300 yards [183–274 m] away, the whole space had been flooded several feet [1–1.2 m] deep by the wave." The 20-foot [6-m] contour line cuts through the town (1955 USGS map). "The agglomeration of wreckage was astounding. Looking strangely out of place among the ruins of shops and houses, were fishing boats or a schooner, and many cases of cargo washed out of ships, together with the masts and riggings from the unfortunate vessels. Under the verandah of the hotel, a huge red buoy lay atop the wreckage, heaped in almost inextricable confusion. During the whole night, the earthquake and subterraneous rumbling continued at intervals sometimes of 2 to 3 minutes, sometimes of half-an-hour, as if one continuous shock all night long; a louder roll announced a heavier shock, and the whole place vibrated beneath our feet. No less than 70 distinct shocks were felt by 8 a.m., without counting the lesser vibrations, the most severe at 10 p.m. and 4 a.m. Unless the earth opened up, we were tolerably safe. The earth did open in one place on Main Street! With daylight, the force of the earthquake appeared to abate. The shocks, occurring at intervals of an hour, seemed not so severe. The heat, fierce, stagnant, pitiless; the shore covered for miles with a mass of sweltering filth, exhaling under that scorching sun horrible incentives to fever and disease. Later, on our way across to the *La Plata,* to steer through the complicated forest of masts and spars and masses of wreckage strewn in all directions was most difficult. The *La Plata* was moored alongside the remains of the Royal Mail Company's coal-wharf, sheltered by a jutting point of land. Yet the tidal wave had worked its destruction here, too. Even in

1867 11 18 *[continued]*			that sheltered corner, the storekeeper reported that the water had rushed 16 feet [4.9 m] deep and more over everything. What remained untouched was the mountain of coal. We were glad that only a few hours would be needed to draw our supply, and we would leave that fetid den for Jamaica that evening, but again we were doomed to disappointment. The workers on shore, though offered exorbitant wages, refused to work. All energy seemed to have deserted them. We had to wait until the cooler part of the evening allowed our own crew to get the coal into the ship themselves. Night brought no cooling and as darkness closed in, the oppression seemed to increase. We were surrounded by black, stagnant, almost putrid water, with enough festering abominations, to generate the most fearful epidemics. The earthquake continued. Night seemed to have renewed its strength. From dark to daylight the unceasing trembling continued as during the night before. At intervals came the shocks, and as each rolled like thunder beneath us, its terrors were supplemented by the shrieks of the wretched inhabitants on shore reaching us like the lamentations of souls in agony. With the steady fortitude of British seamen in an emergency, throughout the night our crew labored; as a stronger shock made the whole ship vibrate, it required all their courage to continue the work. The next day as we saw the last hawser cast off, a sense of security came over us for the first time in 50 hours. The crewmen used their sounding lines carefully, lest a new reef had been thrown up by the convulsion. Passing Little Saba about an hour later, we discovered and picked up one of our boats. She had been torn away by the waves and dashed on that erupting island. Little Saba, itself, was rent and torn in all directions, its precipices almost shattered, and mounds of fallen rocks had piled up on its shoreline. In one place, a large crater gaped, and grey lava was still coming down from it to the sea. We were told by those who knew Little Saba well, that the whole shape of it was changed. It is probable that had not this small island acted as a safety valve and provided an escape for the imprisoned elements, the earth shocks at St. Thomas would have been far more severe. Not one stone would have been left on another, and few of us aboard the *La Plata* would have lived to tell the tale, were it not for this off-shore safety valve. Even now, as we lay off Little Saba, waiting for our crewmen to

DATA	LOCATION	(m)	DESCRIPTION OF TSUNAMI AND SEISMIC EFFECTS
1867 11 18 *[continued]*			pick up our boat, one more earth shock came, as if it were a parting salute. It was severe enough to make the rocks roll furiously down the cliffs of the little island and to cause a fresh rush of lava to pour from its crater." The *Avis* reported tidbits of news that arrived from New York, "By the Gulf Cable," that said, "The sea rose 50 feet [15.2 m], and a volcano vomited forth lava," and that "the English, Danish, and Dutch consuls on the island [St. Thomas] saved nothing." Another tidbit reported that "Little Saba was submerged for a short time, after being visited by earthquake and volcanic eruptions simultaneously." Reid said people confused volcanic activity with landslides, yet the distinct differentiation between flowing lava and rock falls seems to negate this, as do reports of smoke plumes, gases, or steam from volcanic openings. Disputes about volcanic eruptions appear to stem from the idea that a volcanic eruption could be the tsunamigenic origin. Latimer had suggested a submarine volcano between St. Thomas and St. Croix as a possibility. About Little Saba Latimer noted, "It is historical that some 75 years ago the small island of Saba was an active volcano, and on the 18th and 19th of Nov. last, emitted smoke; thus showing the volcanic action was not distant." Örsted reported that Little Saba "presents a crater formed of lava." The death toll at St. Thomas tallied 16 to 17 victims from the tsunami, though many more were presumed drowned. Perishing from the tsunami were 10 or 11 of the workers who had been coaling the *La Plata*, one person picked up by the *Susquehanna,* and two others noted in obituaries: a young man trying to rescue his father and a man aboard an English steam launch. The "two or three lady passengers and children" imperiled in the dangling *La Plata* lifeboat that "somehow survived," should not be confused with Palmer's account of three female passengers washed up ashore on Hassel Island, which lay many km away on the opposite side of Water Island. From fellow passengers, who had gone ashore earlier, it was learned that "loss of life was probably great, as numbers of inhabitants, after the first great earthshock, had put off in boats to flee they knew not whither. The tidal wave destroyed most of these frail craft, and the great majority of their occupants must have drowned. Also, vessels in the harbor had been driven ashore with the loss of many seamen." In a young woman's memoirs, were noteworthy

1867 11 18 *[continued]*			observations: "The first shock happened suddenly. The weather was exceptionally pleasant and things seemed perfectly normal. The only unusual happening was the behavior of Mrs. Gad's dog. The animal howled constantly and seemed very restless. It was asked that the dog be put into the next room. No sooner had the door been shut on the dog, than the first earthshock took place." Looking toward the approaching "tidal wave, high as a tower, covering the complete width of the ocean, people all around us went wild with fear crying, 'St. Thomas is sinking! God help us!' At that moment we believed the ocean would swallow the island and everything on it. In fright from the aftershocks, we camped outside on the hill. Sick people who had abandoned the hospital lay all around. The jail had been opened and some of the prisoners allowed to run away were here too. For more than a month, we slept fully clothed, never knowing when we would have to flee the [borrowed] house." Remarking about the days of anxious waiting on the hills above town, she recounted: "Small children were baptized in a hurry." **St. Croix:** At St. Croix, the earthquake was nearly as severe as at St. Thomas. According to Reid, "The waves broke upon the north and west coasts with great violence." The *Saint Croix Avis* reported that Christiansted, on the north coast, experienced "one of the most dreadful visitations that has ever befallen this island. The oldest inhabitant here knows no parallel to the sad calamity that happened here about 1:50 p.m. Two very severe shocks of Earthquake, one immediately after another, occurred, these lasting over 3 minutes. The frightened people rushed out of their houses; and then beheld the troubled sea, which had receded soon after the two violent shocks, coming furiously, mountain high, and dashing on the shore as far as 100 yards [91 m] in some places. The greatest damage was in Gallows Bay, where upwards of 20 small cottages were upturned or completely demolished from the violence of the encroaching sea. Not an estate in the country remains scathless. Where the Negro-village did not receive damage, either the dwelling-house, the mill, or the works were injured." Aftershocks rang church bells and tumbled walls. During the next few days, "Whatever chimnies and walls that were cracked and remained standing, have now either fallen or received more damage. Ever

DATA	LOCATION	(m)	DESCRIPTION OF TSUNAMI AND SEISMIC EFFECTS
1867 11 18 *[continued]*			since the Earthquakes, the water from the wells in town and elsewhere is salt—though not quite so salt as the sea. What is more remarkable, wells in Eastend that were formerly giving no water, are now springing abundantly." Frederiksted [MMI ≅ IX], on the west coast, was punished. The *Avis* continued, "This chastisement from God has been dealt out also to the people of Frederiksted, who suffered, as we understand in far greater degree. The heavy sea having no barrier to stop its impetuosity; no reef to break its force, boiled up and swept everything before it, shaking the foundations of large buildings and removing the smaller ones, carrying them no short distance from where they stood." The ocean withdrew and returned as a wall of water 25 to 30 feet [7.6–9.1 m] high and swept 76 m inland in immense sheets, according to the *Avis,* and "damage along the Bay Street was very great. Wooden buildings had been tossed about like corks and floated inland for considerable distances. Many substantial stone houses were completely destroyed." Reid reported that "The U.S. warship, *USS Monongahela,* had a most thrilling experience in the harbor." The tale was reported in a monograph by van Housel, who had been aboard the *Monongahela.* Watlington commented, "Housel's non-scientific, anecdotal report unsuspectingly presents the historical and social circumstances of the people of the Danish West Indies at a significant turning point in their history." The midshipman's narrative powerfully describes the trials of the *Monongahela* surviving the conflagration at St. Croix: "After the first severe shocks near 3 p.m., lasting 30 seconds, the receding sea quickly changed direction, and drove the *Monongahela,* lying at 7 Fathom [12.8 m], rapidly towards the beach, the shore approaching at railroad speed. Our vessel advanced and receded from the shore with the waters until, as if some great power had raised up the bottom of the bay, the sea rapidly closed in on the town, filling the houses and covering the street running along the beach to a depth of 24 feet [7.3 m]. Our ship, following the current, took a course toward the south end of the town, until over the edge of the street it swung her bow toward the north and was carried along, smashing a frame storehouse, and breaking down a row of shade-trees." A frantic effort was made to hoist the jib in hopes of keeping off the town. With the rush of the backwash, they were carried out

1867 11 18 *[continued]*			450 m, "when our vessel grounded [in 4.6 m of water] and careened over onto her port beam's ends." The bottom was visible for 800 m beyond. Usually 55 to 73 m deep, "that immense body of water, which had covered the bay and town, was reforming—with the whole Atlantic Ocean as an ally—for a tremendous charge upon us and the shore. This was the supreme moment of catastrophe. As far as the eye could reach to the north and to the south was a high threatening wall of green water. It seemed to pause, as if marshaling its strength, then on it came in a majestic unbroken column. The suspense was terrible. Our noble vessel seemed as a tiny nutshell to withstand the mighty rushing Niagara advancing upon us. Many a hasty prayer was muttered by lips unaccustomed to devotion." The tsunami struck the ship with gigantic force, "making every timber shiver, but the decks were not engulfed. Being rather flat-bottomed [drawing about 5 to 6 m], the first effect of the blow was to send her over on her starboard beam's ends, giving the water an opportunity of getting well under her before righting, when she was buoyed to the crest of the wave and carried broadside to the shore, finally landing on the edge of the street in a cradle of rocks that seemed prepared for her reception. Here she rested with her decks inclined at an angle of 15°. A small Spanish brig was carried bodily inland across the cane fields, landing in the midst of the King's highway." In a letter cited by the *Avis* it was stated, "My reliable witness [as opposed to a previous account accused of inaccuracies], described the tidal action in this way. In the beginning the sea went out gradually a short distance, then flowed back slowly towards the land. The waters receded again beyond the Man-of-War, *Monongahela*, lying at anchor. Just enough water was left to prevent the *Monongahela* from keeling over into the mud. Outside and beyond the warship, the sea rose majestically to a series of heights like a chain of hills of graduated heights, swelling to gain velocity to overwhelm everything, when its forward movement began. The *Monongahela* seemed doomed, lying as it was, at the base of the towering wave. Suddenly, to the joy of the beholder, the ship began to rise by degrees. Up, up went the *Monongahela*, so high as to be seen plainly from where people had run for safety. There was the *Monongahela* floating on the top of the great incoming wave, on an apparent level with

DATA	LOCATION	(m)	DESCRIPTION OF TSUNAMI AND SEISMIC EFFECTS
1867 11 18 *[continued]*			the tops of the houses in Bay Street. The warship landed in an open area of Bay Street, but did not stay there. The powerful receding waters that carried away lumber sheds and other buildings on the waterfront, deposited the warship with architectural precision on a ledge of reef on the shore's edge." Commodore Bissell recounted that, at 3 p.m., "A terrible rumbling and vibration of the vessel "as if violently blowing off steam from the boilers: it was a severe shock of an earthquake that lasted about a minute. Then a quiet occurred until 2 minutes past 3, when the water receded from the shore between Butler's Bay and Sandy Point with astonishing rapidity, and the island seemed to be approaching us. In a few seconds a reflux tide set in from seaward towards the land with such velocity as to carry the vessel and two others toward the beach. Such was her velocity that the anchor chain was torn from the deck and rushed entirely out the hawse hole. Thus adrift, she was washed in by a heavy swell over the beach, bumping injuriously several times on the rocks; but more water coming in, she remained yet afloat. Sails were instantly hoisted to pay her around to seaward. At this moment, a curling wall of water about 15 feet [4.6 m] high came in from seaward, struck the port broadside, rolled her deeply to starboard, and, receding, left her high and dry on the ledge of the rocky wharf where she now lies, parallel to the shoreline, careened 15° to port, which side looks seaward, her port bilge resting on the very edge of the rocky wharf: this wharf descends perpendicularly 2 feet, from which the bottom slopes gradually seaward, so that if she were to fall over it would be completely on her beam ends." From the time the water and when the second wave returning as "a wall of water 25 to 30 feet high [7.6–9.1 m]. Had the ship been carried out a few feet farther over the edge of the reef, it would have been a total loss." In van Housel's account, the executive officer ordered the men over the side: "the waters again retreated and assumed such a threatening appearance, fearing another tidal wave (which would have dashed us against the stone houses or against the walls of a Danish fort just ahead of us)." Two prisoners in double irons hobbled on deck and begged to have them removed, but the keys were ashore with a fleeing officer. Someone took pity on them. "The chains being cut, they were left to escape with the bracelets still on wrists and ankles." Many

1867 11 18 *[continued]*			running to safety "were beset by a rush of water thrown far up into the town, seeking its way back to the sea. We were soon in water waist deep, contending with the strong current as best we could. This water bore on its surface all manner of debris gathered from the yards and houses in its course—chairs, cradles, bedsteads, broken fences and doors, together with flocks of ducks and geese quacking and gabbling." Two American ships, a brigantine and the sloop, *Dauntless,* were carried far inland. Four of the *Monongahela*'s crew drowned, and one, whose legs had broken while descending the ropes, died from internal injuries suffered in the fall, dreadfully, after the agony of one leg being amputated, "but before he had undergone the ordeal of the second, saving him further unnecessary suffering," according to van Housel, who helped carry the injured seaman to a makeshift infirmary in a nearby church. Two other fatalities were reported at St. Croix. A Moravian Brother at a parish in Friedensthal reported, "A poor blind helper sister of this congregation died the day after the shock, in consequence of the inundation." The Brother also said, "A woman gathering wood with her son was drowned. The boy saved himself by clinging to a deer, which had been washed into the sea, and which swam to shore, where it shook off the boy." Details of this story were undoubtedly the ones confused with another story, in which a Marine grabbed a horse to "rescue a Negro girl from her perch on a fence." Heading to higher ground, there were "two or three 'Blue Jackets' towing astern, hanging on the horse's tail." Mixed-up reports said the boy had clung to the deer's tail, understandably considered spurious. In all, 7 victims of the tsunami had been reported at St. Croix. The reported fatalities for both islands numbered at least 23. Paiewonsky said that the 3 ships-of-war, the *Susquehanna, De Soto,* and *Monongahela,* comprised a squadron of naval vessels that brought U.S. officials to the Danish West Indies to consummate the purchase of the islands from Denmark. The *La Plata* had arrived with news from King Christian IX, agreeing to the "Terms of Sale" pending approval by a popular vote. Undertaken by Secretary of State, Seward, remembered for the purchase of Alaska, "Seward's Folly," the negotiations finally failed, due not only to the 5 deaths and damage to 2 of the 3 warships sent there, but also to growing doubt about the advisability of purchasing islands

DATA	LOCATION	(m)	DESCRIPTION OF TSUNAMI AND SEISMIC EFFECTS
1867 11 18 *[continued]*			prone to such havoc as the devastations by the hurricane, the earthquake, the tsunami, and the yellow fever epidemic that followed. (A cholera epidemic rampaged the island exactly one year before.) Yellow fever carried off Rear Admiral Palmer, Commander of the U.S. fleet, 3 weeks later. A 50-year interruption in the process ensued. *Harper's Weekly* spoke of "the city and bay of St. Thomas, the chief town and principal harbor of our new possessions in the West Indies," which was reported prematurely, since the U.S. Senate did not ratify the purchase. Brete Harte added his parody about the situation as a poem about the "demise" of St. Thomas entitled: "Saint Thomas, A Geographical Survey," beginning: *Very fair and full of promise* *Lay the island of St. Thomas:* which ends: *So the Mountains shook and thundered,* *And the Hurricane came sweeping,* *And the people stared and wondered,* *As the Sea came on them leaping:* *Each, according to its promise,* *Made things lively at St. Thomas.* *Till one morn, when Mr. Seward* *Cast his weather eye to leeward,* *There was not an inch of dry land* *Left to mark his recent island.* *Not a flagstaff or a sentry,* *Not a wharf or port of entry,—* *Only—to cut matters shorter—* *Just a patch of muddy water* *In the open ocean lying,* *And a gull above it flying.*

1867 11 18 *[continued]*			A vote of the populace in Jan. 1868, overwhelmingly approved cession to the U.S., with 205 unanimous votes at St. John and 22 opposing out of 1,039 at St. Thomas. Objections to being separated from her brother islands by St. Croix took form as songs and letters of protest. These laments, in the island patois, consisted of correspondence back and forth between Sissy Shanna (Christiansted), on behalf of sister Frederika (Frederiksted) and herself, Buddy Tomas (St. Thomas), and Buddy Jonny (St. John). A sample from the *St. Croix Avis* (1868): "But de wust ting of all me dear Buddy is de King Letter wha' he write for say dat he sell you to de Yankees and he no sell a' we too." Watlington said the strategy of presenting potentially controversial social commentary in the voice of a native dialect-speaker survives since then. **St. John:** As the *Tidende* reported, "St. John suffered enormously." Brother Warner, a Moravian Brother writing from Emmaus, "St. Jan," stated, "At 10 minutes before 3 p.m. we were all startled and driven out of our house by a fearful shock of an earthquake. We had scarcely made good our escape, when a second shock, almost as severe as the first, again shook the house to its foundations, and rocked it fearfully. We fully expected it would fall, but it was once more [as with the terrible hurricane of Oct. 29] graciously preserved from harm. The people living at the station having soon gathered around us, we knelt down, and committed ourselves, for life or death, into the hands of our merciful Saviour." In the midst of "continued rumblings and constant shocks." they spent the night in the yard, "with no other shelter than the vault of heaven. Above, all was clear and beautiful, but beneath all was in ferment." Brother Warner proceeded to comment about the tsunami: "The unusual ferment in the sea struck terror into many hearts, and brought some desperately wicked men upon their knees crying for mercy." An immense tsunami would have arrived at St. John, 3.2 km east of St. Thomas, only minutes after its arrival at St. Thomas, but details are lacking. **Puerto Rico:** "The same phenomena of earthquake and tsunami were produced," at Puerto Rico [MMI ≅ IX], causing damage throughout the island. The shocks were very severe at Vieques and Culebra, the islands closest to the epicenter. San Juan suffered damage to major buildings. At Ponce, the force of the earthquake damaged sugar estate works and 70 out of 80

DATA	LOCATION	(m)	DESCRIPTION OF TSUNAMI AND SEISMIC EFFECTS
1867 11 18 *[continued]*			factory chimneys collapsed. A violent shock of an earthquake was felt at Mayagüez. The largest waves were observed at Vieques, where on the south coast, immediately after the shocks, "a high wave broke on the shore." Later waves washed its northern shore, but runups are not known. According to Latimer, "The shocks extended to Porto Rico, and were felt throughout most of it; but the force of, and damage done by them to buildings, sugar estate works, and chimneys, was chiefly at the east end, the north side as far down as Arecibo, and the south side as far down as Ponce. Below those places, neither on the north or the south side, was any damage done to buildings. The greater rolling wave passed into the harbor and over beaches at the east end, and down the south side of the island of Porto Rico, but it did not extend to the north side or west end." Waves of 6.1 m were seen at Puerto Rico, principally along the eastern and southern coasts. Latimer also said, "At Arroyo, the sea had the same characteristics as at St. Thomas." Dramatic accounts of these characteristics should have been available, although, unfortunately, details are lacking. Rivers rose 0.9 to 1.5 m, and waves inundated 40 m from the shore. Waves were also very high at Salinas. At Yabucoa, the sea receded 137 m, and then advanced an equal distance over the land. Smaller waves were observed to the northeast at Fajardo, and up to 96 km along the southern shore. Latimer also reported that "The great wave, which soon followed the first heavy shocks was seen for some time rolling on towards St. Thomas and Porto Rico, from the south to the north, while at the same moment another similar wave (perhaps even larger than the first) rolled on towards St. Croix, from the north to the south; thus showing that the force which caused them occurred somewhere between those two islands, and sent the water in both directions." **British Virgin Islands:** At Virgin Gorda, large chunks of granite were "wrent" during the first shocks and, at Peter Island, the sea rose to such an alarming height that the inhabitants took to their boats and rowed over to Tortola, according to Rumbold. Parsons noted that "On the north side, in Virgin Gorda sound, the effect was very slight, the wave rising but a few inches [10–15 cm], whereas on the south and southwest parts it fell 7 feet [2.1 m] and rose 5 feet [1.5 m]. The reefs were left out of water as walls, and the sea took some time to subside."

1867 11 18 *[continued]*			Consequently, the full excursion of the wave would have been about 3.6 m. Rumbold said that at Road Harbor (Road Town), Tortola, "Immediately after the first shock had ceased, the sea receded or rather sunk about 4 feet [1.2 m], and immediately rose between 4 and 5 feet [1.2–1.5 m] above its original level." The trough-to-peak was then about 2.4 to 2.7 m. The ocean flooded inland 9 to 12 m, "submerging the whole of the lower part of the Town, sweeping before it nearly all of the smaller dwellings which had not been entirely destroyed by the late hurricane and had been replaced in their original positions. On the sea rising, an excessively strong current set in, first towards the land, then seaward. The first tide and current threw up a quantity of deposit from the sea with a strong sulphurous *[sic]* smell. From the velocity with which the old lumber and other matters were taken along by the current, first in one direction and then in another, he estimated it as running at the rate of at least 12 knots [22 kph]. The inhabitants of the Town betook themselves for safety from the overflowing tide," scrambling up the hillsides. Some found shelter in the ruins of the houses earlier damaged by the hurricane and then the earthquakes. Parsons reported that at Road Harbor, the sea abruptly receded, "leaving shallow grounds dry; it very quickly came back, and disturbed all the mud at the head of the bay; some boats broke adrift (but were not lost), and traversed the bay with the various currents. The debris of fallen buildings was all close to the shore, as the street skirts the bay, and is only 2 to 3 feet [0.6–0.9 m] above high water; consequently, when the wave came in, a great deal of this washed about." Rumbold said, "There is not a stone building of any description in this Island which has not been injured in some degree. With regard to the out islands, at Salt Island, as in Tortola, the earth was observed to open on the hills above Tortola and in several places near the seashore where water rushed upwards from the openings." **Leeward Islands:** Damaging waves inundated St. Martin, Saba, and St.-Barthélemy, and a very high wave was reported at Saba. At 2:50 p.m., the 2.1-m-deep St.-Barthélemy Carenage was left dry. Vessels were abruptly deposited on the ground and several were set adrift by the return wave. A schooner rushed about with the currents, at a rate of 10 knots, and was not

DATA	LOCATION	(m)	DESCRIPTION OF TSUNAMI AND SEISMIC EFFECTS
1867 11 18 *[continued]*			able to be secured until 8 p.m. The wharves were submerged by 1 m of water. On the west coast of Barbuda, the sea receded and rose 1.4 m, and then subsided as at other islands. Small waves were reported at Sint Eustatius. At Basseterre, St. Kitts, the ocean receded, then rose quite high, inundating 2.4 to 3 m. At St. John's, Antigua, the tsunami rose 2.4 to 3 m, washing 1.5 m over the wharves, and entered the town. The reflux receded quietly, followed by another wave. At 3 p.m. in English Harbour, the current surged about in the bay at 5 to 7 knots, becoming quite turbid, the commotion not ceasing until 5 p.m. The water rose and fell 0.6 m as the eddies changed and overflowed the dockyard. The shock was felt at 3 p.m. at Montserrat. Movement by the sea was not observed. On the other hand, according to letters reported by Sainte-Claire Deville, things were different at Guadeloupe. Îles des Saintes, situated just south of Guadeloupe, "experienced with some violence this displacement of the sea. All of Fond-du-Cure was submerged; the water *envahi* [invaded] the houses to a height of 1 m. The inhabitants ran away before the looming sea." Saint-Claire Deville cited that, at Basse-Terre, "a very singular phenomenon occurred," and that "a *phénomènes extraordinaire* [extraordinary phenomenon] was produced here in the roadstead. The sea swelled without any outward sign: no waves, no backwash. It rose 1 m in the span of 5 minutes, floating boats tied on the bank. It then dropped in the same span of time, retiring 8 to 10 m, leaving places that were always submerged, revealed. The same phenomenon was produced again, and then the sea went back to normal. Having seen all phases of this spectacle, it can be said that the difference in the extremes of the water levels was not less than 2 m. Several were sure they felt an earthquake at nearly the same time." *Le Commercial* reported, "We have felt, at the Pointe-à-Pitre, one or even two shocks of an earthquake," which were said to have been quite weak, lasting 2 minutes or less. The wave at Pointe-à-Pitre was "hardly more than a swell." Saint-Claire Deville said, "it was completely otherwise on the north side of the isle." He cited a resident of Sainte-Rose who wrote, "Towards 3 p.m. [3:18 p.m. in another account], the ocean suddenly withdrew more than 100 m from the shore. This retreat was preceded by slight oscillations from the earthquake, which lasted perhaps an estimated 5 or 6

1867 11 18 *[continued]*			seconds. Then all at once, a first wave, of 60 *pieds* [feet] [19.8 m] in elevation, rising to the north, approximately 3 *milles* [nautical miles] [5.6 km] broad, rolled violently towards the land, where it crashed, immersing the whole shore, and swamping the houses that were there. Boats drawn up on shore, fishing nets, wood, and other materials—everything was left adrift. A second and third of these enormous waves rolled in from north to south, one after another, at short intervals, and like the first one, overwhelmed everything along the way." On the northwest shore at Deshaies, a tsunami ravaged the town and smashed nearly all the houses. Said one resident, "There's no more bread! Everyone has taken refuge in the church." The immediate arrival and the massive height of the Sainte-Rose and Deshaies tsunami compared to other areas in Guadeloupe, suggest that a local submarine landslide generated this immense tsunami, which was much higher than the tsunamis experienced at many islands closer to the tsunamigenic source. **Windward Islands:** Parsons, designated to survey the effects of the "earthquake wave" for the West India Survey, reported, "Overall, the waves were lesser on the east coasts of both The Leewards and The Windwards. Generally, the waves came on as long swells of the ocean with the advancing ridge of water having a smooth appearance, without a breaking crest unless meeting an obstacle; an overcasting wave would have soon destroyed itself." At Prince Rupert's Bay, Dominica, the sea receded from the shore 4 feet [1.2 m] perpendicular at 4 p.m. Parsons said the 4 p.m. time must have been incorrect. Possibly, use of local solar times could account for time irregularities. As it came back in, it appeared as a wall of water 10 feet [3 m] high, but did not overfall. Several waves followed; the sea did not subside until 6 p.m. Effects were seen elsewhere, but not on the east shore. The larger waves on the north shore of Dominica may have traveled to Dominica from the 19.8-m tsunami that deluged Sainte-Rose, Guadeloupe, then continuing southward. This may explain the greater waveheights at such great distance to the northwest, from islands that were closer to the tsunamigenic source. Waves were noted at Martinique. Other islands in the Windward Islands chain had waves along the west and south shores, perhaps due to some tectonic release, as at Grenada. At St.

DATA	LOCATION	(m)	DESCRIPTION OF TSUNAMI AND SEISMIC EFFECTS
1867 11 18 *[continued]*			Lucia on the west coast, fishermen at Layon and Anse-la-Raye reported that their nets had been overturned by waves of 3 to 4 feet [0.9–1.2 m]. Near Rivière Dorée on the south coast, people observed "the ocean leave the shore, and return about the same amount, and rocks that had never before broken the surface or shown near the water's edge appeared." At St. Vincent a little before 5 p.m., fishermen in Cumberland Bay on the west coast, saw a small wave not over 2 feet [60 cm] perpendicular that receded and advanced several times on the shore. A shallow channel between the shore and Bottle-and-Glass Rocks became dry. Bequia, in The Grenadines, had 3 great slow waves that rose 1.8 m, lasting above 40 minutes. Parsons reported that at Admiralty Bay, "the wave assumed distinct features; it receded from the shore a little before 5 p.m., but soon making a ridge about the entrance, it returned and overflowed the bay to a 6-feet [1.8-m] perpendicular depth—or 160 yards [146 m] up the slope," which swiftly flooded houses near the shore to a depth of 1.2 m. "It receded again, with another very decided wave and after that subsided, with several minor waves and circular eddies about the bay." The reefs and shoals were left dry, "looking like walls coming out of the sea, due to the steep sides and dark coral." The inhabitants described this wave as quiet, but with quick recess and inflow. "The ridge was seen, but it did not overfall or break in the advance, but reached the bay like a smooth undulation. All along these islands, the wave came in 3 times, with the second wave generally the highest, and then several smaller occurred whilst the water was reducing itself to its usual level." At Canouan, a slight wave was seen on the southwest coast, and on the southwest coast of Carriacou, the wave was noticed by fishermen in Tyrrell's Bay about 5 p.m. The sea went out, leaving the reefs dry, then came in again in several small waves. At Grenada, the southern-most of The Windwards, the chief effect was seen on the west shore at St. George's Carenage and at Gouyave. Parsons said, "The wave presented alarming features that frightened the inhabitants." At 4:45 p.m. (5:20 in another account), "the sea suddenly dropped 6 feet [1.8 m], leaving much of the harbor dry, the reef opposite rising from the water as a dark wall. The lagoon poured out its water, and all the retreating streams made a head at the entry to the Carenage. In a few minutes the ridge quickly

1867 11 18 *[continued]*			advanced to the shore, but did not overfall, and flowed over the wharves to a height of 5 feet [1.5 m]. A sloop of 30,000 tons was placed on the wharf by this wave, and carried off with the next. It was said the second wave, occurring 1 minute after the first, was of superior elevation; a minor third followed, then the waters of the Carenage obtained their normal condition, after a series of slight undulations and eddies, which lasted an hour." According to Reid, "The wave action was repeated 6 times, then all was quiet." Steamers at the West India coal wharf were left dry on the bottom, a perpendicular drop of 12 feet [3.7 m]. A wharf was demolished, and its 90-kilogram stones were moved 3 m by the reflux of the wave. Several small houses were carried 15 m inland together "with their inmates." All loose things on the shore were sent floating about the Carenage. At about 5 p.m. at Gouyave, further up the west coast, the incoming wave was "quite equal to that at the Carenage." Vertical oscillations of 6.1 m trough-to-peak were reported. The tsunami damaged boats and buildings in the town. On the south shore at Clarke's Court Inlet and also at Westerhall Bay, Parsons said, "a moderate wave came in, not more than 2 feet [60 cm], while on the east coast at La Baye, the wave was just discernible, and rose about 1 foot [30 cm]. No lives were lost." While some noted they "felt the earthquake itself distinctly, others said there was nothing of the kind; many inhabitants declared that a sulphurous stream was emitted from the middle or eastern arm of the Carenage." The *Grenada Chronicle* detailed the destruction occasioned by "the visitation," and reported that "A long-hidden volcanic crater had come into action and discharged clouds of volcanic gas, by which the waters of the Carenage boiled up and rose to a fearful height." Robson indicates an old crater on the edge of the Carenage. Refuting the possibility of any volcanic activity, Parsons commented, "The mud of the shallows being so disturbed that nasty vapors tainted the air for some time, but sulphur gases never existed at the moment." Parsons did report, however, that "On Nov. 14th, the Central American volcanic mountains showed signs of more than usual activity, and discharges occurred constantly from that date until beyond the occurrences in the islands." Volcanoes in Central America, both new and old began to erupt. Becker referred to a new volcano, Cerro Negro, in Nicaragua,

DATA	LOCATION	(m)	DESCRIPTION OF TSUNAMI AND SEISMIC EFFECTS
1867 11 18 *[continued]*			that erupted for the first time on Nov. 14. The thread of volcanic activity continues with a report of Mount Vesuvius in Italy, erupting from Nov. 14 to the 21st. These phenomena may render plausible the reports, otherwise entirely dismissed, of volcanic activity associated with the earth convulsions at Little Saba and Grenada. They may help to clarify the great height of waves at St. George's and Gouyave, and their arrival primarily on the south and west coasts of The Grenadines. In the *Jamaica Daily Gleaner,* it was reported that "The whole world appears to be suffering from violent convulsions of Nature. Hurricanes in the West Indies, Gulf of Mexico, and Hong Kong—Earthquakes at St. Thomas, Porto Rico, Jamaica, and Tortola and Santa Cruz [St. Croix]—Volcanic eruption in Grenada Harbour two hours after the Earthquakes at St. Thomas—Volcanic eruption in Iceland—Volcanic eruption of Mount Vesuvius, in Italy from 14–21 November." Worthy of note is drop in the level of the Aegean Sea near Izmir, Turkey, reported by Soloviev, who cited Schmidt for Nov. 19 at 21:00 UT, "The sea level was 1.1 m lower than usually." When the seabed collapsed in the Anegada Deep, the resultant drop would cause the sea to rush in to fill the hollow. The drop in sea level at the far end of the Mediterranean may have been a distant effect of waters pulled into the swell of the ensuing tsunami. Such worldwide effects are not without precedent, as exemplified by the worldwide tsunami resulting from the Krakatoa eruption in 1883 and the cataclysm that produced the vast Lisbon teletsunami in 1755. The Aegean is almost due east along the fault-line described by Myles as "the crustal boundary that extends to the Azores–Gibraltar Ridge." Myles said the subduction by the Eurasian Plate of the African Plate, in 1755, from Syria to the Mid-Atlantic Ridge, triggered the massive earthquake. The ensuing tsunami propagated throughout the North Atlantic, reaching the West Indies with 5- to 10 m-waves. Aristides Rohas, 1867; Backshall (no date); Baratta, 1901; Becker, 1869; Brooks, 1919; Campbell, 1991; Chapman, 1867; Dixon, 1869; Fuchs, 1886; *Grenada Chronicle,* 1867; Hall, 1917a&b; *Harper's Weekly,* 1868; Harte, 1868; Harvey, 1876; Housel, 1878; *Jamaica Daily Gleaner,* 1867; Lander & Lockridge, 1989; Lander & O'Loughlin, 1997 & 1999b&e; Latimer, 1868; *Le Commercial,* 1867; Maskell, 1868; Montadon, 1962; Myles, 1985;

1867 11 18 *[continued]*			O'Loughlin, 2001; Örsted, 1867; Paiewonsky, 1979; Parsons, 1868; Perrey, 1870; Piddington, 1852; Porter, 1867; *Port-of-Spain Gazette,* 1867; Raupach, 1867; Reid & Taber, 1920; Riise-Baumann & Baumann, 1910; Robson, 1964; Robson & Tomblin, 1966; Rumbold, 1867; *Saint Croix Avis,* 1867 & 1868; Sainte-Claire Deville, 1867; *Sanct Thomæ Tidende,* 1867; Schmidt, 1881; Schubert, 1994; *Scribner's Monthly,* 1878; Shepherd & Lynch, 1992; Soloviev *et al.,* 2000; Taber, 1922b; Verbeek, 1885; Warner, 1876; Watlington & Lincoln, 1997. V4.
1868 Venezuela	Maiquetía, Venezuela		A wave was reported at Cabo Blanca, Maiquetía. No precise date was given. One of the continuing aftershocks from the Nov. 1867 event, culminating in the strong Mar. 17 earthquake and tsunami, may have been associated with this event. Singer *et al.,* 1983. V1.
1868 03 17 [06:45 LT] 18.1N 65.1W Puerto Rico	Puerto Rico: Arroyo, Naguabo & Vieques Danish West Indies (U.S. Virgin Islands): St. Thomas: Charlotte Amalie Bequia: Admiralty Bay	 0.6 0.9	A sharp 30-second earthquake, heavy all over Puerto Rico [R–F ≅ IX] that felt "nearly as strong as the earlier violent earthquake." Reid (1919a) noted reports from 18 locations, but lamented the incompleteness of the information. "The eastern part of Porto Rico suffered most in the earlier earthquake; the western part suffered most in the later earthquakes." Ships in San Juan Harbor distinctly felt the shock, and along the southeast coasts at Arroyo and Naguabo a small flux and reflux of the sea occurred with flooding. At Vieques, the sea receded and a small wave was noticed. Reid (1920) said the distribution of intensities and the sea-wave indicated that it originated along the north edge of the Anegada Deep, about 48 km southwest of the previous submarine source. The *Sanct Thomæ Tidende* reported waves 2 feet [60 cm] in height, with recession and flooding at Charlotte Amalie. The sea dropped 1.4 m, with a 10-minute period. The only other tsunami report was by Browne for Bequia in The Grenadines: "On the 17th of March, a steady strong wave was seen bearing down on Admiralty Bay; it had no perceptible crest, and was 3 feet [90 cm] in height; it encroached upon the land varying distances from 70 to 400 feet [21–122 m]. A second smaller wave followed. No shock of an earthquake was felt." Note the extensive area between report areas. The 1868 [no month or year] wave at Maiquetía, Venezuela, may have been associated with this tsunami. Browne, 1869; Campbell, 1991; Heck, 1947; O'Loughlin, 2001; Long, 1999; Reid & Taber, 1919a & 1920; *Sanct Thomæ Tidende,* 1868; Taber, 1922a&b. V4.

DATA	LOCATION	(m)	DESCRIPTION OF TSUNAMI AND SEISMIC EFFECTS
1868 08 13 [11:37 UT] 10.7N 63.8W Venezuela Peru	Venezuela: Juan Griego, Isla de Margarita & Río Caribe Tide gauge: San Juan, Puerto Rico	 0.6	Singer reported tsunamis at Juan Griego, Isla de Margarita, and at Río Caribe, along with extensive fault activity. Waves registered as far away as the tide gauge at San Juan, Puerto Rico [60 cm]. The event is notable because tsunamis occurred at the same time in both the Pacific and the Caribbean. Silgado reported tsunamis reaching throughout the Pacific beginning the night of the 13th, caused by a massive earthquake in Peru that devastated the coasts of Peru, Chile, and Bolivia. Waves up to 21 m at Chile, carried the *USS Watertree,* anchored at Arica, 4.8 km up the coast and 805 m inland, where it still lies today. Billings, 1915; Montessus de Ballore, 1912; Schubert, 1994; Silgado, 1985; Singer *et al.,* 1983. V4.
1873 10 14 [18:05 LT] 10.2N 80.0W Panamá	Panamá: Colón Harbor & Aspinwall		Camacho reported a tsunami at Colón Harbor, Panamá. A shock [MMI ≅ V] was strongly felt aboard ships. In Aspinwall, there was "much fright and fear of the tidal wave added to the excitement." Arce *et al.,* 1998; Camacho & Víquez, 1993; Molina, 1997; *Panamá Star and Herald,* 1873; Rojas *et al.,* 1993; Víquez & Toral, 1987. V3.
1874 05 or 08? Trinidad Venezuela	Trinidad		Singer reported terrain uplift in May or Aug., described as an *isla de lodo efímera* [short-lived mud island] that appeared south of Trinidad. Shepherd reported a shock on Aug. 17 at eastern Venezuela [MMI ≅ VIII] and at Trinidad [MMI ≅ IV], which may have been associated with the appearance off Trinidad. The transient islet also appeared in 1853, 1911, 1928, 1934, and as in 1911, waves may have occurred. Centeno-Graü, 1940 & 1969; Fiedler, 1961; *Port-of-Spain Gazette,* 1874; Robson, 1964; Shepherd & Lynch, 1992; Singer *et al.,* 1983. V1.
1874 03 11 [04:30 LT] 15.5N 61.5W St. Thomas	Dominica & St. Thomas		At Dominica, the *Corsica* reported a series of heavy rollers in the harbor near 5 a.m., lasted ½ hour, and rendered communication with shore impossible. Reid said a submarine shock southeast of St. Thomas, accompanied by noise, shook the island and especially ships in the harbor, which had been perfectly calm. Suddenly it appeared "turbid as though clouded by sand and mud. A little later, strong ripples from the south, agitated the water surface." The turbidity may have been from seismic activity agitating the bottom. Reid said the waves at Dominica suggested a local source and were probably not connected with the St. Thomas shock. Perhaps the shock provided the trigger-action for a submarine disturbance near Dominica. Berninghausen, 1968; Palgrave, 1874; Reid & Taber, 1919a. V4.

1880 01 04 [11:00 LT] 15.2N 61.2W Dominica	Roseau, Dominica		The Valley of Desolation Volcano (Grand Soufrière) erupted. Mud and sand rained on Roseau for 10 minutes. Ash spewed in a swath 4 km wide at the coast 10 km away. Floods of muddy water ran down the Point Mulatre River and Rivière Roseau, which rose 3.7 m above its usual level. A 12-foot rise in the river level from mudflows would probably give rise to a significant wave upon entering the sea. Bert, 1880; Daubrée, 1880; Robson & Tomblin, 1966. V2.
1881 08 12 [05:20 LT] 19.6N 75.5W	Kingston, Jamaica	0.46	A well-marked double-shock [Hall ≅ II] was reported with a 1-second vertical shock, then in 3 seconds, a 4-second horizontal shock. A wave was reported on the north coast. At Kingston Harbor, the water rose 46 cm 6 hours after the earthquake. Berninghausen felt that this wave was not caused by the earthquake, but does not give any reason for his conclusion. A hurricane passed near Cuba on this date. The storm may have caused the wave or a true tsunami triggered by an earthquake generated by the immense pressure on the seafloor from the hurricane. Berninghausen, 1968; Brooks, 1919; Dunn & Miller, 1964; Hall, 1907; Hebert *et al.*, 1993; Piddington, 1852; Taber, 1920; Tomblin & Robson, 1977. V1.
1882 09 07 [09:18 LT] 10.0N 79.0W Panamá	Panamá: Archipiélago de San Blas & Miguel de la Borda Tide gauge at: Colón	3.0 0.62	A tsunami "3 m high or more" was cited by Camacho, which swept the San Blas coast. The tsunami, with an estimated magnitude between I–I ≅ 1.0 [3 m] and I–I ≅ 2.6 [5–6 m], washed out most of the archipelago, which remained submerged for several minutes. Between 75 and 100 people perished, most of them drowned. The ocean receded a great distance, and upon returning, swept away villages built on the beaches of different islands and the mainland. Mendoza said the outflow and inflow occurred 4 times, the greatest damage from the third wave. The marigram of a distinct 62-cm train of waves from the French Canal Company tide gauge, at Colón, was lost. Near the coast at Río Indio, two freshwater lagoons were left dry. A "seaquake" was felt 3½ km from the coast aboard the *Honduras*. No visible waves appeared on the surface, but the vessel "began to move, as if at full speed, and ran aground on a bed," according to Montessus de Ballore. Sieberg reported that "The earthquake was accompanied by tsunami waves, which were observed from Panamá to Guayaquil." The tsunami was generated by an Ms ≅ 7.9 earthquake at 7:50 a.m. that was observed at Panamá; Rivas and Greytown, Nicaragua; Buenaventura and Cartagena, Colombia; Maracaibo, Venezuela; and

DATA	LOCATION	(m)	DESCRIPTION OF TSUNAMI AND SEISMIC EFFECTS
1882 09 07 *[continued]*			Guayaquil, Ecuador. It was felt strongly by vessels offshore at Panamá City, but no anomaly was observed at the tide gauge on Isla Naos (Pacific side). Sarasola reported a hurricane with a "good number of victims," at Cuba, from Sep. 5–6, which could have served to trigger the shock, as discussed by Brooks. The cable between Colón and Jamaica broke, which suggests a submarine landslide. Submarine slides triggered by earthquakes resulted in submarine cable breaks in the Gulf of Corinth, Greece, where it was established by Forster that submarine landslides were responsible for the majority of "seismic sea waves" accompanying local earthquakes. Arce *et al.*, 1998; Brooks, 1919; *Bulletin du Canal Interocéanique,* 1882; Camacho & Víquez, 1994; Fernández *et al.*, 1999; Forster, 1890; Fuchs, 1883; Grases, 1974; Hall, 1907; Iida *et al.*, 1967; Leeds, 1974; Mendoza & Nishenko, 1989; Milne, 1912; Molina, 1997; Montessus de Ballore, 1888; Nelson, 1889; *Panamá Star and Herald,* 1882; Ramírez, 1969; Rappaport & Fernández-Partagas, 1995; Rockwood, 1883; Rojas *et al.*, 1993; Sarasola, 1928; Sieberg, 1932; Solov'ev & Go, 1984a; Tabor, 1967; Víquez & Toral, 1987. V4.
1883 08 27 [10:00 LT] 05.8S 106.3W Krakatoa, Sunda Straits, Java	St. Thomas Elsewhere: English Channel Table Bay, Cape of Good Hope, Africa	 0.36 0.36	The *Monthly Weather Review* reported, "A tidal wave occurred here on Aug. 27th," at St. Thomas. The sea receded 3 times from the shore. The eruption (30 billion tons of magma) of Krakatoa in the Sunda Straits near Java is considered one of the world's worst historical disasters, which created 30 to 40 m waves at Java and other islands, causing 36,000 fatalities. The explosion created air pressure waves that caused tsunamis recorded at Australia, Japan, South Sandwich Islands, Hawai'i, California, and Alaska. The English Channel and Table Bay, Cape of Good Hope, Africa, had 36-cm waves. Berninghausen said that it could be assumed that similar small waves occurred throughout the Atlantic. It has been debated whether changes in barometric pressure from the explosive eruption or energy transmitted directly through the water propagated the tsunamis that traveled the world's oceans. Possibly enhancing the effect at St. Thomas, is its location nearly antipodal to Krakatoa. Forel reported a subterranean noise heard at *l'Îlot Caiman-Brac* [Cayman Brac Islet] in the Cayman Islands. Berninghausen, 1964; Evans & Wharton, 1888; Ewing & Press, 1955; Forel, 1885; *Monthly Weather Review*, 1883; Taber, 1922b, Sigurdsson, 1996; Verbeek, 1885. V4.

1887 09 23 [06:55 LT] 19.5N 74.4W Hispaniola Jamaica	Haiti: Môle St.-Nicolas, Jérémie, Anse d'Hainault & Golfe de la Gonâve		An earthquake, originating in the Bartlett Trough southwest of Môle St.-Nicolas, was strongly felt throughout Haiti and in the Northern Depression of Hispaniola. The disaster struck Môle St.-Nicolas [MMI ≅ IX] the heaviest. Môle St.-Nicolas was covered in ruins from the collapse of nearly all the houses. According to Scherer, "Springs of hot water burst forth in the middle of town and the La Gorge River, which had been dry since 1878, reappeared in all its force. The sea, at the moment of the shock, drew back a great distance from the city. At Cap-Haïtien, the shocks were numerous; 88 houses were cracked, and the few which had resisted in 1842, fell." Taber said, despite the severity, it was not as intense as the 1842 Cap-Haïtien disaster. The tsunami was observed at Môle St.-Nicolas, and significant waves were reported near Golfe de la Gonâve. Scherer said that at Jérémie, "The sea drew back 20 m, leaving numbers of fine fish out of water, and the people were still disputing about these when the water returned with a rush upon the beach. The same movement of the sea made itself felt in all the neighboring ports, particularly at Anse d'Hainault." One-minute shocks came from the northeast at Gonaïves [MMI ≅ VII], where the population bivouacked outdoors. The havoc was considerable at Plaisance, Gros Morne, Marmelade, St.-Marc, and Port-de-Paix [MMI ≅ VII]. At Port-au-Prince [MMI ≅ VI], oscillatory movement lasted 2½ minutes. Strong shocks were felt during several minutes at Jérémie [MMI ≅ VI], and the inhabitants abandoned their dwellings. Santo Domingo [MMI ≅ VII] "was much damaged." Taber said the distribution of intensities was consistent with Scherer's assessment that the earthquake evidently had its epicenter to the south of Môle St.-Nicolas, "for the accompanying sea-wave was practically limited to the shores of the Windward Passage and the Golfe de la Gonâve, being of no importance along the north coast of Haiti." It was also felt at Jamaica [Hall ≅ III]. At Great Inagua [MMI ≅ VII], Montadon noted a *forte répercussion* [strong seismic repercussion]. The lighthouse-keeper at Inagua reported a very severe shock at 7 a.m. and another at 8:10 p.m. Woodring reported the same areas affected as those shaken by the event in 1842, and Santiago de Cuba [MMI ≅ VIII] had severe 30-second oscillations followed by three shorter shocks. Rudolph said the earthquake was felt on the south coast of Cuba at 7.

DATA	LOCATION	(m)	DESCRIPTION OF TSUNAMI AND SEISMIC EFFECTS
1887 09 23 *[continued]*			a.m. The Captain of the *Alps* said, "A shock, one from a submarine earthquake, lasted some 45 seconds; the *Alps* was left shaken everywhere." The *Tamora* reported 3 distinct shocks in a half-hour. The ship jolted as if it had hit ground, and the masts and rigging were shaken about heavily. The *San Francisco* reported similar effects observed north of Cuba, 20 *seemeilen* [nautical miles] [37 km] from land. Shocks at 7 a.m. were felt aboard the British, *Daisy,* and American barquentine, *Miranda.* At sea between Haiti and Cuba, the American barque, *Evan 1,* felt severe shocks at 6:47 and 6:54 a.m. In the next 3 hours, she felt 30 more, plus another 100 more through noon the next day. Heck mistakenly identified the area as the Philippines. Berninghausen, 1968; Hall, 1907; Heck, 1947; Milne, 1912; Montadon, 1962; Rudolph, 1895; Scherer, 1912; Shepherd & Lynch, 1992; Taber, 1922b; Woodring, 1924. V4
1893 02 19 [01:30 LT] 17.6N 76.7W	Kingston, Jamaica		At Kingstown, a short sharp shock occurred. A pronounced ripple crossed the harbor from south to north. It was said that fish jumped out of the water onto the wharves and jetties. The fish may have been stranded by the backwash of the wave. Tomblin & Robson, 1977. V1.
1896 02 16 [13:00 LT] 17.6N 76.7W	Kingston, Jamaica		A sharp shock was felt at Kingston. It was felt lightly throughout the whole island. It was followed by heavy roaring of the surf that lasted half-an-hour. Whether the heavy roaring was from ocean waves or was a sound reverberation is not clear. (See account of a "booming" sound set up on Jan. 15, 1907.) Such a roaring noise, if due to a disturbance of the sea, can serve as a natural tsunami warning signal. Hall, 1907; Tomblin & Robson, 1977. V1.
1897 11 29 16.4N 62.1W Montserrat	Montserrat		A report in the *Journal of Geology* on the frequent occurrence of earthquakes at Montserrat, said, "After the large tsunami attacked the island on Nov. 29, 1897, a number of earthquakes were felt. Particularly severe were those on Feb. 15, 18, and 20, 1898. As the island is volcanic, there might be a possibility that serious eruption of volcanoes could occur in the near future." Shepherd cited an account about Feb. 15 that said, "A terrific earthquake between 11 a.m. and noon occurred here today, the severest in the island for many years, and which dwarfs into insignificance the shocks we have been having since the flood (Nov. 29, 1896 *[sic]*)." Despite uncertainty as to which year, a large tsunami "attacked" Montserrat. Murphy, 1964; Shepherd & Lynch, 1992; *Journal of Geology,* 1898. V4.

1897 12 29 [06:32 LT] 19.3N 71.5W Hispaniola	Port-au-Prince, Haiti, Puerto Playa, Dominican Republic		A powerful 1½-minute shock at Port-au-Prince, Haiti [R–F ≅ V], was recorded instrumentally at Cecchi, and world-wide at Rome, Ischia, and Catania, Italy; Shide, England; Toronto, Canada; and Nicolaiew, Russia. It was felt forcibly at Mirebalais and Jacmel, Haiti. At Santo Domingo (Dominican Republic), ground subsided and great cracks formed, but where houses had been built on pilings none tumbled down in spite of R–F ≅ IX–X intensities. The submarine cable break east of Puerto Playa may have been triggered by a submarine landslide, according to Taber. He said major "sea-waves" were not reported for the disasters of 1564 or 1783, nor for this event. Nevertheless, a submarine collapse that broke underwater cables may have produced a tsunami. Forster established that, in the Gulf of Corinth, Greece, submarine landslides triggered by earthquakes, which resulted in submarine cable breaks, were responsible for the majority of "seismic sea waves" accompanying local earthquakes. Forster, 1890; Scherer, 1912; Taber, 1922b, Woodring, 1924. V1.
1900 10 29 [04:42 LT] 11.0N 66.4W Venezuela Trinidad	Venezuela: Puerto Tuy, Macuto, Río Neveri & Puente de Hierro	10.0	A 10-m tsunami occurred at Puerto Tuy associated with an earthquake, according to Singer, who noted that a tsunami might have occurred at Macuto also. Centeno-Graü placed the epicenter between Isla La Orchila and the coast at Macuto, citing a ruinous 50-second temblor [MMI ≅ IX], killing 25 at Guarenas. Charallano, San Casimiro, and Cúa were destroyed. Great damage with thousands of victims occurred at Caracas, Guatire, Higuerote, Tacarigua, Río Chico, Brochico, Macuto, and Curiepe. Schubert said this was one of only two very large "sea-waves" reported in Venezuela, the other in 1726. An islet at the mouth of Río Neveri disappeared. At Río Chico, railway and telegraph service to Carenero was interrupted; 20 shocks cracked the railway, streets, and edifices at Caracas; and a road split and shifted 300 m at Petaquire. Puente de Hierro was devastated, with part of the village subsiding under water. Fiedler estimated the convulsion affected a million square km. Singer reported numerous phenomena: fault activity, liquefaction, surface ruptures at Los Roques; a channel altered at Puente Colorado; a landslide at Farallón La Tortuga; and collapse at Islas Borrachas. Very severe shocks were reported at Trinidad. Bolt *et al.,* 1975; Centeno-Graü, 1969; Fiedler, 1961; Grases, 1971; *Port-of-Spain Gazette,* 1900; Robson, 1964; Singer *et al.,* 1983. V4.

DATA	LOCATION	(m)	DESCRIPTION OF TSUNAMI AND SEISMIC EFFECTS
1902 05 07 13.2N 61.1W St. Vincent	St. Vincent The Soufrière		Schubert called this one of the volcanism-produced tsunamis in the West Indies (also citing Mont Pelée, 1902). The Soufrière began to erupt on Apr. 30, killing 56 inhabitants, continuing until May 7, when an estimated 1.4 cubic km of pyroclastic flow erupted; 1,565 perished. Mud and water from the crater lake were thrown out above the rim. Then vast mudflows 6½ hours later descended to the mouth of the Rabacca River on the east coast. Tsunamis were produced when these large mudflows entered the sea. For events from May 7–8, Woodworth cited the *Elizabeth Daily Journal,* "Wallibou and Richmond plantations and villages on the leeward coast were wrecked. Wallibou was partly under water, which had swept in from the sea by a tidal wave. Five other plantations were gone." Richmond plantation remained under water. On the Richmond estate near the seashore, "a new crater had formed, from which fire and ashes were being vomited." The Carib Indians completely disappeared. Only a few Carib individuals were left on St. Lucia and Dominica. Woodworth commented, "It would be extremely important to see if we can find out if most, or all, of the Caribs were killed by the ocean, or other results of the eruption. If the ocean, this would be a very rare tsunami indeed; one that basically wiped out a branch of the human race." The explosions were heard at Trinidad, 280 km to the south, and 672 km to the northwest in the Virgin Islands. Woodworth (2001b) reported that the Wallibou estate was destroyed, at least in part, by the tidal wave there. "The strange color and great depth of the new inlet of the sea on the Wallibou estate imply that an eruption occurred there and that a new crater has been formed which extends from the land into the sea." Robson said The Soufrière erupted from its "old" crater, previously active in 1718 and 1812. The "new" crater, first active in 1812, became inactive from 1902 onwards. Heilprin said portions of the sea bottom adjacent to St. Vincent to the north, "sank incident to the eruption of The Soufrière." Heilprin, 1903; Robson & Tomblin, 1966; Woodworth, 2001a&b. V4.
1902 05 08 14.5N 61.1W	Saint-Pierre, Martinique		Mont Pelée violently erupted sending a *nuée ardente* [blazing cloud or pyroclastic flow] that destroyed Saint-Pierre, devastated 52 square km and killed 38,000 people (only one person survived—a prisoner in the depths of a jail). The blast was tornadic in the violence of its

1902 05 08 *[continued]* Mont Pelée, Martinique			sweep. An eyewitness, who saw the flow coming down the mountain from his ship: "The fiery cloud reached the ship, then another roar, and with it all the water in the harbor seemed to gather up and rush upon the shipping. Every craft heeled over to the great tidal wave and seemed to careen and sink, according to Woodworth." Afterwards, "The sea was rolling like the heaviest kind of ground swell." When this witness, severely injured, looked at Saint-Pierre, the town was gone. "All around us the ships were sunk or aflame, and between us and them and the shore dead bodies floated singly and in groups." Many ships were capsized by the tornadic blast of pyroclastic debris, described by Captain Freeman, of the *Roddam,* as "an enormous black cloud, like a wall with patches of fire in it, approaching the sea from the land, according to Woodworth (2001b). "With it came an immense tidal wave of boiling water, accompanied by a loud and terrible noise. Immediately the steamer was caught and tossed over on her side, almost capsizing." The wall of fire swept over the town and bay, struck the *Roddam* broadside with terrific force, and covered her "from stem to stern with tons of powdered lava," which perilously retained its heat for hours. An eye witness from another ship said, "The terrific tidal wave which had swept over the *Roddam* and nearly capsized her had parted her cable, and set the vessel adrift." Chief-Officer Scott, of the less-fortunate *Roraima,* said the firestorm caused "a rain of burning ash that was directly responsible for burning most of the shipping in the roadstead." Following the wave, "The sea was a confused mass of boiling mud." Centeno-Graü, 1969; Heilprin, 1903; Robson & Tomblin, 1966; Woodworth, 2001a&b. V4.
1902 05 09 14.0N 61.0W St. Lucia	St. Lucia		At Hodder's Submarine Mount, lying 8 km west of St. Lucia, two bubbling white oval patches appeared 8 m apart, visible from 8:30 a.m. to 1:30 p.m. The turbulent areas, 91 m × 55 m and 91 m × 137 m, then ceased. Such commotion from a volcanic vent may have produced waves. Heilprin reported disturbances in the ocean noted 11 m west of St. Lucia, and that "sea-captains remarked on a material change in the course of currents sweeping along the coasts of Martinique." This series of eruptions in May comprises a three-day period of intense volcanism. Anderson & Flett, 1903; Heilprin, 1903; Robson & Tomblin, 1966. V1.

DATA	LOCATION	(m)	DESCRIPTION OF TSUNAMI AND SEISMIC EFFECTS
1902 05 20			Woodworth said eruptions of Mont Pelée and The Soufrière, St. Vincent, continued, "and, yes, tsunamis were produced." The *Elizabeth Daily Journal* of May 21, 1902, "Tidal Wave Destroys A Village," cited by Woodworth for Martinique, said "it was reported that a tidal wave had destroyed a portion of the village of Le Cabet." He said this tidal wave was on May 20, and "it evidently extended to Fort-de-France, where, as mentioned in the May 22 article, "Fort-de-France on Tuesday was covered with ashes, stones were falling, and a tidal wave added to the terror of the population." Further investigation is needed to confirm whether the continuing pyroclastic flows of the period produced tsunamis. Woodworth, 2001b. V3.
1902 08 30 [21:00 LT] 14.5N 61.1W Martinique	Fort-de-France, Martinique	1.0	At 9:25 p.m., the sea retreated from Fort-de-France, followed by a more rapid rise of 1 m, which covered the quays and inundated to the border of the *savane* [savanna], according to Heilprin. Earlier, at 1 p.m., "a great flocculent volcanic cloud flowed from the crater of Mont Pelée, becoming dark gray in the center and turning white on the border," halting about half-way to Fort-de-France. A violent eruption occurred at 9 p.m., comparable to the May eruption. With "intense lightning and electric manifestations," it advanced rapidly almost to Fort-de-France, accompanied by a light fall of ashes and small stones. Heilprin reported "simultaneous scintillant and zigzag discharges, producing an uninterrupted crepidation, from time to time effaced by a great glow, which illuminated the clouds." A strong odor of ozone was perceptible and the barometer depressed 3 mm and rose 4 mm in a 10-minute period. Schubert mentioned volcanism-produced tsunamis in the West Indies taking place "in 1902 at Mont Pelée and St. Vincent." Either of the 1902 Mont Pelée events could have been intended—this event, since it occurred in concert with The Soufrière eruption, or the Aug. 30 eruption, which specifically noted a tsunami—possibly both were meant. Dudley & Min 1998; Heilprin, 1903; Robson & Tomblin, 1966; Schubert, 1994. V4.
1904 12 20 [05:42 UT] 09.2N 82.8W 1904 12 20	Bocas del Toro, Panamá Limón, Costa Rica		An Ms ≅ 7.8 earthquake may have generated a tsunami at Bocas del Toro. At anchor aboard the Man-of-War, *USS Dixie,* the crew felt the shock severely. Part of Sapodilla Islet subsided. At Limón, after the earthquake that was felt strongly along the *Litoral Atlántico* [Atlantic seaboard], uplift of submerged flat reefs made them visible. The thrust-type tectonic source

1904 12 20 *[continued]* Panamá Costa Rica			was within the North Panama Deformed Belt. Ambraseys & Adams, 1996; Arce *et al.*, 1998; Bermúdez León *et al.*, 2000; Boschini & Montero, 1994; Camacho & Víquez, 1993; Feldman, 1984; González Víquez, 1910; Molina, 1997; Oddone, 1907; Pacheco & Sykes, 1992; Rojas *et al.*, 1993; Víquez & Toral, 1987. V2.
1906 Venezuela	Cabo Blanca, Maiquetía, Venezuela		Singer reported a tsunami at Cabo Blanca, Maiquetía [no day or month], noting an association with an earthquake, reported separately from the immense earthquake and tsunami on Jan. 31. Schubert listed Maiquetía for Jan. 31st. Schubert, 1994; Singer *et al.*, 1983. V2.
1906 01 31 01.0N 81.5W Venezuela	Venezuela: Cumaná, Carúpano, Río Caribe, Nueva Esparta		Tsunamis occurred at Cumaná, Carúpano, Río Caribe, and Costas Nueva Esparta. The death toll from tsunamis was estimated at 500 victims. An Ms ≅ 8.2–8.9 earthquake produced 3- to 5-m tsunamis on the Pacific coasts of Ecuador, Panamá, and Costa Rica—a rare report of tsunamis occurring in both the Pacific and Atlantic oceans. Grases, 1971; Molina, 1997; Rudolph & Szirtes, 1906; Schubert, 1994; Singer *et al.*, 1983; Solov'ev & Go, 1984a. V4.
1907 01 14 [15:30 LT] 18.2N 76.7W Jamaica	Jamaica: Port Royal Gun Cay, Kingston, Buff Bay, Hope Bay, Orange Bay, Sheerness, St. Ann's Bay, Annotto Bay, Port Maria, Ocho Ríos, Port Antonio	 2.5 2.5 2.5 2.5 2.5 2.2 2.2 2.5	An eye witness account reported by Hall said the "great earthquake" of Kingston [MMI ≅ IX], "came without any warning, and in the space of 20 seconds, Kingston Town was shaken into a mass of shapeless ruins." It caused the immediate death of 1,000 and injured another 1,000, many of whom later died; 90,000 were left homeless. The only seismograph was broken by the earthquake. "Shocks from warped a road as if it was a ship's deck in a heaving sea." An eye witness said in one place the railway "buckled up like a miniature switch-back—sleepers, cement, roadway, and all—some 20 or 30 feet [6–9 m], forming a curve under which an ordinary-sized man could walk. At other places, the rails formed a sharp curve, still parallel, but the rails were curved like a 'U' as if forged by a blacksmith." Hall, himself a witness, said that at Kingston, "When the cloud of dust lifted, it was seen that the walls of houses and stores had fallen across the narrow streets, burying the foot-passengers, horses and carriages, and here and there an electric car. In other cases, the walls of the stores had fallen inwards, burying buyers and sellers alike." Fire prevented victims beneath the rubble from being rescued; "That night made a clean sweep of the ruined business part of Kingston." Hall quoted a report, "We had not left the land end of the pier more than a few yards [1.8–2.7 m]

DATA	LOCATION	(m)	DESCRIPTION OF TSUNAMI AND SEISMIC EFFECTS
1907 01 14 *[continued]*			before the awful sounds of the earthquake began. Looking round, I saw the covered pier totter to its fall, and instantly the ground of the garden over which we were walking heaved up beneath us, causing us to reel and stagger, and fissures in the garden paths and grass appeared, from which sprang upwards, volumes of sea water. The garden was studded with coconut palms, and to these we clung as we were thrown about. The earth wave passed under the hotel in front of us and the building instantly fell down within a few yards of us." Subsidence occurred in Kingston Harbor, at Port Royal. The Military Hospital built on "thick brick arched piers sank bodily into the earth 4 to 5 feet [1.2–1.5 m]." The Victoria and Nelson batteries, "sank 1.2 m into the ground, which appeared to have sunk as many feet." The Point was submerged under 2.4 m of water southeast of the forts. Mangrove bush that had been previously above water, at the eastern edge of the Palisadoes, was measured at 4 fathoms [7.3 m] deep. A wide strip of beach [46–55 m] subsided and disappeared; "the lagoon and sea are one." Many furrows formed in the deep dry sand, 27 m × by 4.5 m, and 2.7 m deep. At places, sand had poured into the fissures. At others, mud and water had been ejected up through the fissures. Those who saw the sea immediately after the shock said it "was mountain high." The line of seaweed that remained visible confirmed the high wave. Unfortunately, no runup height was estimated. Berninghausen said seiches of 2.5 m were set up in Kingston Harbor. Hunt, a pilot of Port Royal, who was sailing his sloop of 5 tons, 3 km southeast of Port Royal, reported that he and his crew observed "a sort of half misty cloud coming from the SSW over the surface of the water, resembling a spray of steam or a sort of boiling appearance." They felt the boat rock and toss as if it was bumping on the ground and in a few seconds, they witnessed "the huge wave strike Gun Cay and completely cover it from view, and almost simultaneously the [Palisadoes] Point was struck and obscured from view. An enormous cloud of dust was seen to rise from town." The same phenomenon was seen when it struck Kingston. The submarine cable was broken in three places. Great damage was done to surrounding areas. Intensities of MMI ≅ VIII were noted at Holland Bay, St. James Bay, and Port Maria, where it twisted a suspension bridge. Buff Bay was destroyed by a

1907 01 14 *[continued]*			violent tremor, "and the sea withdrew some distance from the land." Near Buff Bay, two people were killed by falling rocks. A tsunami pounded the northern coast with waves up to 2.5 m at St. Ann's, Sheerness, Buff, Orange, and Hope Bays. At Ocho Ríos, the sea withdrew 69 m. At Port Maria, the sea receded 25.6 m, returning as waves of 1.8 to 2.4 m. At Annotto Bay, about 3 minutes after the shock, the sea retreated 73 to 93 m, dropping 3 to 3.7 m. The wave then rising 2.2 m above normal (trough-to-peak of 5.2–5.9 m) swept into the lower parts of town, destroying many dwellings. On the higher land, it inundated 7.6 to 9 m. At Port Antonio, the strong backwash moved a small building towards the beach. A passenger aboard a launch in Port Antonio reported, "Suddenly the launch reversed ends and the pilot seemed to lose all control of it," not understanding what was happening until he observed "boys running along the hills staggering and the trees shaking," along the shore. Waves of lesser significance were reported along the south coast of Jamaica. The recession of the sea and short time period after the earthquake indicated a local submarine landslide. Hall placed the epicenter 5 km south of the Palisadoes, between Plumb Point Lighthouse and Harbour Head. Hall determined that, since the intervening mountains would have attenuated those shocks, a second focus lay between Buff Bay and Enfield, formed below the geological fault in the St. George district. The seismometer, when restored to operation, recorded aftershock traces revealing the two foci. The first bearing cut through Harbor Head, and the second cut the north shore between Annotto and Buff Bays. Hall noted that a well-marked shock was felt at Guantánamo, Cuba [Hall ≅ II], at a distance of about 175 miles [282 km]. Based on barometric measurements, Hall reported, "It thus appears that in some cases, two small atmospheric waves pass over Kingston about the time of an earthquake: that the first wave passes about 8 hours before the shock: that the second wave passes about 16 hours after the shock: and that the shock itself takes place during the small depression between them. For this event a wave passed at 7 a.m., 8 hours before the shock, and the second wave passed at 3 p.m. on the 15th, 24 hours after the shock. The depression at the time of the shock was much larger than with the small shocks usually experienced in Jamaica." Most earthquakes at

DATA	LOCATION	(m)	DESCRIPTION OF TSUNAMI AND SEISMIC EFFECTS
1907 01 14 *[continued]*			Jamaica are probably due to submarine landslides due to its location atop the ridge between two great "deeps," the major strike-slip faults having little or no vertical displacement. Jamaica is on a foundation of a submarine ridge that juts 805 km out from Cabo Gracias a Dios on the coast of Honduras. The ridge lies between the Bartlett Deep on the north side and the deep southeast of Jamaica. From this ridge, 2000 fathoms [3.7 km] above the bed of the ocean and 800 fathoms [1.5 km] below the surface, elevations rise forming the Rosalind, Gorda, and Pedro Sand Banks. The ridge culminates on Jamaica's Blue Mountain, at 2,256 m, the highest point along the ridge. Many of the earthquakes and the attendant tsunamis have similar characteristics because of the shape and features of Kingston Harbor. The great 1692 Port Royal tsunami had much in common with this event. Berninghausen, 1968; Forster, 1890; Hall, 1907; Heck, 1947; Lynch & Shepherd, 1995; Murty, 1977b; Rubio, 1982, Taber, 1920 & 1922b; Tomblin & Robson, 1977. V4.
1907 01 20 [23:00 LT] 18.2N 76.7W Jamaica	Buff Bay, Jamaica		Hall cited an unnamed correspondent to the *Daily Telegraph,* who recorded details of the Jamaica events at Buff Bay for Jan. 14, 15 and 20. He listed details about the shocks, atmospheric conditions, and rumblings: "The noise apparently comes from the sea." For the 20th he wrote: "I visited the Race Course between 4:30 and 6 p.m. and still heard rumbling and felt vibration which seemed only apparent in this locality." He reported two persons killed from falling rocks, and then noted: "The sea withdrew some distance from the land." This withdrawal may denote a tsunami, but there is only the one report. Hall, 1907. V2.
1907 02 12 18.2N 76.7W Jamaica	Port Royal Harbor, Jamaica		The eyewitness reports of Kirkpatrick were cited by Hall, including one for a disturbance of the waters "about a month later." These events were located "outside the Palisadoes" as well as "inside the Harbour." Aftershocks had continued daily since the major Jan. 14. Then, "On Feb. 12th, at 9 a.m., there was an unusual commotion in the harbor; waves were dashing on the shore from a belt of disturbed water, in the middle of the harbor, about half-a-mile in length [805 m]. At a point where the greatest ebullition appeared, the depth of the water is about 60 feet [18.3 m]. A gentle breeze was blowing at the time." Although the waves may have been due to a seaquake, a small tsunami might have been generated. Reports of "agitation of

1907 02 12 *[continued]*			the waters" are not uncommon in the historical literature, but such a vivid picture yields details otherwise often lacking. Kingston Harbor is across the narrow Palisadoes Sandspit from the epicenter of the main shocks. Hall, 1907. V2.
1910 05 11 18.3N 70.7W Hispaniola	Bahía de Ocoa, Dominican Republic		A severe earthquake occurred with a distribution of intensities that indicated an origin along the faults connected to the Swan Island–Jamaica–South Haiti Trough, the same fault zone that produced the tsunami that overwhelmed Azua in Oct. 1751. Scherer reported that "the shock was strongest in Bahía de Ocoa [MMI ≅ VII], where the sea wall was broken." Due to the geophysical situation in the bay reported by Taber for 1751, and the severity of the shock, waves much as the ones in 1751 may have been generated. The shock was "equally strong" at Barahona and Cristóbal [MMI ≅ VII]. Forceful shocks were felt at Santo Domingo and in Haiti at Cap-Haïtien, Port-de-Paix, and Port-au-Prince. Lynch & Shepherd, 1995; Scherer, 1912; Taber, 1922b. V1.
1911 11 03 10.0N 61.4W Trinidad	Trinidad		Following the volcanic explosion of a mud island volcano off Trinidad, "extraordinary waves" were observed. Singer reported terrain uplift of the beach 5 days later, on Nov. 8, described as an *isla de lodo efímera* [short-lived mud island]. Similar *islas de lodo efímera* appeared, also south of Trinidad, in 1853, 1874, 1928, and 1934, possibly with waves also. Arnald and Macready, 1956; Berninghausen, 1968; Centeno-Graü, 1940; Singer *et al.,* 1983. V4.
1916 04 25 [20:45 LT] 09.2N 83.1W Panamá Costa Rica	Panamá: Bocas del Toro, Isla de Carenero, Careening Cay, Isla Colón & Isla Bastimentos	 0.6 1.2 1.3 0.5	A severe earthquake [Ms ≅ 7.3] caused considerable damage at Bocas del Toro and Almirante [MMI ≅ VIII–IX], disrupting electric and water service. With its epicenter on the *Litoral Atlántico* [Atlantic seaboard], according to Bermúdez León, it caused major damage at Sixaola, Costa Rica. Houses were shifted from their supports, small buildings tumbled down, and a reinforced three-story building swayed 2 feet [60 cm] from the perpendicular. Fissures opened and closed and fresh water flowed from cracks at Bocas del Toro and at Almirante. Reid reported the submarine telephone cable linking the two areas pulling apart in two places, "the strange thing being that the wires were drawn to the size of a coarse hair" before they became separated. A small tsunami [I–I ≅ 0] flowed knee deep over [60 cm] the land at Bocas del Toro and carried canoes and debris 200 m inland. Large tanks washed away and a

DATA	LOCATION	(m)	DESCRIPTION OF TSUNAMI AND SEISMIC EFFECTS
1916 04 25 *[continued]*			factory "came to grief." An eye witness, debarking at Bocas del Toro, said that, at 8:45 p.m., "just as we arrived, a violent earthquake registered in a terrifying way, causing a huge panic in the entire city. We were on board during the earthquake, but the quaking was felt as if we were on land. The boat was lifted by the waves and swept by strong currents. The lights of vessels anchored in the harbor could be seen slowly gliding a great distance, impelled by the strange sea currents." The pier was raised up by the waves and pulled along by a strong current "like that of a river." Isla de Carenero, situated at the mouth of the port, was inundated with 4 feet [1.2 m] of water. Careening Cay was severely affected by the earthquake and overwhelmed by 1.3 m of water; parts of it were completely submerged. The place where one observer had been standing was plunged entirely under the waves. At Old Bank, Isla Bastimentos, "The sea flowed beyond the usual limit for some time." The tsunami flooded Isla Colón, reaching 50 cm in height. The shock was strongly felt on a ship. Eastern Hispaniola [MMI ≅ IX] experienced the shock, as did Puerto Rico [MMI ≅ VI]. Ambraseys & Adams, 1996; Arce *et al.,* 1998; Bermúdez León *et al.,* 2000; Berninghausen, 1968; Camacho & Víquez, 1993 & 1994; Feldman, 1984; Forster, 1890; Grases, 1990; Güendel, 1986; Heck, 1947; Kirkpatrick, 1920; Lynch & Shepherd, 1995; Molina, 1997; *Panamá Star and Herald,* 1916; Reid, 1917; Rojas *et al.,* 1993; Víquez & Toral, 1987. V4.
1916 11 12 10.3N 67.8W	Ocumare de la Costa, Venezuela		A wave was reported at Ocumare de la Costa. Singer noted an association with an earthquake. A shock was reported at Puerto Rico at 5:20 p.m. This wave may or may not have been a tsunami. Reid & Taber, 1919a; Schubert, 1994; Singer *et al.,* 1983. V2.
1918 10 11 [10:14 LT] 18.5N 67.5W Puerto Rico	Puerto Rico: Punta Agujereada, Punta Borinquen,	 5.5–6.1 4.6	One of the most severe earthquakes known to strike Puerto Rico originated in the Mona Canyon, 24.2 km west of Punta Higüero [MMI ≅ IX], probably triggered by thrusting into the trench near the Brownson Deep in the Fault Zone. Generally reported as magnitude 7.5, according to the map in Reid 's report to the U.S. Congress, Punta Borinquen [MMI ≅ IX], closest dry land to the epicenter, experienced an Ms ≅ 9 earthquake. A "great sea wave" lashed the northwest coast, which had the highest runups of 6.1 m. With the trough preceding the crest, the tsunami was observed almost immediately following the shock. Observers noted,

1918 10 11 *[continued]*	Aguadilla & at the head of the bay, Río Culebrinas, Punta Higüero & 1 km SE, Mayagüez, Isla Mona Bahía de Boquerón Gúanica & 1 km SE Playa de Ponce, Cayo Cardona, Isla Caja de Muertos, Río Grande, Río Grande de Loíza, Isabella Puerto Arecibo Tortola	2.4–3.7 3.4 4.0 5.5 2.7 1.5 4.0 1.1 0.43 0.5 0.75 1.5 0.01 1.0 2.0 0.6 0.70	"The ocean first withdrew from the land, in places exposing reefs and stretches of bottom never before visible during the lowest tides, then the water returned reaching heights equally high above normal. At some places the great wave was followed by one or more smaller ones, and especially in sheltered bays, the water continued to ebb and flow for some time." The tsunami caused extensive damage, especially to villages established in a flood plain. At Punta Borinquen, the 15-m tall lighthouse solidly built on rock was badly shattered and would have fallen but for the support of the iron stairway, and later abandoned. Reid said the keeper reported waves reaching 4.6 m above sea level, "not counting the wash of the wave." The surge penetrated 100 m into a coconut grove. With 5.5- to 6.1-m runups along the base of 100- to 120- m limestone cliffs near Punta Agujereada [MMI ≅ IX], the tsunami drowned 8 people, turned a narrow strip of beach used for pasture into a sandy waste, and uprooted several hundred coconut palms. At Aguadilla [MMI ≅ VIII], waves drowned 32 and destroyed 300 huts. Runup heights reached 3.7 m, but varied in different parts of the city. Reid said evidence from stained wallpaper enabled fairly accurate measurements: "In no place were the measurements less than 2.4 m above sea level, and near the head of the bay the crest of the wave must have been 3.4 m in height." At Río Culebrinas, 4 m southwest of Aguadilla, 1,000-kilogram blocks of limestone from the Columbus Monument wreckage were carried 45 and 75 m inland by 4-m waves. Two minutes after the sea receded at Punta Higüero (Jigüero) lighthouse, 5.5-m waves uprooted palm trees and stranded fish between the rails of railroad tracks 5.2 m above sea level; southeast about 1 km waves were 2.8 m. Though 2½ times as far from the epicenter, Añasco [MMI ≅ VIII], built on alluvial material, felt a considerably greater intensity than Rincón [MMI ≅ VI], built on rock. An eyewitness, 4 km southwest of Rincón, noticed surface waves that moved along the ground as fast as a train with a 1-m wavelength and an amplitude of 15 cm. At Mayagüez [MMI ≅ VIII], 23 minutes after the shock, seawater entered the lower floors of waterfront buildings and destroyed a few native huts along the beach. A small house carried seaward by the backwash was left stranded near the shore. Watermarks to 1.5 m above sea level were left on houses. Debris made streets

DATA	LOCATION	(m)	DESCRIPTION OF TSUNAMI AND SEISMIC EFFECTS
1918 10 11 *[continued]*	St. Thomas: Krum Bay Charlotte Amalie Río Ozama, Dominican Republic Tide gauge at: Atlantic City, New Jersey	1.2 0.45 0.70 0.10	impassable, impeding fire trucks and rescue efforts. More than 100 deaths occurred in Mayagüez. Boats anchored 300 to 400 m from shore were not affected. At Isla Mona, the receding water bared the reef and the 4-m tsunami washed away a pier and flooded a cistern. At El Boquerón, the sea rose 1.1 m. An observer said the ocean "withdrew about an hour after the earthquake, the water going out gradually during a period of 20 minutes." A small boat at in Bahía de Boquerón anchored 50 m from shore, where the water was normally 1.5 m, rested on the bottom for a few minutes. The sea returned more rapidly than it retired. Runups reached 43-cm at Gúanica, and 1 km southeast near the entrance to the bay, 50 cm. Water movements were small near Playa de Ponce. The sea rose 75 cm at Cayo Cardona. At Isla Caja de Muertos, the water rose to 1.5 m in height and covered 15 m of the beach. A 10-cm bore went up the Río Grande. After 24 minutes, the water receded, rising 1 m at Río Grande de Loíza. At Isabella, the water rose 2 m and 60-cm waves were observed at Puerto Arecibo. At Ponce [MMI ≅ VII], the great earthquake killed 4 people. Most towns reported buildings damaged. Damage to property for all of Puerto Rico exceeded $4 million. Submarine cables from San Juan to Kingston and St. Thomas to Puerto Plata were broken in several places. Earthquake effects with lessening intensities were felt at St. Thomas, Tortola, St. Martin, and Saba, and in the Dominican Republic at Santo Domingo, Puerto Plata, La Romana, Samaná, Santiago, and San Pedro de Macorís, and in Haiti, at Port-au-Prince, Gonaïves, Cap-Haïtien, Bayeux, and Port-de-Paix. Disturbances of the sea up to 70 cm in height were reported at Tortola. At Krum Bay, St. Thomas, waves rose 1.2 m during a ½-hour period. Oscillations continued for 3 hours. At Charlotte Amalie, the water rose and fell up to 45 cm several times. In the Dominican Republic, the waters of Río Ozama fell 70 cm, remaining out for 10 minutes and then rose as high above normal (the full excursion of the wave was 1.4 m). The water continued to rise and fall for 2 to 3 hours with a 40-minute period. The tide gauge at Atlantic City, 2,200 km north of the origin, registered a 10-cm wave, 4 hours and 14 minutes later; oscillations lasted several hours, with a 10- to 15-minute period. Reid said, "Waves set up by earthquakes are rarely noticed at sea; it is only when they approach shallow water that they

1918 10 11 *[continued]*			become dangerous. The ship, *La Rigolette,* in sight of Vieques, felt the shock severely but did not report the wave." This may inform other reports of sharp shocks felt at sea about possible co-incident tsunamis. The official death toll was 116 persons. Since more than 100 perished at Mayagüez, plus 4 at Ponce, plus 40 tsunami deaths at northwestern Puerto Rico, the tally would have been more than 116. The greatest number of fatalities was said to be due to the tsunami. Mercado reported an account by an "old, very well known gentleman in Aguadilla," who said, "deaths should have exceeded 100; the reason was that many persons were unaccounted for—therefore, not included in the official toll." Dillon reported 91 killed by the tsunami in northwestern Puerto Rico. Berninghausen, 1968; Campbell, 1991; Canals, 1919; Dillon *et al.,* Forster, 1890; 1999; Lander & Lockridge, 1989; Lander & O'Loughlin, 1997 & 1999e; Lynch & Shepherd, 1995; Mercado & McCann, 1998; Mercado Irizarry, 1997b; Reid & Taber, 1919a&b; Robson, 1964; *Seismological Notes,* 1918; Taber, 1922b. V4.
1918 10 24 [23:43 LT] 18.5N 67.5W Puerto Rico	Mona Passage, Puerto Rico Tide gauge at: Galveston, Texas		Considered a terrific aftershock of the Oct. 11 event, both this and another earthquake on Nov. 12 had epicentral intensities of R–F ≅ VIII to IX. In marked contrast to the vertical vibrations of the initial disturbance, these were characterized by horizontal oscillations. A small wave was recorded at the Galveston, Texas, tide gauge. Taber said these two shocks were not accompanied by "sea-waves," although it has been suggested that, at that time of night, a small wave may have escaped notice. Submarine cables were again cut, possibly indicating submarine slumping. The steamer, *Mariana,* 11 km southwest of the Isla Mona Lighthouse, plunged and rolled heavily. Heck reported Oct. 25. Forster, 1890; Heck, 1947; Lander & Lockridge, 1989; Lander & O'Loughlin, 1999e; Lynch & Shepherd, 1995; Mercado Irizarry, 1999; *Port-of-Spain Gazette,* 1918; Reid and Taber, 1919a&b; Taber, 1922b. V4
1922 05 02 [20:24 UT] 18.4N 64.9W	Tide gauge at Galveston, Texas	0.64	A wave with an amplitude of 64 cm was reported on the tide gauge at Galveston. A train of three waves with a 45-minute period was followed in 8 hours by a 28-cm wave in a similar train of smaller waves. Parker associated it with an earthquake felt 4 hours earlier at Vieques, Puerto Rico. According to Campbell, the 2-second shock was slight and unlikely to have been the tsunamigenic source. Campbell, 1991; Lander & Lockridge, 1989; Parker, 1922. V4.

DATA	LOCATION	(m)	DESCRIPTION OF TSUNAMI AND SEISMIC EFFECTS
1928 09 13 Venezuela	Carúpano, Venezuela		Singer reported a wave at Carúpano, but the absence of any association with an earthquake. A hurricane from Sep. 12–17 caused 1,800–2,500 deaths at Martinique, Guadeloupe, Puerto Rico, and Grand Turk. Though the path of the hurricane was 1000 km north, it may have produced waves near Carúpano. Another possibility is that the immense pressure changes affecting the sea floor from a hurricane served to trigger an earthquake and resultant tsunami. Douglas, 1958; Flament & Martin, 1980; *Monthly Weather Review,* 1971; Rappaport & Fernández-Partagas, 1995; Salivia, 1972; Singer *et al.,* 1983; Snow, 1952; Soulan, 1994. V1.
1928 12 Trinidad	Trinidad		Centeno-Graü reported a small island elevated to 8 m higher than sea level appearing off Trinidad. Singer termed the terrain uplift an *isla de lodo efímera* [short-lived mud island]. Waves may have accompanied the emergence as in 1911. Other such islands were reported in 1853, 1874, and 1934, located south of Trinidad. Centeno-Graü, 1940; Singer *et al.,* 1983. V1.
1929 01 17 [6:52 LT] 10.6N 64.6W Venezuela	Venezuela: Cumaná, Manicuare, El Salado, Puerto Sucre & El Barbudo		A powerful earthquake [Ms ≅ 6.9] destroyed Cumaná [MMI ≅ IX], killing 50 people and injuring 800, and was felt in Caracas and Barcelona. Tsunamis followed causing great damage at Cumaná, Manicuare, and El Salado, Puerto Sucre. At El Barbudo, the tsunami caused two 5-ton launches to be washed ashore and stranded and a steamer off shore was endangered by a huge wave. Many sailboats and dwellings were wrecked by the tsunami. Singer reported the rupture of an active fault, landslides, liquefaction, settlement and collapse at Pointe Guzmán Blanco at Río Manzanares, and other earthquake-related phenomena throughout the area. Berninghausen, 1968; Centeno-Graü, 1969; Heck, 1947; Lynch & Shepherd, 1995; Paige, 1930; Robson, 1964; Schubert, 1994; *Seismological Notes,* 1929; Singer *et al.,* 1983. V4.
1931 12 01 22.0N 84.0W Cuba	Playa Panchita, Rancho Veloz, Cuba	1.0	A tsunami was reported at Playa Panchita, Rancho Veloz, Las Villas. Waves beat on the beaches; the sea inundated the jetty and 1 m into houses along the coast; furniture was damaged. A hurricane passed north of Cuba on Nov. 25, becoming a tropical storm by Nov. 30. No earthquake or storm was reported locally. Neumann *et al.,* 1988; Rubio, 1982. V1.
1932 02 03 [06:16 UT] 19.8N 75.8	Santiago de Cuba		A strong earthquake [Ms ≅ 6.7], which affected 80% of the structures at Santiago de Cuba [MMI ≅ VIII], killed 8 people and injured 300 more. Hardly a building was left undamaged and a few were totally destroyed. A wave was also observed from aboard a North American

1932 02 03 *[continued]* Cuba			ship. Marigram records from different points in the Caribbean confirmed that the tsunami was relatively minor. Rubio placed the epicenter about 25 km from Santiago de Cuba. *Barbados Advocate,* 1932; Berninghausen, 1968; Hess, 1932; Lynch & Shepherd, 1995; Rubio, 1982. V4.
1932 11 04	Cumaná, Venezuela		A wave at Cumaná was reported by Singer with an uncertain link to an earthquake. From Nov. 4–10, a hurricane caused 2,500 fatalities in Cuba, Jamaica, and The Caymans. It may have produced waves at Cumaná or, as Brooks indicated, the storm may have provided the necessary trigger-action to set off an earthquake and the resulting wave. Brooks, 1919; Clark, 1988; Rappaport & Fernández-Partagas, 1995; Seon (no date); Singer *et al.,* 1983. V1.
1934 08 30 Trinidad	Trinidad		Singer reported terrain uplift of an *isla de lodo efímera* [short-lived mud island] south of Trinidad. Waves were a possibility as happened following the volcanic explosion in the 1911, where "extraordinary waves" were observed. Other mud islands erupted in 1853, 1874, and 1934, also off the south coast of Trinidad. Centeno-Graü, 1940; Singer *et al.,* 1983. V1.
1939 07 24 [12:02 LT] 12.5N 61.5W The Grenadines	The Grenadines: The Sisters Kick-'em-Jenny Submarine Volcano Grenada: Sauteurs		Since the Kick-'em-Jenny Submarine Volcano, located near The Sisters, north of David Point, Grenada, was first noted in 1939, "It has erupted at about 5-year intervals. The 1939 and 1990 events may have generated very small tsunamis of local extent," according to Sigurdsson. The first reported eruption of the Kick-'em-Jenny Submarine Mount was observed from Sauteurs, Grenada. Devas reported that light shocks were felt from 8 a.m. at frequent intervals in the northern part of Grenada followed by 7 heavy shocks from 12:02–1:15 p.m. The most prolonged shock of 12 seconds "was accompanied by the discharge of a column of black smoke from a point in the sea estimated as 5 miles [8 km] distant bearing 7½° true," and that "The eruption cloud rose 150 to 200 feet [45.7–61 m] and contained 'solid, dark-colored bodies of considerable size, which fell back into the sea'. At 6:04 p.m. on the same day, a second discharge of brown smoke from the same locality was projected rapidly to 900 feet [274 m] above sea level." From 8 a.m.–7 p.m., 1- to 3-second period *T*-phase tremors were recorded at the Montserrat Volcano Observatory. Sigurdsson said, "The eruptions have been small, but have generated underwater acoustic signals or *T*-phase, probably due to the collapse

DATA	LOCATION	(m)	DESCRIPTION OF TSUNAMI AND SEISMIC EFFECTS
1939 07 24 *[continued]*			of steam bubbles from the reaction between the erupting magma, at over 1000° C, and the sea water above, a process known variously as hydro-eruption, fuel-coolant interaction, or phreatomagmatic explosion." In Oct. 1943, similar acoustic signals were recorded at stations in Martinique, Guadeloupe, and Dominica, and again in Oct. 1953. In Oct. 1965, from Puerto Rico to Trinidad, *T*-phase were recorded with ground tremors felt at Île de Ronde [MMI ≅ V], Grenada [MMI ≅ IV], St. Vincent, and Martinique, and in May 1966, in the eastern Caribbean. From Aug. 3–6, 1966, 168 *T*-phase events were recorded along with 2 earthquakes that were felt at Sauteurs on Aug. 3. Soundings in 1962, taken from the survey ship, *Vidal,* placed the depth of the vent between the contour lines at 50 and 100 fathoms [91–183 m]. In 1999, *The Sun* carried a report that "There could be an eruption any time now," at Kick-'em-Jenny. "The summit of the volcano is now around 130 to 140 m below the surface and the pressure of the water dampens the explosions, but once it reaches 110 m, eruptions would become more violent and a massive eruption will trigger a tsunami. Grenada is expected to be hardest hit with waves around 150 feet [45.7 m] in 15 minutes in a worst case scenario, while Barbados is expected to be least affected—by 12-foot [3.7-m] waves in 40 minutes. These waves will be traveling at 300 to 500 mph [483–805 kph]." Sigurdsson said that the alternative, much smaller eruption scenario by Smith and Shepherd, was more realistic, given the rate of Kick-'em-Jenny's growth since 1939. "The theoretical tsunami runup height from this size eruption ranges from 8 meters on Grenada to 1 meter in the Virgin Islands." Devas & Macadam-Sherwin, 1939; González, 1999b; Molard, 1947; Robson & Tomblin, 1966; Sigurdsson, 1996; Smith & Shepherd, 1993; *The Sun,* 1999; *West Indies Pilot,* 1942. V2.
1939 08 15 [01:45 LT] 22.2N 79.1W Cuba	Cuba: Cabo Francés		At Cabo Francés, movement of the sea awoke sailors on a barque and a yacht. The *Barbados Advocate* reported a moderate 2-second tremor [Mb ≅ 5.6] at Santa Clara [MMI ≅ VII–VIII]; the Remedios hospital was damaged and many injured. Rubio said the shock caused the tsunami, and that "This 'tsunami' was a II on the Rudolph scale[an earthquake intensity scale]; the epicenter was localized in the sea." *Barbados Advocate,* 1939; Bodle, 1941; Lynch & Shepherd, 1995; Robson & Tomblin, 1977; Rubio, 1982; *Seismological Notes,* 1939. V2.

1946 08 04 [17:51 UT] 19.3N 68.9W Hispaniola	Dominican Republic: Northern coast Matancitas, Julia Molina Cabo Samaná Tide gauges: San Juan, Puerto Rico Bermuda Daytona Beach, Florida Atlantic City, New Jersey	 4.6 2.4 4.0–5.0 0.66	Two violent earthquakes laid waste to the Dominican Republic in 1946 on Aug. 4 and Aug. 8. The huge earthquake [Ms ≅ 8.1] on Aug. 4, originated 64 km offshore northeast of Julia Molina in the Mona Passage, and on Aug. 8, a powerful Ms ≅ 7.5 aftershock continued the destruction. According to Lynch and Bodle, evidence showed that the epicentral region was very extended. Effects extended into Haiti, and shook many other islands, including Puerto Rico [MMI ≅ VI] where some damage occurred on the west coast. A 4.6-m tsunami pounded the northeast coast of Hispaniola. The greatest loss and devastation were caused by a 2.4-m tsunami at Matancitas (Matanzas), [MMI ≅ X]. Continuing shocks and tidal waves battered the coastal towns for months. According to Lynch and Bodle, Matanzas on Bahía Escocesa was so totally destroyed that it had to be abandoned. "Loss of life, approaching a total of 100, was greater than at any other place in the Republic." In 1998, an important revision of this death toll was communicated by Herridge de Guerrero, Dominican Disaster Mitigation Committee Coordinator, which greatly increases the total fatalities due to tsunamis in the Caribbean. She related that "Just off the northern coast, a wave was formed by a sudden disturbance in the ocean floor. The ocean had receded from the Matancitas coast and the people went to collect the fish exposed. The tidal wave entered almost a kilometer inland, and swept the city and several villages into the ocean, killing approximately 1,790. The great tragedy is that the Dictator, Trujillo, silenced the story for political reasons and the population was never taught to be ready for earthquakes and tsunamis." Trujillo had commissioned the earthquake study at the time to allay people's fears about the continuing earthquakes, according to Lynch and Bodle, who recommended against rebuilding at Matanzas "on anything beyond a picnic-ground scale." Though only 3 km northwest of Matanzas, Julia Molina [MMI ≅ IX] suffered little from the tsunami, apparently protected by a high sandy beach. Near the mouth of Río Boba, the river parallels the shore 200 to 300 m inland. Lynch and Bodle said the tsunami was reported "to have risen above the high sandy shore to join with the river. This would have required a rise of some 4 to 5 m. When the beach was flooded to the north, observers noted that it was bare to the south. Their impression was that

DATA	LOCATION	(m)	DESCRIPTION OF TSUNAMI AND SEISMIC EFFECTS
1946 08 04 *[continued]*			the crest-and-trough effect moved toward the south." On the northern escarpment of the Península de Samaná [MMI ≅ IX], great rock slides collapsed into the ocean. At Samaná [MMI ≅ IX], buildings near the beach were damaged "a little short of severely." No damaging effects from the waves were reported at Bahía de Samaná, although abnormal ebb-and-flow effects were noted. At Sabana de la Mar [MMI ≅ VIII], no wave effects were noted. The pier at Sánchez [MMI ≅ IX] was damaged heavily and railroad tracks were warped into an S-shaped curve where the pier met the shore. At Cabo Samaná, several ebbs and flows were observed, but caused no damage. Quicksand and mud were ejected in places from small cracks along the banks of the Río Yaque del Norte, Yuma, and Boba rivers. Lynch and Bodle reported a shift in the low-tide level at Ciudad Trujillo, illustrating it with a photo of a person by the old low-tide perimeter, many meters distant from the low-tide limit afterwards. Other evidence of possible terrain changes was reported by an inhabitant who said a swimming hole that was blasted out of the rocks by his father years before, near the sandy shoreline, was "left high and dry even at high tide." A small tsunami was observed along the western and northern coasts of Puerto Rico. It was recorded as 66 cm on the San Juan tide gauge, 36 minutes after the earthquake. Waves were recorded with travel times of 35 min. at San Juan; 2 hr. 07 min. at Bermuda; 3 hr. 59 min. at Daytona Beach; and 4 hr. 49 min. at Atlantic City. Much damage from the earthquake was also reported at Cabrera, San Francisco de Macorís, and Castilla [all MMI ≅ X]; Puerto Plata, Moca, and Cabo Cabrón [all MMI ≅ IX]; El Seibo, Santiago, Ciudad Trujillo, San Pedro de Macorís, La Romana, and many other places. Berninghausen, 1968; Bodle & Murphy, 1948; Campbell, 1991; *Earthquake Notes,* 1946; Heck, 1947; Herridge de Guerrero, 1998, Lander & Lockridge, 1989; Lynch & Bodle, 1948; Lynch & Shepherd, 1995; Murty, 1977b; Small, 1948. V4.
1946 08 08 [13:28 UT] 19.7N 68.5W	Dominican Republic: Sabana la Mar		A major aftershock [Ms ≅ 7.9] occurred with nearly as great as the Aug. 4 earthquake, but located 100 km northwest of the original epicenter. It shook Samaná Bay [MMI ≅ VIII], and generated a tsunami at Sabana de la Mar, smaller than the one on the 4th. Rubio reported 75 perishing from the tsunami at Santo Domingo. The tsunamis and earthquakes left 20,000

1946 08 08 *[continued]* Hispaniola Puerto Rico	Cuba: Santa Fé, Guanabo, Marianao, Jaimanita & Baracoa Puerto Rico: San Juan, Aguadilla & Mayagüez Tide gauges: San Juan Bermuda Daytona Beach, Florida Atlantic City, New Jersey	0.6	inhabitants homeless. Lynch listed as badly damaged many of the same towns as on the 4th, including Matanzas, Cabrera, Salgado, Puerto Plata, San Francisco de Macorís, Moca, Santiago, Julia Molina, and many others. At Sabana de la Mar, a mild recession was seen, but not high water. A small tsunami was observed at Mayagüez, Puerto Rico, where the sea withdrew 76 m before returning. A wave was also observed at Aguadilla where the sea drew out 24 m and returned. At San Juan [MMI ≅ VI], 35 minutes following the shock, the tsunami registered 60 cm on the tide gauge. Power and communication facilities were temporarily disrupted and tidal fluctuations caused no serious damage. In addition, slight shocks were felt at the Virgin Islands. The tsunami propagated from east to west at Cuba, with swells reported at the beaches of Guanabo, Marianao, Santa Fé, Jaimanita, and Baracoa. Bathers and those in pleasure boats were alarmed, but no damage was reported. Singer reported the collapse of mines at Santa María, Naricual, Venezuela, for the same date of Aug. 8. Waves were recorded with similar travel times to the Aug. 4 tsunami: 35 minutes after the earthquake at San Juan; 2 hr. 02 min. after the shock at Bermuda; 4 hr. 02 min. after the shock at Daytona Beach; and 4 hr. 42 min. after the shock at Atlantic City. Berninghausen, 1968; Bodle & Murphy, 1948; Lander & Lockridge, 1989; Lynch & Bodle, 1948; Lynch & Shepherd, 1995; Rubio, 1982; Schubert, 1994; Singer *et al.*, 1983. V4.
1950 08 03 10.6N 69.5W Venezuela	Puerto Cabello, Venezuela		Waves were reported at Puerto Cabello. Although Singer noted an uncertain link to an earthquake, the possibility of a tsunami cannot be disregarded. For that date he reported a widespread seismic activity such as numerous ground movements, surface displacements and ruptures at Cerro de El Paraíso, Maracaibo; San Rafael, Sanare; and La Culebrina, Humocaro Bajo; rockfalls at La Aguada; El Peñón, Humocaro Bajo; and Las Adjuntas, El Tocuyo, with 100 victims; landslides at La Boca, Anzoátegui, and Barrio El Atlántico, Caracas; *Deliziamentos rotacionales* [rotational landslides] at Humocaro Alto and Curamato, Guárico; a mudslide at Laguna La González, Chabasquén; and the Lagunas del Catire overflowed in Caserío Providencia, Chabasquén. Subsequently, 3 days later, on Aug. 6, volcanic activity commenced at Volcán de Humo, Sanare. Murphy & Ulrich, 1952; Singer *et al.*, 1983. V1.

DATA	LOCATION	(m)	DESCRIPTION OF TSUNAMI AND SEISMIC EFFECTS
1953 05 31 [19:58 UT] 19.7N 70.4W	Tide gauge: Puerto Plata	0.06	A 6-cm wave was recorded on the Puerto Plata tide gauge at the Dominican Republic. An earthquake [Ms ≅ 7.2] was felt moderately there and at Puerto Rico. Its epicenter was placed 565 km from Puerto Rico. Campbell, 1991; Murphy & Cloud, 1955. V4.
1955 01 18 11.3N 69.4W Venezuela	La Vela, Venezuela		A tsunami wrecked 4 ships and a number of waterfront buildings in La Vela. A minor earthquake [Mb ≅ 5.5] off the coast of Panamá at the time, is a possible source. Marigraph records may exist. Berninghausen, 1968; Murphy 1964; *Seismological Notes,* 1955. V4.
1961 06 16 [10:31 UT] 8.9N 73.4W Colombia	Lago de Maracaibo, Venezuela		An *irrupción* [bursting forth; surging up] of waves occurred at Lago de Maracaibo. The towns south of the lake were partially flooded. Singer confirmed an association with an earthquake [Ms ≅ 6.5], and reported landslides at Altamira and Calderas. The epicenter was located in northern Colombia. Lander & Cloud, 1963; Singer *et al.,* 1983. V4.
1967 12 04 Puerto Rico			High waves battered the northern coast of Puerto Rico, especially Levittown, Catano, Arecibo, and Fajardo. In Old San Juan, where people live right at the seashore, 80 dwellings were swept into the Atlantic, leaving 300 homeless. At La Perla, the labyrinthine seaside section between the San Geronimo Fort and the shore, one of the most exposed locations, huge waves caused the heaviest damage. No weather warnings had been issued, nor was an earthquake reported. Mercado reported in an unpublished report about swells affecting Puerto Rico that a destructive storm was referenced along the northern coast, due to an extra-tropical storm, or nor'easter, a kind of winter storm, occurring typically from Oct. to Feb. Hoose, 1969; Mercado Irizarry, 1999; *The New York Times,* 1967. V0.
1968 09 20 [06:09 UT] 10.7N 62.7W Venezuela Trinidad	Venezuela		Waves following a violent 30-second shock [Ms ≅ 6.2] at Venezuela [MMI ≅ VII] were reported. One person was killed and 50 others were injured. Coffman said, "Reports of a tsunami following the earthquake near the coast of Venezuela could not be verified," but gave no details. Singer did not mention a tsunami, but reported settlement and collapse at Península de Paria, on the coast at Güiria, and destructive landslides at Río Caribe and Tunapuy. The earthquake killed 2 people in Trinidad and severely damaged Diego Martin and Port-of-Spain. It was felt at Tobago, Guyana, and throughout Grenada [MMI ≅ IV], and Colombia. Coffman & Cloud, 1970; Lynch & Shepherd, 1995; Singer *et al.,* 1983. V1.

1969 12 25 [21:32 UT] 15.8N 59.7W	Barbados, Antigua & Dominica	0.46 0.30 0.12	An earthquake [Ms ≅ 7.7] occurred that generated a small tsunami that was recorded with amplitudes of 46 cm at Barbados, 30 cm at Antigua, and 12 cm at Dominica. *Preliminary Determination of Epicenters*, 1969; Von Hake & Cloud, 1971 V4.
1976 02 04 [09:01 UT] 15.2N 89.2W Guatemala	Golfo de Honduras Tide gauge at: Puerto Cortés, Honduras	0.45	A major earthquake [Ms ≅ 7.5] at a depth of 5 km occurred at Guatemala with a left-lateral surface break 230 km long at the Motagua Fault. A tsunami [I–I ≅ −0.5] that affected the Golfo de Honduras was recorded at the Puerto Cortés tide gauge with a maximum amplitude of 45 cm. Ambraseys & Adams, 1996; Arce *et al.*, 1998; Buckman *et al.*, 1978; Espinoza, 1976; Feldman, 1984; Grases, 1990; Kanamori & Stewart, 1978; Molina, 1997; Plafker, 1976; Rojas *et al.*, 1993; Víquez & Toral, 1987; Young *et al.*, 1989. V4.
1979 09 03 Venezuela	Puerto Cumarebo, Venezuela		The pier at Puerto Cumarebo was destroyed by a wave. Singer reported the absence of an association with an earthquake. Singer also reported landslides at Caracas and San Agustín del Sur; roads obstructed; and 3 dwellings destroyed at San José, with 24 people perishing. Hurricane *David* killed 2,063 victims at both Dominica and the Dominican Republic from Aug. 29 to Sep. 5. Singer reported the hurricane affecting Isla de Aves, a tiny Venezuelan Dependency, located at sea on a level between Îles des Saintes and Dominica. The hurricane caused damage and a *partición* [separation] of Isla de Aves a week earlier on Aug. 27. The pier at Puerto Cumarebo may have been destroyed by a storm wave due to the hurricane (depending on its path and its duration), or by a collapse of the seabed generating a submarine-landslide tsunami that may have been triggered by the immense pressures exerted on the ocean floor by a powerful hurricane. Brooks stated that "stress from the violence of a storm" causes "a load effect upon an unstable sea bottom that often provides the necessary trigger-action to set off an earthquake and the tidal wave that follows." Brooks, 1919; Clark, 1988; Rappaport & Fernández-Partagas, 1995; Schubert, 1994; Singer *et al.*, 1983. V1.
1985 03 16 [14:54 UT] The Leewards	Tide gauge at: Basse-Terre, Guadeloupe	0.1– 0.12	At Guadeloupe [MMI ≅ VI], a moderate earthquake [Ms ≅ 6.3] injured 6 people. Feuillard reported that 4 of the injuries occurred at Point-à-Pitre. The earthquake was felt at Nevis and Montserrat [both MMI ≅ V–VI]; Martinique, Dominica, St. Kitts, Antigua, and Puerto Rico also experienced the shock. A several-centimeter [0.10- to 0.12-m] tsunami was recorded at

DATA	LOCATION	(m)	DESCRIPTION OF TSUNAMI AND SEISMIC EFFECTS
1985 03 16 *[continued]*			Basse-Terre, Guadeloupe. Lynch said this event triggered a landslide at Redonda. Two earlier events in Feb. 1843 and Apr. 1690, which had striking similarities to each other, both described masses of rock plunging into the sea, giving rise to a wave at Redonda. Feuillard, 1985; Lynch & Shepherd, 1995; *Preliminary Determination of Epicenters*, 1985. V4
1989 11 01 [10:25 UT] 18.9N 63.8W Puerto Rico	Cabo Rojo, Puerto Rico		A minor earthquake [Ms $\cong$ 5.2] occurred in the Mona Passage, which generated a small tsunami. The Puerto Rico Civil Defense reported a notable augmentation of the sea level in the area of Cabo Rojo, Bahía de Boquerón. Mercado received a report from eyewitness, Luis Padilla, who had observed a small tsunami at Cabo Rojo. *El Nueva Día,* 1989; Mercado Irizarry, 1997a; *Preliminary Determination of Epicenters,* 1989. V4.
1990 12.5N 61.5W The Grenadines	Grenada: The Grenadines: The Sisters Kick-'em-Jenny Submarine Volcano		The Kick-'em-Jenny Submarine Volcano at The Sisters, in The Grenadines, near Île de Ronde, north of David Point, Grenada, has erupted at about 5-year intervals, since it was first observed in 1939. The most recent occurrence was in 1990, according to Sigurdsson, who said, "The 1939 and 1990 events may have generated very small tsunamis of local extent. Most important for Kick-'em-Jenny is the probability of eruptions larger than those observed in the historical period. Typically, eruption frequency varies inversely with eruption magnitude." Active volcanoes generally have small eruptions, and volcanoes with long periods of dormancy are likely to have catastrophic, large-magnitude eruptions at the initiation of an eruptive cycle. The vent of the Kick-'em-Jenny growing submarine mount has remained below the surface, although based on its rate of growth in the period from 1939, "the runup height of a theoretical tsunami (projected by the smaller of the Smith and Shepherd scenarios) ranges from 8 m at Grenada to 1 m in the Virgin Islands. Sigurdsson stated, "Of great importance to Kick-'em-Jenny, is the depth of the water above the crater, which is now approaching maximum efficiency for tsunami generation in terms of water depth," but as the water depth above the vent decreases, "the explosive volcanic energy will be increasingly partitioned into the atmosphere. Only time will tell, if the shallowing of the crater in the next decades will result in larger tsunami generation potential." González, 1999b; Sigurdsson, 1996; Smith & Shepherd, 1993. V2.

1991 04 22 [21:56 UT] 09.6N 83.2W Costa Rica Panamá	Costa Rica: Limón, Cahuita– Puerto Viejo, Boca del Río Matina, Bocas del Río Pantano, Río Estrella, Río Moín, Moín, Canal de El Matina, Westfalia, Cahuita Reef, Puerto Viejo Final de Cocles, Punta Uva, Manzanillo & Gandoca Panamá: Bocas del Toro:	2.0– 3.0 2.0 3.0 2.0– 3.0 2.0– 3.0	A strong earthquake [Ms ≅ 7.6], caused damage along the whole *Litoral Atlántico* [Atlantic seaboard] from Bocas del Toro, Panamá, to Batán and Turrialba, Costa Rica, according to Bermúdez León, with landslides in the Cordillera de Talamanca and minor damage in Valle Central, causing 50 deaths. Maximum intensities assigned were MMI ≅ VII–IX. Arce reported, "A seismic seawave 2 m high was recorded at San Cristóbal. This wave affected the whole coast from north of Limón to Panamá. A tsunami with wave heights of 2- to 3-m was observed along the coast from Cahuita, Costa Rica, to Isla Bastimentos, Panamá. This wave flooded from 10 to 200 m inland." The magnitude I–I ≅ 1.0 tsunami was generated locally in the North American–Caribbean Plates Boundary Zone tectonic environment. In Molina's detailed account waves were reported up to 3 m and "runups" to 180 m. It may be assumed that the word "runup" is used to define the distance that the sea flowed inland. "Runup" is used as reported. At Costa Rica, the ocean receded from 200 to 500 m immediately following the earthquake. At Boca del Río Matina, the sea drew back 400 m, and inundated the beach with runups up to 180 m. Bocas del Río Pantano experienced a 100-m withdrawal; the sea then returned within 3 minutes without flooding. At Moín, the water receded 200 m, and a "sea wave overtopped 3-m dikes in Río Moín," according to Molina. Two people were drowned at Canal de El Matina, 12 *millas* [miles] from Moín. The sea withdrew 200 m at Westfalia, followed by movement and shifting of the sands. At Cahuita Reef, the sea retreated rapidly and after 1 hour returned slowly with no flooding. At Puerto Viejo, the waters receded 100 m, followed seconds later by a large wave. The sea withdrew again and returned with runups of 30 to 70 m. The seawater retreated 20 m from the cliff and beach of Final de Cocles, and returned slowly [5 to 6 minutes] to the previous level. The sea withdrew 300 to 400 m from the beach at Punta Uva. The water receded 400 to 600 m at Manzanillo, and returned in 5 minutes. The ocean inundated 30 to 70 m, with movement of the sand. The seawater retreated 300 m from the beach at Gandoca, and it returned 5 minutes later with a 70-m runup. At Panamá, the initial recession began 10 to 15 minutes after the shock. The recession of the sea varied from 100 to 400 m for a period of from 5 to 45 minutes. At Bocas

DATA	LOCATION	(m)	DESCRIPTION OF TSUNAMI AND SEISMIC EFFECTS
1991 04 22 *[continued]*	Las Delicias Sand Bank, Isla de Carenero, San-San Refuge, Playa de Julio Abrego, Río Changuinola, Punta Tiribibi, Isla Colón: Lime Point, West Knapp Hole, Ground Creek, Punta Cauro, Boket Bay, Nancy Key, Isla San Cristóbal, Puertobelo & Isla Bastimentos	0.6 2.0 2.5–3.0 2.0 0.6	del Toro, less than 10 minutes after the earthquake, Las Delicias Sand Bank, which was normally covered by 60 to 90 cm of seawater, "emerged to the surface like a submarine," where it remained above the surface of the water for 5 to 7 minutes, according to Camacho. Afterwards several waves entered the bay from the northeast with great force, flooding 50 to 100 m into the flat northern part of the town. As the fury of the sea continued, people evacuated to the surrounding hills. The perturbation of the sea lasted several hours and the waters were reported as dirty and muddy. At Isla de Carenero, 15 to 20 minutes after the earthquake, the seawater withdrew and remained out for a quarter of an hour. The inhabitants fled in their canoes to more protected neighboring islands. Only 10 to 20 minutes after they departed, a violent wave flooded the beaches with runups of 100 to 150 m, which finished the destruction of their dwellings. At San-San Natural Refuge, waves deposited 1 m of sand, which created dunes and covered turtle nests. The seawater receded minutes after the earthquake from Playa de Julio Abrego 3 to 6 times with runups to 100 m. A high-energy current was observed at Río Changuinola and sand was deposited at Punta Tiribibi. On Isla Colón at Lime Point, the sea withdrew 6 times, exposing the reefs; water flooded the coastal road. Minutes after the shock, great turbulence and a strong current were observed in the bay at West Knapp Hole. The beach had runups of 5 to 10 m. At Ground Creek, water in the bay receded and returned as a strong current 60 cm in height with a 10-m runup. At Punta Cauro beach, next to the cliff, the sea retreated 5 times. The final wave was 2 m high. At the cove in Boket Bay, waves of 2.5 to 3 m arrived 10 to 15 minutes after the earthquake. On the point at Nancy Key, the water withdrew several times and returned as a strong current. The sea receded several meters for 45 minutes at Isla San Cristóbal. Many fish were trapped and people ran out onto the exposed area of the beach to catch the fish. The returning sea was a slow wave. Tsunamis were observed at Puertobelo [60 cm] and Isla Bastimentos, where the ocean withdrew from the reef 10 to 15 minutes after the earthquake, and then returned. Nishenko reported two principal types of runup behavior observed following the initial retreat. First, in some places, a wave 2 to 3 m in height advanced rapidly to the beach and

1991 04 22 *[continued]*	Tide gauges: Coco-Solo, Colón Isla Magueyes, Puerto Rico Limetree, St. Croix	 0.08 0.07 0.07	flooded from 10 to 100 m inland, according to location and elevation. This type of runup behavior was reported at the mouths of major rivers, as at Río Matina, Bocas del Río Pantano, Río Moín, and Río Estrella. Local subsidence of poorly consolidated sediments amplified the effects, according to Nishenko. Second, in other places along the coastline, the runup behavior consisted of a gradual increase in the water level about an hour after the initial retreat. This occurred without the formation of a large wave, as the runup behavior was experienced near Moín, Limón, Punta Uva, Cahuita, in Costa Rica, and at Archipiélago de Bocas del Toro Islas Bocas del Toro and Carenero, in Panamá. "Along some coastal areas, fringing reefs provided protection against runup effects by damping tsunami waves. Co-seismic coastal uplift also helped minimize the impact of flooding," according to Nishenko. The closest tide gauge to the earthquake epicenter was at Coco-Solo, Colón, 320 km east of the source zone. At Coco-Solo, a maximum amplitude of 8 cm was recorded an hour after the earthquake. About 3½ hours later, tide gauges at Isla de Magueyes, Puerto Rico, and at Limetree, St. Croix, both recorded maximum amplitudes of 7 cm. Based on eyewitness observations and the tide gauge records, the ocean disturbance along the coastline from Costa Rica to Panamá lasted approximately 4 to 5 hours. Ambraseys & Adams, 1996; Arce *et al.*, 1998; Barquero & Rojas, 1994; Bermúdez León *et al.*, 2000; Boschini & Montero, 1994; Camacho, 1994; Denyer *et al.*, 1992; Molina, 1997; Nishenko *et al.*, 1992; Plafker & Ward, 1992; *Preliminary Determination of Epicenters*, 1991; Rojas *et al.*, 1993. V4.
1997 07 09 [19:25 UT] 10.6N 63.5W Venezuela	Venezuela: Isla de Margarita, Cariaco Cumaná & Sucre Tobago		An earthquake [$M_s \cong 6.8$] occurred off the coast of Venezuela, near Isla de Margarita, which caused extensive damage and landslides in the Cariaco–Cumaná region. At least 7 people perished and 522 others were injured. Rodríguez reported that, although official tsunami data were not available, "some people from the FUNVISIS Seismology Department and other departments were in the region and received information about the sea level dropping or retiring along the coasts in the state of Sucre. Mercado reported an eye witness observation of a wave coming ashore on the south coast of Tobago a few minutes after the earthquake. Marigraph records may exist for this wave. Mercado Irizarry, 1997c; Rodríguez, 1998. V3.

DATA	LOCATION	(m)	DESCRIPTION OF TSUNAMI AND SEISMIC EFFECTS
1997 12 26 [03:00 LT] 16.4N 62.1W Montserrat	Montserrat	1.0	A volcanic debris avalanche of approximately 25 million cubic meters slid into the ocean on Dec. 26 in the White River valley of Montserrat, named the Boxing Day Collapse. The slope failure at 3 a.m., which lasted 15 minutes, was the first major eruption of this volcano since 1646. The failure caused a 44-million-cubic-meter caldera collapse. Calder said, "This collapse represents the highest-magnitude, and perhaps the most intense volcanic activity yet at the volcano." During the Boxing Day event, the main pyroclastic flows were largely confined to the White River valley although some material did spill out over the prominent bend and travel down towards Morris. On the night of the eruption, a wave inundated to a maximum distance of 80 m inland at Old Road Bay, 10 km from the landslide site. Calder reported that a small tsunami moved inshore near the jetty. The wave was estimated to have been about 1 m higher than the level of the road which lies approximately 2 m above water level and to have moved a maximum distance of 80 m inland. A variety of objects, including a small wooden boat, the roof of a shelter, and a stone table were displaced several meters inland, and a large log was carried inland even farther by the wave. Impact marks from the inundation of up to 1 m high could be seen on palm trees on the side facing the sea. The orientation of the grass indicated the flow of the backwash. An observer reported seeing the sea move out and then back in, typical of a landslide-generated tsunami. According to Calder, the tsunami was probably generated by the steep flow front of the debris avalanche, possibly assisted by the pyroclastic flows as they entered the sea at the mouth of the White River. The wave was refracted around the coastline of Montserrat and achieved considerable runup in Old Road Bay. It was determined that the focusing of the wave at Old Road Bay could be attributed to the abrupt change of coastline there. The wave flowing parallel to the shoreline would have met the shore at Old Road Bay head-on, causing the runup. The shallow offshore bathymetry and onshore topography in the area also aided the extended inundation. Models developed for the Montserrat Volcano Observatory determined that a debris slide of 25 million cubic meters, entering the ocean at 50 m per second, should produce a source waveheight of 5 m. The model conditions appear consistent with estimates for the mouth of

1997 12 26 *[continued]*			the White River soon after 3 a.m., according to Calder. "The coastal wave would decay at about the square root of the distance," so a wave should be a height of about 1 m at 10.7 km distance at Old Road Bay. A discrepancy appears to exist in the waveheights. If the runup as stated was 1 m higher than the road, and the road was reportedly at an elevation of 2 m above sea level, the wave would have needed approximately 3 m to reach 1 m above the road. Since July 18, 1995, when the Soufrière Hills first began erupting, several debris slides reached the ocean. Although waves may have been produced, no wave reports are known. Perhaps, they simply have not been looked for or observations were not reported. Calder *et al.*, 1998; Long, 2000; Mangeney *et al.*, 1998; Robson and Tomblin, 1966. V4.

Tsunamis are the wave of the future !

BIBLIOGRAPHY

Primary and Secondary Sources Referenced in the Text

Though quite extended, the following is far from an exhaustive listing of relevant works. Many websites are available as well.

Acosta, Fr. J. de (1792) *Natural and Moral History of the Indies by Father Joseph de Acosta,* 6th Edición, Madrid (in Spanish).

Acts of the Privy Council of England, Colonial Service (1690) Public Records Office, London.

Affleck, Capt. (1757) An account of the agitation of the sea at Antigua, November 1, 1755, By Capt. Affleck of the *Advice* Man-of-War, Communicated by Charles Gray, Esq., F.R.S., in a Letter to William Watson, F.R.S., Letter CIII, *Philosophical Transactions* **49**, Part ii, Royal Philosophical Society of London, 668–670; also in *Philosophical Transactions* **11**, Abridged, London, 9–10 (1809).

Agostinho, J. (1935) Notice sur quelques raz-de-marée aux Açores, *Annales de la Commission de l'Étude de Raz-de-Marée* (5), Union Géodésique et Géophysique Internationale, 21–24 (in French).

Allemange Konst en Letterbode (1825) Severe earthquake damage September 20 at Demerara, Guyana, December 2 (in Dutch).

Ambraseys, N.N. and Adams, R. (1996) Large magnitude Central American earthquakes, 1898–1994, *Geophysical Journal International* **127**, 665–692.

Anderson, J. (1997) "Danger of tsunamis potent in vulnerable Caribbean," *Maine Sunday Telegram,* July 6, 3C.

Anderson, T. and Flett, J.S. (1903) Report on the eruptions of the Soufrière in St. Vincent, in 1902, and on a visit to Montagne Pelée in Martinique, Part I, *Philosophical Transactions* **200**, Series A, Royal Philosophical Society of London, 353–553.

Annales de Chémie et de Physique (1823) Tidal wave reported for November 30, 1823, at Saint-Pierre, Martinique **22**, 411 (in French).

Annales de Chémie et de Physique (1824) Rise and fall of the sea reported for November 30, 1824, at Saint-Pierre, Martinique **30**, 412 (in French).

Annales de Chémie et de Physique (1828) Ships thrown upon the strand November 30, 1827, at Saint-Pierre, Martinique **33**, 408 (in French).

Annual Register (1781) October 3, 1780 report; given as authority by Thomas Southey, 292, 35.

Annual Register (1842) Shocks felt aboard ships May 7 at Cap Haytien, 109.

Antigua Herald (1843) Report from French warship of a wave rising up on February 8 at Redonda.

Antigua Weekly Register (1843) The sea rose 1.2 m February 8 at St. John's, Antigua, February 9.

Arce, M.F., Molina, E., Havskov, J., and Atakan, K. (1998) Tsunamis in Central America, *Technical Report II 1–12, Reduction of Natural Disasters in Central America, Earthquake Preparedness and Hazard Mitigation, Phase II,* Institute of Solid Earth Physics, University of Bergen, Norway, 34 pp.

Archive des Découvrirs (1824) Rise and fall of the tide September 13 at Plymouth, 215 (in French).

Ardouin, B. (1853) Histoire d'Haïti, in *Études sur l'Histoire Haïti* **9**, 8°, Paris (in French).

Aristides Rohas, D. (1867) An opinion on the maritime disasters of the Antilles, *Federalista,* Caracas (November 28, 1867); reprinted in *Smithsonian Annual Report, 1867,* 466–467.

Arnald, R. and Macready, G.A. (1956) Island forming mud volcano in Trinidad, British West Indies, *Bulletin of the American Association of Petroleum Geologists* **40**(11), 2756.

Bacardí, E. (1925) *Crónicas de Santiago de Cuba,* Reimpression, Santiago de Cuba (University of the West Indies Library, Mona, Kingston) (in Spanish).

Backshall, P. (no date) The great hurricane of 1867—looking back after more than 100 years, St. Thomas, 27–32.

Baralt, R.M. and Díaz, R. (1841) *Resumen de la Historia Antigua de Venezuela,* Paris, 448 pp. (in Spanish).

Baratta, M. (1901) *Istoria Terremoti d'Italia,* Fratelli Bocca (editori), Torino, 949 pp. (in Italian).

Barbadian (1825) Smart shock on September 20 at Barbados, September 23.

Barbadian (1827) Report of alarm on November 30 at Fort Royal, Martinique, December 12.

Barbadian (1843) Columns of water 100 feet in height observed at Guadeloupe, February 15 & 18.

Barbados Advocate (1932) Earthquake effects for February 3 reported at Santiago de Cuba, February 5.

Barbados Advocate (1939) Tremor felt August 15 at Santa Clara province, Cuba, August 16.

Barquero, R. and Rojas, W. (1994) Sismicidad inducida por el terremoto de Limón, *Revista Geológica de América Central,* Volumen Especial: Terremoto de Limón, San José, 111–120 (in Spanish).

Bates, R. and Jackson, J.A. (1987) *Glossary of Geology,* American Geological Institute, Alexandria, Virginia, 788 pp.

Becker, Capt. (1869) *Voyage Round the World of H.M.S. Sulphur, 1836–1842* **1**, 2 Volumes (Records from Acapulco, Mexico), 147–149.

Beckford, W. (1790) *A Descriptive Account of the Island of Jamaica* **1**, London, 91–94.

Belcher (1756) A letter concerning the extraordinary motion of the waters in Lake Ontario, *Philosophical Transactions* **49**, Royal Philosophical Society of London, 544.

Benzoni, G. (1565) *Istoria del Mondo Nuovo* **3**, in tre librii (in Italian).

Bermúdez León, K., Chinchilla Montes, S., Franco Zapata, M.G., and Murillo Madrigal, E.L. (2000) *Temblores y Terremotos Importantes en la Historia Escrita de Costa Rica,* Observatorio Vulcanológico y Sismológico de Costa Rica-Universidad Nacional (OVSICORI-UNA), *http://www.ovsicori.una.ac.cr/histsis.html* (in Spanish).

Berninghausen, W.H. (1964) Tsunamis and seismic seiches reported from the eastern Atlantic south of the Bay of Biscay, *Bulletin of the Seismological Society of America* **54**(1), 439–442.

Berninghausen, W.H. (1968) Tsunamis and seismic seiches reported from western North and South Atlantic and the coastal waters of northwestern Europe, *Informal Report 68–85,* U.S. Naval Oceanography Office, Washington, 41 pp.

Bert, L. (1880) Éruption et chute de poussières volcaniques, le 4 janvier 1880, à la Dominique (Antilles anglaises), *Comptes-Rendus de l'Academie des Sciences de Paris* **90**, 622–624 (in French).

Bevan, J. (1997) The deadliest Atlantic tropical cyclones, 1492–Present, update to E.N. Rappaport and J. Fernández-Partagas, *NOAA Technical Memorandum NWS–NHC–47* (1995), National Hurricane Center; at web-site *http://www.nhc.noaa.gov/pastdeadlyal.html.*

Bevis, J. (1757) Phænomena of the great earthquake of November 1, 1755, in various parts of the globe, *The History and Philosophy of Earthquakes–1757,* Tomas, London, 280–334, Shelf mark 18–m–11 (British Museum Manuscript).

Bewick, B. (1809) Of the earthquakes at Cadiz, Nov. 1, 1755, *Philosophical Transactions* **10**, Royal Philosophical Society of London, Abridged, 662.

Billings, L.G. (1915) Some personal experiences with earthquakes, *National Geographic Magazine* **27**(1), 57–75.

Blackford, M.E. (1996) Tsunami warning systems: An American perspective, *Summary Report of the UNESCO IOC IOCARIBE Tsunami Warning Workshop* (United Nations Educational Scientific and Cultural Organization, Intergovernmental Oceanographic Commission, IOC Sub-Commission for the Caribbean and Adjacent Regions), UNESCO International Conference on Tsunamis, St. John, U.S. Virgin Islands (May 1996), Paris.

Blainville, C. de (1843) Sur un volcan qui a fait éruption entre la Guadeloupe et Marie-Galante, le 17 mars 1843, *Comptes-Rendus de l'Academie des Sciences de Paris* **16**, 1083 (in French).

Bochet, E. (1843) Lettre sur la plus grande fréquence des tremblement de terre ressentis depuis quelques années aux Antilles, *Comptes-Rendus de l'Academie des Sciences de Paris* **16**, 1084 (in French).

Blocksma, M. (1989) *Reading the Numbers: A Survival Guide to the Measurements, Numbers, and Sizes Encountered in Everyday Life,* Penguin Books, New York.

Bodle, R.R. (1941) Summary of instrumental epicenters–1939, *United States Earthquakes, 1939,* Serial No. 637, Coast and Geodetic Survey, Washington, 40.

Bodle, R.R. and Murphy, L.M. (1948) Noninstrumental results–Puerto Rico, and also Tidal disturbances of seismic origin, *United States Earthquakes, 1946,* Serial No. 714, Coast and Geodetic Survey, Washington, 22–23.

Bois, C. (1928) Chronique Séismologique, *Materiaux pour l'Étude des Calamités* (16), Genève (in French).

Boletín Servicio Sismológico (1933) Universidad de Chile (23) (in Spanish).

Bolt, B.A. (1993) *Earthquakes, Newly Revised and Expanded,* W.H. Freeman, New York.

Bolt, B.A., Horn, W.L., Macdonald, G.A., and Scott, R.F. (1975) *Geological Hazards,* Springer-Verlag, New York, 138.

Borlase, W. (1756) An Account of the Agitation of the Waters on 1st November 1775, *Philosophical Transactions* **49**, Royal Philosophical Society of London, 373.

Borlase, W. (1762) Some Accounts of the Extraordinary Agitation of the Waters in Mount's Bay and other Places on 31st March 1761, *Philosophical Transactions* **52**, Part ii, Royal Philosophical Society of London, 418–433; also in *Philosophical Transactions* **11**, Abridged, London, 601 (1809).

Boschini, I. and Montero, W. (1994) Sismicidad Histórica e Instrumental del Caribe de Costa Rica, *Revista Geológica de América Central,* Volumen Especial: Terremoto de Limón, 65–72 (in Spanish).

Boscowitz, A. (1885) *Les Tremblements de Terre,* Paris, 135–155 (in French).

Brière de l'Île (1839) Lettre à M. Guillory, sur le tremblement de la Martinique, le 11 janvier 1839, *Bulletin de la Société Industrial d'Angers* **10**(2), 189–191 (in French).

Brière de l'Île (1843) Lettre à M. Guillory, sur le tremblement de terre de la Martinique, le 8 février 1843, *Bulletin de la Société Industrial d'Angers* **14**(2), 193 (in French).

Brigham, W.T. (1871) Volcanic manifestations in New England: Being an enumeration of the principal earthquakes from 1638 to 1869, *Memoirs Boston Society Natural History* **2**, 1–28.

Brooks, C.F. (1919) Stresses placed on the ocean floor by a hurricane, *Monthly Weather Review,* U.S. Weather Bureau.

Browne, G.M. (1869) On the floods in the Island of Bequia, *Quarterly Journal of the Geological Society* **25**, London, 12–13.

Buckman, R., Plafker, G., and Sharp, R. (1978) Fault movement following the Guatemala earthquake of February 4, 1976, *Geology* **6**, 170–173.

Bulletin du Canal Interocéanique (1882) Troisième Année, Imprimerie de la Société de Publications Périodique, Issues 76–78, Paris, 647–648, 663–664 (in French).

Bulletin of the Global Volcanism Network (1997) **5**, 1980–1997, Smithsonian Institution.

Buros, L.M. (no date) *Descripción del Terremoto del 26 de Marzo de 1812, del Cual Fui Testigo,* 336–337 (in Spanish).

Calais, E., Bethoux, N., Mercier de Lepinay, B. (1992) From transcurrent faulting to frontal subduction: A seismotectonic study of the northern Caribbean Plate Boundary from Cuba to Puerto Rico, *Tectonics* **11**(1), 114–123.

Calder, E.S., Young, S.R., Steve, R., Sparks, J., Barclay, J., Voight, B., Herd, R.A., Luckett, R., Norton, G.E., Pollard, L., Ritchie, L., Robertson, R.D.A., and the Montserrat Volcano Observatory Staff (1998) *Special Report 06, The Boxing Day Collapse, December 26, 1997,* at *http://www.geo.mtu.edu/volcanoes/west.indies/soufriere/govt/specrep/specrep06.html.*

Camacho, E. (1994) El tsunami del 22 Abril de 1991 en Bocas del Toro, Panamá, *Revista Geológica de América Central,* Volumen Especial: Terremoto de Limón (in Spanish); also in *Tsunami Newsletter* **25**(1), 6–7 (1994).

Camacho, E. and Víquez, V. (1993) Historical seismicity of the North Panama Deformed Belt, *Revista Geológica de América Central* **15**, 49–64 (in Spanish).

Camacho, E. and Víquez, V. (1994) El terremoto de San Blas del 7 Septiembre de 1882, *Revista de Humanidades* (2), Tercera Época (Abril 1994), Panamá, 49–64 (in Spanish).

Campbell, J.B. (1991) *Earthquake History of Puerto Rico, Seismicity Investigation,* Appendix I, Part A, Aguirre Nuclear Power Plant, Weston Geophysical Research, Weston, Massachusetts.

Canals, J.A. (1919) El terremoto ultimo, informe preliminar acerca de los Daños causados a las iglésias de la Mitad Occidental de la isla por el terremoto del 11 de Octubre de 1918, *Boletín Éclesiástico de la Diócesis de Puerto Rico* (15), 2[d] Serie (in Spanish).
Capadose, H. (1845) *Sixteen Years in the West Indies,* 2 Volumes, Newby, London.
Carr, M. and Stoiber, R. (1977) Geological settings of some destructive earthquakes in Central America, Appendix I: Destructive earthquakes in Central America, *Bulletin of the Geological Society of America* **88**, 151–156.
Castellanos, J. de (1955) *Elegias de Varones Ilustres de Indias* **1**, Edición de Bogotá, 4 Volúmenes, Elegy XIII, Canto 2, Edición primitiva impresa en Madrid (1589) (in Spanish).
Centeno-Graü, M (1940) *Estudios Sismológicos de Venezuela,* Litografía del Comercio, Caracas (in Spanish).
Centeno-Graü, M. (1969) *Estudios Sismológicos* **VIII**, Segunda Edición, augmentada y corregida, Biblioteca de la Academia de Ciencias Físicas, Matemáticas y Naturales, Caracas (in Spanish).
Chapman, D.J. (1867) Our southern summer storm, *U.S. National Weather Service Report,* Norfolk.
Charlevoix, Fr. P.-F.-X. de (1730a) *Histoire l'Isle Espagnole ou de Saint-Domingue; Écrite Particulièrement des Mémoires Manuscrits du Jean-Baptiste Labat le Pers, Jésuite, Missionaire à Saint-Domingue, et sur les Pièces Originales, qui se Conservent au Dépôt de la Marine* **1**, 6 Tommes, 12°, Paris, 225 pp. (British Museum Library) (in French).
Chocque, H. (1843) Note sur tremblement de terre de la Pointe-à-Pitre, le 8 février 1843, *Comptes-Rendus de l'Academie des Sciences de Paris* **17**, 352–356 (in French).
Chuy, T. (1982) Actividad sísmica de la Provincia Holguín, *Investigaciones Sismológicas en Cuba* (2), La Habana, 20–45 (in Spanish).
Chuy, T. and O. Pino (1982) Datos macrosísmicos en la provincia Santiago de Cuba, *Investigaciones Sismológicas en Cuba* (2), La Habana, 46–136 (in Spanish).
Clark, G. (1988) Hurricanes of the Caribbean Sea (1974–1987), National Hurricane Center, American Meteorological Society, Miami, 101–125 (unpublished notes).
Cochrane, J.D. and Arthur, R.S. (1948) Reflection of tsunamis, *Journal of Marine Research* **7**(3), New Haven, 239.
Codrington, C. (1901) Dispatch to the Lords of Trade and Plantations, Documents on British territories administration, 1689–1692 (Colonial), *Calendar of State Papers,* Public Records Office, London.
Coffman, J.L. and Cloud, W.K. (1970) Tsunamis, *United States Earthquakes, 1968,* Environmental Science Services Administration, Coast and Geodetic Survey, Washington, 73.
Coffman, J.L., Von Hake, C.A., and Stover, C.W. (1982) *Earthquake History of the United States,* plus supplement, U.S. Department of Commerce and U.S. Department of the Interior, 208.
Cotte, le P. (1807) Tableau chronologique des principaux phénomènes météorologiques observés en différents pays depuis 33 ans de 1774 a 1806—et comparés avec les températures correspondantes de Paris, *Journal de Physique* **65**, Paris 161–168, 250–264, 329–365 (in French).
Cruz, G. and Wyss, M. (1983) Large earthquakes, mean sea level, and tsunami along the Pacific Coast of Mexico and Central America, *Bulletin of the Geological Society of America* **73**, 553.
Currejolles (1770) Note about the depression of a barometer at Cap-Français, Haiti, before the earthquake of 1770, *Journal de Physique* **56**, 106, 292 (in French).
Curtis, G.D. and Pelinovsky, E.N. (1999) Evaluation of Tsunami Risk for mitigation and warning, *Science of Tsunami Hazards* **17**(3), 187–192.
Daubrée, M. (1880) Examen des poussières volcaniques tombées le 4 janvier 1880, à la Dominique et de l'eau qui les accompagnait, *Comptes-Rendus de l'Academie des Sciences de Paris* **90**, 624–626 (in French).
Davison, C. (1936) *Great Earthquakes,* Thomas Murby & Company, London, 286 pp.
Davy, Rev. C. (1756) *Letters Addressed Chiefly to a Young Gentleman, upon Subjects of Literature, &c.* **2**, A. Pereira, London, 12–60, 114–127.
Delpech, M. (no date) *Descripción del terremoto del 26 Marzo de 1812, del Cual Fui Testigo* (in Spanish).

Denver Rocky Mountain News (1998) "Landslides blamed for tidal waves; Scientist finds cause of Puerto Rico disasters," May 31, 49A.

Denver Rocky Mountain News (2000) La Palma volcano collapse will generate a giant tsunami across the Atlantic, October 9, 35A.

Denyer, P., Arias, O., and Personius, S. (1992) Generalidades sobre los efectos geológicos del terremoto de Limón, *Revista Geológica de América Central,* Volumen Especial: Terremoto de Limón, 29–38 (in Spanish).

Devas, R.P. and Macadam-Sherwin, T. (1939) Manifestations of volcanic activity observed July 24, 1939 from Gunton Estate House, Grenada, dated July 25, 1939, 2 pp. (typescript).

Diario de la Habana (1853) Apuntes para la historia de los terremotos en Santiago de Cuba, particularmente del 11 de Junio de 1766, June 24 & 25 (in Spanish).

Dillon, W., Acosta, J., Uchupi, E., and Brink, U. ten (1998) Joint Spanish–American Research Uncovers Fracture Pattern in Northeast Caribbean, *EOS Transactions*, American Geophysical Union (July 14, 1998), 309–310.

Dillon, W., Brink, U. ten, Frankel, A., Mueller, C., and Rodríguez, R. (1999) Seismic and Tsunami hazards in Northeast Caribbean, *EOS Transactions* **80**(28), American Geophysical Union, 309–310.

Dillon, W.P., Edgar, N.T., Scanlon, K.M., and Coleman, D.F. (1996) A review of the tectonic problems of the strike-slip northern boundary of the Caribbean Plate and examination by GLORIA, in J.V. Gardner, M.E. Field, and D.C. Twichell (eds.), *Geology of the United States' Seafloor: The View from GLORIA,* Cambridge University Press, Cambridge, 135–164.

Dillon, W.P., Edgar, N.T., Scanlon, K.M., and Klitgord, K.D. (1987) Geology of the Caribbean, *Oceanus* **30**(4), 42–52.

Dixon (1869) The St. Thomas Treaty, Articles to the *Boston Daily Advertiser* from Washington Correspondent, Dixon, of the *Boston Daily Advertiser,* New York.

Dorel, J. (1981) Seismicity and seismic gap in the Lesser Antilles arc and seismic hazard in Guadeloupe, *Geophysics Journal of the Royal Astronomical Society* **67**, 679–696.

Douglas, M.S. (1958) *Hurricane,* Rinehart & Company, New York, 393 pp.

Dudley, W.C. and Min, L. (1998) *Tsunami!,* Second Edition, University of Hawai'i Press, Honolulu.

d'Ulloa, Don A. (1756) An account of the earthquake of November 1, 1755, in a letter transmitted to the Royal Philosophical Society, *Philosophical Transactions* **49**, London.

d'Ulloa, Don A. (1809) Of the earthquakes at Cadiz, Nov. 1, 1755, *Philosophical Transactions* **10**, Royal Philosophical Society of London, Abridged, 662–663.

Duchassaing (1843) Observations sur le tremblement de terre de la Guadeloupe, du 8 février 1843, *Comptes-Rendus de l'Academie des Sciences de Paris* **27**, 190–191 (in French); also in *Bulletin de la Société de Géologie de France* **14**, 611–612 (in French).

Dunbar, P.K. and Whiteside, L.S. (1994) Earthquake coupling with other natural disasters, Poster presentation at the American Geophysical Union Meeting (December 1994), San Francisco.

Dunn, G.E. and Miller, B. (1964) *Atlantic Hurricanes,* Louisiana State University Press, Baton Rouge, 377 pp.

Dutrouleau et Gonenet (1844) Relation médico-chirurgicale du tremblement de terre de la Pointe-à-Pitre, le 8 février 1843, *Annuaire Maritime et Colonial,* 8°, 51 pp. (in French).

Earthquake Notes (1946) Important Earthquakes of 1946: Dominican Republic Earthquake of August 4, 1946, *Eastern Section of the Bulletin of the Seismological Society of America* **18**(1/2), 2–4, 11–12.

El Nueva Día (1989) Small tsunami November 1 at Cabo Rojo, Puerto Rico, November 2 & 11 (in Spanish).

Espinoza, F. (1976) The Guatemalan Earthquake of February 4, 1976, A Preliminary Report, *Professional Paper 1002,* U.S. Geological Survey, Washington.

Evans, F.I. and Wharton, W.L. (1888) On the seismic sea waves caused by the eruption of Krakatoa, 26th and 27th August, 1883, and subsequent phenomena, *Philosophical Transactions,* Part iii, Royal Philosophical Society of London, 89–151.

Evans, J. (1848) Hurricanes, 1493–1848, *Nautical Magazine,* London.

Ewing, M. and Press, F. (1955) Tide-gauge disturbances from the Great Eruption of Krakatoa, *EOS Transactions* **36**(1), American Geophysical Union, 53–60.

Eyriès, J.-B.-B. (1824) *Abrégés des Voyage modèrnes depuis 1780 jusqu'à nos Jours,* 14 Tommes, 8°, Paris, 1822–1824 (in French).

Eyriès, J.-B.-B. (1825) About the destruction of Mérida and Caracas during the 26th March 1812 earthquake, *Nouvelles Annales des Voyages* **36**, 58 (in French).

Fayers, R.-J. (1843) Letter concerning the Earthquake in the West Indies, *Jameson's Journal* **35**(69), 180.

Feldman, L.H. (1984) A catalogue of historical documents pertaining to the earthquake damage in Panamá and Costa Rica, 54 pp. (unedited manuscript).

Feldman, L.H. (1993) *Mountains of Fire, Lands That Shake: Earthquakes and Volcanic Eruptions in the Historic Past of Central America (1505–1899),* Labyrinthos Press, Culver City.

Fernández, M., Havskov, J., and Atakan, K. (1999) Destructive tsunamis and tsunami warning in Central America, *Science of Tsunami Hazards* **17**(3), 173–185.

Fernández, M., Peraldo, G., Rojas, W., and Flores, R. (1993) Tsunamis en Centroamerica, *Tecnología en Marcha* **12**(2), Costa Rica, 17–30 (in Spanish).

Férussac (1824) Saint-Pierre, Martinique, November 30, *Bulletin des Sciences et Mathématiques* **vi**, 17, and **viii**, 303 (in French).

Feuillard, M. (1982) La séismicité historique de la Guadeloupe et la Martinique, Ph.D. thesis, Observatoire Volcanologique de Guadeloupe (in French).

Feuillard, M. (1985) *Macroséismicité de la Guadeloupe et de la Martinique* (in French).

Fiedler, G. (1961) Areas afectadas por terremotos en Venezuela (1791–1814), *Boletín Geología* **4**, Publicacion Especial 3, Memorias del III Congreso Geológico Venezolano, Ministeria de Minas e Hidrocarburos, Caracas (in Spanish).

Flament, P. and Martin, R. (1980) Le cyclone Allen—du 29 julliet au 12 août 1980, *L'Association Météorologique de la Martinique,* avec le concours, du Conseil General de la Martinique et du Service Météorologique Antilles–Guyane, 32 pp. (in French)

Forel, A. (1885) Bruit souterrains entendus le 26 août 1883 l'îlot Caiman-Brac, *Comptes-Rendus de l'Academie des Sciences de Paris* **100**, 266 (in French).

Forster, W.G. (1890) Earthquake origin, *Transactions of the Seismic Society of Japan* **15**, Yokohama, 77.

Fowke (1755) *Letter to Mr. J. Fowke, from his Brother near Lisbon,* November, Dublin, 8 pp.

Fuchs, C.W.C. (1883) *Die vulkanischen Ereignisse des jahres* 1882 **5**, *Jahresbericht, Mineralogie und Petrographie Mitteilungen, Neue Folge,* Wien, 339–381 (in German).

Fuchs, C.W.C. (1886) *Statistik der Erdbeben von 1865 bis 1885* **92**, Heft iii, *Sitzungs berichte der Kaiserlichen Akademie Wissenschaften,* Wien, 215–625 (in German).

Gazette de France (1755) Ebb and flow November 18 at St. Martin, December (in French).

Gazette de France (1767) Extrait d'une lettre du M. Mentelle, ingénieuer géographe du roi, des secousses de avril 24 à Surinam et Martinique, July 17; September 4 & 21 (in French).

Gazette de France (1771) Violent shock September 3 at la Jamaïque, December 18 (in French).

Gentleman's Magazine (1755) **25**, 534–536, 541–542, 554–564, 587–594.

Gentleman's Magazine (1756) **26**, 7–8, 67–72, 99–100.

González, F.I. (1999a) Tsunami!, *Scientific American* **280**(5), 56–65.

González, F.I. (1999b) Email from P. Abbott on the threat from the Kick-'em-Jenny Sub-Sea Volcano, personal communication, March 22.

González Víquez, C. (1910) *Temblores, Terremotos, Inundaciones y Erupciones Volcánicas de Costa Rica, 1608–1910,* Tipográfica Avelino Alsina, San José, 200 pp. (in Spanish).

Grases, J. (1971) *La Sismicidad Histórica del Caribe,* Investigaciones, Documentos de Trabajo, Informe que presenta a la Junta Directiva del Comité Conjunto del Concreto Armado de Venezuela (Octubre), FUNVIS, Caracas (in Spanish).

Grases, J. (1974) *Relación Cronológica de Sismos Destructores de América Central, lapso 1526–1900* **2**, Universidad de Venezuela, Caracas (in Spanish).

Grases, J. (1990) *Terremotos Destructores del Caribe, 1502–1990,* UNESCO–RELACIS, Caracas (in Spanish).

Grenada Chronicle (1867) Clouds of volcanic gases observed November 18th at St. George's Carenage, Grenada, November.

Grenada Free Press (1831) December 3rd earthquake rang the church bell at St. George's, Grenada, December 7.

Güendel, F. (1986) Seismotectonics of Costa Rica: An analytical view of the southern terminus of the Middle America Trench, Ph.D. Thesis, University of California, Santa Cruz.

Guyana Chronicle (1825) On September 20th a phænomenon lasting 3–4 minutes ended with a movement resembling waves of the sea, in the Demerara region of Guyana; The sea was agitated by an oscillatory motion analogous to that felt on shore, September 21.

Hall, Capt. D.C. (1917a) The launching to the *USS Monongahela,* in J.W. Greenslade (ed.), *Proceedings of the United States Naval Institute* **43**, 1161–1172.

Hall, Capt. D.C. (1917b) The wreck of the *USS De Soto,* in J.W. Greenslade (ed.), *Proceedings of the United States Naval Institute* **43**, 1151–1160.

Hall, M. Esq. (1907) The great earthquake of January 14th, 1907, and the aftershocks, *Special Issue of the Jamaica Weather Report 337,* Third Report on Earthquakes in Jamaica, Kingston.

Hall, Esq., M. (1913) The Annotto Bay Epicentre, *Jamaica Weather Report 442,* Kingston, 5.

Hall, Esq., M. (1922) (post mortem) *Earthquakes in Jamaica from 1688 to 1919, with Map,* Kingston, 58 pp.

Harper's Weekly (1868) The late earthquake at St. Thomas, with illustration of the Royal Mail Steamer, *La Plata* **12**(578), New York.

Harte, B. (1868) St. Thomas: A Geographical Survey, *The Heathen Chinee: Poems and Parodies by Brett Harte,* R.E. King & Company Limited, London.

Harvey, Br. J.L. (1876) *Periodical Accounts Relating to the [Moravian] Missions of the Church of the United Brethren, Established Among the Heathen (St. Croix, November 21st, 1867)* **26**, Moravian Brethren's Society for the Furtherance of the Gospel Among the Heathen, London.

Heberden, T. (1761) An account of the earthquake felt in the island of Madeira, March 31, 1761, *Philosophical Transactions* **52**, Royal Philosophical Society of London, 155–156; also in *Philosophical Transactions* **11**, Abridged, London (1809), 543.

Heberden, T. (1809) Of the earthquake in the island of Madeira, November 1, 1755, *Philosophical Transactions* **10**, Royal Philosophical Society of London, Abridged, 664.

Hebert, P.J., Jarrell, J.D., and Mayfield, M. (1993) *The Deadliest, Costliest, and Most Intense United States Hurricanes of this Century,* NOAA Technical Memorandum NWS–NHC–31, National Oceanic and Atmospheric Administration, 41 pp.

Heck, N.H. (1947) List of seismic sea waves, *Bulletin of the Seismological Society of America* **37**(4), 269–286; also in *Annales de la Commission de l'Étude de Raz-de-Marée* (4), Union Géodésique et Géophysique Internationale, Paris (1934), 20–41 (in French).

Heilprin, A. (1903) *Mont Pelée and the Tragedy of Martinique: A Study of the Great Catastrophes of 1902, with Observations and Experiences in the Field,* J.B. Lippincott, Philadelphia and London, 333 pp.

Henderson (1852) *Henderson's Jamaica Almanack and Diary for 1852,* 36.

Herrera y Tordesillas, A. de (1726) *Historia General de los Hechos de los Castellanos en las Islas y Tierra Firme del Mar Océano, 1601–1615,* Madrid (in Spanish).

Herrera y Tordesillas, A. de (1740) *History of the Vast Continent and Islands of America* **5**; English translation by Capt. J. Stevens, 2nd Edition, Book iii, Wood & Woodward, London, 54.

Herridge de Guerrero, C.M. (1998) New fatality data for the 1946 tsunami from the Coordinator of the Dominican Disaster Mitigation Committee, personal communication, July 21.

Hess, H.H. (1932) Interpretation of gravity anomalies and sounding profiles obtained in the West Indies by international expedition in 1932, *EOS Transactions* **13**, American Geophysical Union, 26–33.

Holcombe, T.L., Einwich, A.M., Bowles, F.A., and Egloff, Jr., J. (1977) *The Geological Environment West of St. Croix,* NORDAC Report No. 5, Naval Oceanographic Research and Development Activity (March), Bay St. Louis, 82 pp.

Hoose, H. (1969) A study of swells affecting Puerto Rico, 1956–1968 (unpublished report), National Weather Service, San Juan.

Humboldt, Baron A. de (1826) *Essai Politique sur l'Île de Cuba,* Paris (in French).

Humboldt, Baron A. de (1842) *Voyage aux Régions Equinoxiales* **1/2**, 22, 272, 285, 295; **5**, 5, 28; **7**, 228 (in French).

Humboldt, Baron A. de and Bonpland A. (1818) *Personal Narrative of Travels to the Equinoctial Regions of the New Continent, During the Years 1799–1804,* Translated by H.M. Williams, 7 Volumes, London; also printed by AMS Press, New York **1**, 272–280; **2**, 206–231; **3**, 311–335, 401; **4**, 4–11(1966).

Humboldt, Baron A. von (1819) Account of the earthquake which destroyed the town of Caracas on the 26th March 1812, *Journal of the Edinburgh New Philosophical Society* **1**, 272–280.

Iida, K., Cox, D.C., and Pararas-Carayannis, G. (1967) *Preliminary Catalog of Tsunamis Occurring in the Pacific Ocean,* Data Report 5, HIG–67–10, Hawai'i Institute of Geophysics, University of Hawai'i, Honolulu.

Itier, J. (1843) Renseignements sur le tremblement arrivée à la Pointe-à-Pitre, le 8 février 1843, *Bulletin de la Société Géologie* **14**, 610 (in French).

Jamaica Daily Gleaner (1867) "Packet Summary: violent convulsions of Nature and volcanic eruption from November the 14th to the 21st at Mount Vesuvius in Italy, December 9.

Jameson's Journal (1843) Tremblement le 8 février 1843, **35**(69), 179–185 (in French).

Journal of Geology (1898) **10**, Montserrat Island (The West Indies), Tokyo, 568 pp.

Kanamori, H. (1972) Mechanism of Tsunami Earthquakes, *Phys. Earth Planet. Inter.* **6**, 346–359.

Kanamori, H. and Stewart, G. (1978) Seismological aspects of the Guatemala earthquake of February 4, 1976, *Journal of Geophysical Research* **83**, 3427–3434.

Kant, J. (1756) *History and Nature of the Most Remarkable Occurrence of Earthquakes which Convulsed a Great Part of the Earth at the End of the Year 1755,* Königsberg (in German).

Kelleher, J.A., Sykes, L, and Oliver, J. (1973) Possible criteria for predicting earthquake locations and their application to major plate boundaries of the Pacific and the Caribbean, *Journal of Geophysical Research* **78**(14), 2547–2585.

Kendrick, T. (1956) *The Lisbon Earthquake,* Methuen.

Kirkpatrick, R.Z. (1920) Earthquakes in Panamá up to January 1, 1920, *Bulletin of the Seismological Society of America* **10**(2), 121–128.

Kluge, E. (1863) *Über Synchronismus und Antagonismus von Vulkanischen Eruptionen,* Leipzig (in German).

Labat le Pers, Br. J.-B. (1722) *Nouveau Voyage aux Îles de l'Amérique, avec l'Histoire Naturelle de ces Pays, l'Origine, les Mœrs, la Religion et les Évèntments Singuliers* **vii**, 6 Tommes, 12°, Paris, 440 (in French).

Lander, J.F. (1986) *Project THRUST, Final Report*, National Oceanic and Atmospheric Administration, National Geophysical Data Center, Boulder.

Lander, J.F. and Cloud, W.K. (1963) Summary of instrumental epicenters for 1961, *United States Earthquakes 1961,* Coast and Geodetic Survey, Washington, 68.

Lander, J.F. and Lockridge, P.A. (1989) *United States Tsunamis (Including United States Possessions) 1690–1988,* NOAA Publication 41–2, National Oceanic and Atmospheric Administration, National Geophysical Data Center, Boulder, 265 pp.

Lander, J.F. and O'Loughlin, K.F. (1997) Preliminary table of Caribbean tsunamis, Keynote address presented at the Caribbean Tsunami Workshop (June 1997), Mayagüez, Sponsored by the Sea Grant College Program of the University of Puerto Rico.

Lander, J.F. and O'Loughlin, K.F. (1998a) Caribbean tsunamis: Overlooked threat, Long Term Contributions to Tsunami Science Tsunami Society Award keynote presentation at the Seventh International Conference on Natural and Man-made Hazards (May 1998), Chania, Crete.

Lander, J.F. and O'Loughlin, K.F. (1999b) Update to Caribbean Tsunamis: Newly recognized hazard, Presented to the UNESCO IOC User Group Meeting (January 1999), Honolulu.

Lander, J.F. and O'Loughlin, K.F. (1999c) Caribbean Tsunami Warning System Needs, Invited presentation to the International Meeting of the Tsunami Society (May 1999), Honolulu.

Lander, J.F. and O'Loughlin, K.F. (1999e) Tsunami Events in Puerto Rico and the U.S. Virgin Islands, Presentation at the U.S. Geological Survey Workshop on Seismic and Tsunami Hazards in Puerto Rico and the Virgin Islands (March 1999), San Juan.

Lander, J.F., Whiteside, L.S., and O'Loughlin, K.F. (1999) Use of tsunami histories to define the local Caribbean hazard, *Proceedings of the 1998 International Symposium on Marine Positioning*, (December) Melbourne, Florida.

Landre, M. (1843) *Aaardbevingen in Guiana (Surinam, Berbice, Demerary, Essequebo)*, 4°, 6 pp. (in Dutch); also in *Surinaamsche Courrant*, February 10 (1843) (in Dutch).

Langrené de (1752) Extrait d'une lettre du 16 décembre 1751, sur les tremblements de terre arrivées dans l'île de Sainte-Domingue, à la fin de l'année 1751, *Oeconom*, 127–137 (in French).

Las Casas, Fr. B. de (1876) *Historia de Las Indias* **3**, 5 Volúmenes, escrito en 1552–1561, Madrid, 404–405 (in Spanish); also in B.C. Aribau (ed.) *Biblioteca de Autores Españoles desde la Formación del lenguaja hasta Nuestros Días* **95–96**, Madrid (1947) (in Spanish).

Latimer, G.A. (1868) Observations regarding the earthquakes which occurred in St. Thomas and neighboring islands, commencing November 18, 1867, *Smithsonian Annual Report, 1967*, 465–466.

Le Collection Académique (1755) Ebb and flow November 18 at St. Martin **vi**, 634 (in French).

Le Commercial (1867) Sea swell reported at Pointe-à-Pitre, Guadeloupe, November (in French).

Le Constitutionnel (1824) Remarkable rise and fall of tide at Plymouth [Montserrat], November 16 (in French).

Le Constitutionnel (1829) A violent *raz-de-marée* October 26 and 27 on the coast of Martinique, December 16 (in French).

Leeds, D.J. (1974) Catalog of Nicaragua earthquakes, *Bulletin of the Seismological Society of America* **64**, 1135–1158.

Le Journal de la Belgique (1942) Earthquake on May 7 at Cap-Haïtien, June 25 & July 3 (in French).

Le Journal Historique (1755) Ebb and flow November 18 at St. Martin, December (in French).

Le Journal Historique (1767) Shocks at Surinam, Martinique, and Barbados, October 3 & 18 (in French).

Le Moniteur Belge (1802) March 19 shocks along with agitation of the sea, March 24 (in French).

Le Moniteur Belge (1813) Shock on November 11, 1812 at Kingston, January 26 (in French).

Le Moniteur Belge (1829) *Raz-de-marée* at Martinique on October 26 (in French).

Le Moniteur Belge (1842) Extent of the May 7 earthquake effects reported, June 20 (in French).

Leonhard and Bronn (1833) *Neues Jahresbericht für Mineralogische*, 127 (in German).

Leslie, C.A. (1740) *A New History of Jamaica, from the Earliest Accounts, to the Taking of Porto Bello by Vice-Admiral Vernon, in Thirteen Letters from a Gentleman to His Friend* [Sir Hans Sloane, founder of the British Museum], Letter VIII, Printed for J. Hodges at the Looking-glass on London-bridge, London, 255–259, Shelf mark 1052–a–13 (British Museum Library).

L'herminier, F. (1843) Notes sur les tremblements de terre 11 janvier 1839, et recherches sur les rapports de ce *phénomène* avec l'état météorologique de la Guadeloupe, *Comptes-Rendus de l'Academie des Sciences de Paris* **17**, 980–981 (in French).

Long, D. (1999) Information about a tsunami at Bequia, in The Grenadines, on March 17, 1868, personal communication, November 29.

Long, D. (2000) Boxing Day Collapse of December 26, 1997, at Montserrat, personal communication, November 13.

Long, E. (1774) *The History of Jamaica or General Survey of the Ancient and Modern State of that Island*, 3 Volumes, 4°, London.

Ludlum, D.M. (1963) *Early American Hurricanes, 1492–1870*, American Meteorological Society, Boston.

Lugo, M. and Giraldo, C. (1988) *El Terremoto de Octubre de 1766: Posibles Asociaciónes Sísmotectonicas*, Resúmenes, V Congreso Venezolano de Sismología e Ingeniería Sísmica, Asociación Venezolano de Ingeniería, Estructural Cumaná (in Spanish).

Lyell, Sir C. (1832) *Principles of Geology* **1**, 1st Edition, 4 Volumes, Murray, London, 438–440.

Lyell, Sir C. (1875) *Principles of Geology* **2**, 12th Edition, 2 Volumes, Murray, London, 146–154.

Lynch, Fr. J.J. and Bodle, R.R. (1948) The Dominican earthquakes of August 1946, *Bulletin of the Seismological Society of America* **38**(1), 1–18.

Lynch, L.L. and Shepherd, J.B. (1995) An earthquake catalogue for the Caribbean, Part II, The macroseismic listing for the instrumental period 1900–1991, Paper Presented to the Caribbean and Latin America Seismic Hazard Project Workshop (May 1995), Melbourne, Florida, 45 pp.

Mader, C.L. (1997) 1755 Lisbon Tsunami, CD video of travel times model results, Caribbean Tsunami Workshop (June 1997), Mayagüez.

Mairan, D. de (1752) Secousses de tremblement de terre ressentis aux Antilles, du 15 septembre au 21 novembre 1751, *Comptes-Rendus de l'Academie des Sciences de Paris,* 16–17 (in French).

Mallet, R. (1851) *The Facts of Earthquake Phænomena,* 20th Meeting of the British Association for the Advancement of Science Meeting, Edinburgh (1850), John Murray, London.

Mallet, R. (1852) *The Facts of Earthquake Phænomena,* 21st Meeting of the British Association for the Advancement of Science Meeting, Ipswich (1851), John Murray, London.

Mallet, R. (1853) *The Facts of Earthquake Phænomena,* Catalogue of Recorded Earthquakes from 1606 B.C. to A.D. 1850, Part i, 1606 B.C. to 1755 A.D., 22nd Meeting of the British Association for the Advancement of Science Meeting, Belfast (1852), John Murray, London, 176 pp.

Mallet, R. (1854) *The Facts of Earthquake Phænomena,* Catalogue of Recorded Earthquakes from 1606 B.C. to A.D. 1850, Part ii, 1755 A.D. to 1784 A.D., 23rd Meeting of the British Association for the Advancement of Science Meeting, Hull (1853), John Murray, London, 212 pp.

Mallet, R. (1855) *The Facts of Earthquake Phænomena,* Catalogue of Recorded Earthquakes from 1606 B.C. to A.D. 1850, Part iii, 1784 A.D. to 1842 A.D., 24th Meeting of the British Association for the Advancement of Science Meeting, Liverpool (1854), John Murray, London, 325 pp.

Mallet, R and Mallet, J.M. (1858) *The Earthquake Catalogue of the British Association* (From the Transactions of British Association for the Advancement of Science, 1852–1858), Taylor & Francis, London.

Mangeney, A., Heinrich, P., Roche, R., Boudon, G., and Cheminée, J.-L. (1998) Numerical simulation of the December 1997 tsunami generated by a debris avalanche in Montserrat, Lesser Antilles, *Abstracts of the UNESCO International Conference on Tsunamis* (May 1998), Paris, 88.

Marx, R.F. (1983) *Shipwrecks in the Americas,* Bonanza Books, 482 pp.

Maskell, W.M. (1868) *The Tsunami, Earthquake, and Hurricane That Struck the Danish West Indies, November 18, 1867,* in Paiewonsky (ed.), "History Corner," *Virgin Islands Daily News* (1979).

Mason, A. (1762) Extract of a letter from Barbados to Rev. Thomas Birch, D.D. secretary of the Royal Society, from M. Abraham Mason, relating to an extraordinary agitation of the sea at Barbados, 31 March 1761, *Philosophical Transactions* **52**, Part 2, Royal Philosophical Society of London, 477.

Maul, G. (1997) Cited about the potent danger of tsunamis in the Caribbean, Maine Sunday Telegram, July 6, 3C.

Maul, G. (1999) Letter to the editor of *Scientific American,* personal communication, April, 30.

McDonald, F. (1999) French privateer report of tsunami in 1692, personal communication, June 7.

McDonald, F. and Johnson, C. (1913) Isthmian earthquakes (Panamá), *Canal Record* **7**, 144–149.

McPhaul, J. (1998) "'Tsunamis' can occur in hurricane season," *The San Juan Star*, Puerto Rico, June 1, Local News, 7.

Mendoza, C. and Nishenko, S. (1989) The North Panamá earthquake of 7 September 1882: Evidence for active underthrusting, *Bulletin of the Seismological Society of America* **79**, 12649.

Mercado A. and McCann, W.R. (1998) Numerical simulation of the 1918 Puerto Rico tsunami, *Natural Hazards* **18**(1), 57–76.

Mercado Irizarry, A. (1997a) 1989 Eyewitness report of a small tsunami November 1 at Cabo Rojo, Bahía de Boquerón, Puerto Rico, personal communication, October 29.

Mercado Irizarry, A. (1997b) Fatality count at Aguadilla adjusted upwards for the October 11, 1918 tsunami, personal communication, September 11.

Mercado Irizarry, A. (1997c) Possible tsunami on July 9, 1997 along the coasts of Venezuela in the state of Sucre, personal communication, October 14.

Mercado Irizarry, A. (1999) December 1967 waves batter the north coast of Puerto Rico, personal communication, December 29.

Meyer, H. (1996) Notes in preparation for the IOCARIBE Consultation of Experts on Tsunami Hazards, IOC Workshop Draft Summary Report of the IOC IOCARIBE Tsunami Warning System Workshop, Annex III, Appendix VI, UNESCO International Conference on Tsunamis, St. John, U.S. Virgin Islands (May 1996), 50 pp.

Millas, J.C. (1968) *Hurricanes of the Caribbean and Adjacent Regions, 1492–1800,* Academy of Arts and Sciences of the Americas, Miami.

Milne, D. (1841) Times at which the shock of the 1st November 1755 earthquake reached different places, and the principal phænomena there, *Journal of the Edinburgh New Philosophical Society* **31**, 37–40.

Milne, J. (1911) Unpublished notes relating to destructive earthquakes, *British Association for the Advancement of Science,* 55.

Milne, J. (1912) *A Catalogue of Destructive Earthquakes, A.D. 7 to A.D. 1899,* 16th Report of the Committee on Seismological Investigations Researches in Science, 81st Meeting of the British Association for the Advancement of Science, Portsmouth (1911) London, 94 pp.

Mitchell J. (1761) Conjectures concerning the cause and observations upon the phænomena of earthquakes: Particularly of that great earthquake of the first of November 1755, *Philosophical Transactions* **51**, Royal Philosophical Society of London, 566–634.

Molard, P. (1947) Tremblements de terre des Petites Antilles, et manifestations actuelles du volcanisme de l'archipel (1936 à 1943), *Annales Géophysicae* **3**, 19–21, 113–140 (in French).

Molina, E. (1997) *Tsunami Catalogue for Central America, 1539–1996 for Reduction of Natural Disasters in Central America, Earthquake Preparedness and Hazard Mitigation,* Technical Report II 1–04, Institute of Solid Earth Physics, University of Bergen, Norway, 87 pp.

Molloy (1761) An account of the earthquake at Lisbon, 31st March 1761, *Philosophical Transactions* **52**, Royal Philosophical Society of London, 142–143.

Montadon, F. (1962) Les mégaséismes en Amérique, *Révue pour l'Étude des Calamités, Bulletin de la Union Internationale Sécours* (38), Genève, 57–97 (in French).

Montero, W. (1986a) El terremoto de San Estanislao del 7 de Mayo de 1822; Un gran temblor de subducción del sur la Costa Rica?, *Revista de Ciencias y Tecnología* **10**, 11–20 (in Spanish).

Montessus de Ballore, F. de (1888) Tremblements de terre et éruptions volcaniques au Centre-Amérique depuis la conquête espagnole jusqu'à nos jours, *Société des Sciences Naturelles de Saône-et-Loire de Dijon,* 246–378 (in French).

Montessus de Ballore, F. de (1912) Earthquake and tsunami of August 13, 1868, *Histórica Sísmica de los Andes Meridionales* **2**, 77–158 (in Spanish).

Monthly Weather Review (1883) Brig. and Bvt. Maj. Gen. W.B. Hazen (ed.), **XI**(8) (August), U.S. War Department, Washington, 191.

Monthly Weather Review (1971) U.S. Weather Bureau Report 19–98 (1891–1970).

Moreau de Jonnès, A. (1819) Considerations sur les causes météorologiques qui, dans l'année 1812, ont transporté jusqu'à la Barbade, des déjections du volcan de Saint-Vincent, *Bulletin de la Société Philom.,* 95 (in French).

Moreau de Jonnès, A. (1839) Note sur le tremblement de terre du 11 janvier 1839, à la Martinique, *Comptes-Rendus de l'Academie des Sciences de Paris* **8**, 329 (in French).

Moreau de Jonnès, A. (1843) Les anciens séismes ont été fréquents autrefois et désastreux des Antilles Françaises, savoir: La Martinique et les îles de la Guadeloupe, *Comptes-Rendus de l'Academie des Sciences de Paris* **16**, 1153–1154 (in French).

Moreau de Saint-Méry, M.L. (1797) Description Topographique, Physique, Civile, Politique et Historique de la Partie Françoise de l'Isle Saint-Domingue, *Philadelphie,* 2 Tommes, 4° (in French); 2e Édition, 3 Tommes, 8°, Paris (1875–1876) (in French).

Moreau de Saint-Méry, M.L. (1844) *Descripción de la Partie Española de Santo Domingo,* Traducción de A. Rodríguez, Trujillo (in Spanish).

Moreira, V.S. (1993) Historical tsunamis in mainland Portugal and the Azores: Case histories, in S. Tinti (ed.), *Tsunamis of the World,* Kluwer Academic Publishers, Dordrecht, 65–73.

Murphy, L.M. (1964) *Annotated Bibliography on Tsunamis,* Monographie 27, Spaeth, M.G. (ed.), Union Géodésique et Géophysique Internationale, Compiled and edited by U.S. Coast and Geodetic Survey, Imprimé par l'Institut Géographique National (July), Paris, 249 pp.

Murphy, L.M. and Cloud, W.K. (1955) Tidal disturbances of seismic origin, *United States Earthquakes, 1953,* Serial No. 785, Coast and Geodetic Survey, Washington, 29.

Murphy, L.M. and Ulrich, F.P. (1952) Tidal disturbances of seismic origin, *United States Earthquakes 1950,* Serial No. 755, Coast and Geodetic Survey, Washington, 39.

Murty, T.S. (1977a) Deep water signature of a tsunami, *Marine Geodesy* **1**(1), 73–102.

Murty, T.S. (1977b) *Seismic Sea Waves–Tsunamis,* Department of Fisheries and Marine Research, Bulletin of the Fisheries Research Board of Canada (198), Ottawa, 337 pp.

Musson, R. (1999) East-to-west Caribbean hurricanes, personnel communication, December 2.

Myles, D. (1985) *The Great Waves,* McGraw-Hill, New York, 206 pp.

Nelson, W. (1889) *Five Years at Panamá: The Trans-Isthmian Canal,* Belford Company, New York, 287 pp.

Neumann, C.J., Jarvinen, B.R., Pike, A.C., and Ellms, J.D. (1988) *Tropical Cyclones of the North Atlantic Ocean, 1871–1986 (with figures to 1992)* **6**(2), National Oceanic and Atmospheric Administration, National Climatic Data Center, Historical Climate Service, Asheville.

Neumann, F. (1953) The earthquake problem in Haiti, *Earthquake Notes, Eastern Section of the Bulletin of the Seismological Society of America* **24**(3–4), 24–26.

Nishenko, S. *et al.* (1992) The 1991 Limón, Costa Rica tsunami (unpublished manuscript).

Nugent, Lady M. (1934) *Lady Nugent's Journal: Jamaica One Hundred and Thirty Years Ago,* 2nd Edition, F. Cundall (ed.), West India Committee, London, 404 pp.; Reprinted, P. Wright (ed.) (1966).

Oddone, E. (1907) Tremblements de terre ressentis pendant l'année 1904, *Publication Bureau Central de l'Association Internationale Séismique,* Série B, Strassburg (in French).

Odonel Gómez, L. (1998) Concerning the updated 1946 Dominican Republic fatality estimate of 1790 perishing from the tsunami, personal communication, July 27.

Oldmixon, J. (1741) *History of the British Empire in America,* 2nd Edition, 2 Volumes, London (British Museum Library).

O'Loughlin, K.F. (2001) Unraveling the details of the November 18, 1867 tsunami, earthquake, and hurricane in Saint Thomas and the Antilles (unedited manuscript).

Olsen, P. (1988) Archivist for Denmark translation of excerpts from the records of the Danish West Indies and Guinea Company, *Journal det paa St. Thomas passerede* (498), translated from German; and from the Governor of St. Thomas to the Danish West Indian and Guinea Company Board dated "Christiansfort of 8 April, 1690," in *Breve og dokumenter fra Vestindien, 1674–1698* (98), translated from Danish, personal communication, May 9.

Örsted, A.S. (1867) Extract from a Treatise by the Danish Naturalist A.S. Örsted on the "Natural Features of the Westindia Islands," *Sanct Thomæ Tidende,* November 30.

Oviedo y Valdes, G.F. de (1547) *Crónicas de las Indias: La Hystoria General de las Indias, Nuevamente Impresa, Corregida y Emendada; y con la Conquista del Peru* **1**, Libro XIX, Capítulo VII, Salamanca (in Spanish).

Pacheco, J. and Sykes, L.R. (1992) Seismic moment catalog of large, shallow earthquakes, 1900–1989, *Bulletin of the Seismological Society of America* **82**, 1306–1349.

Pacine, M. (1839) Lettre à M. Martins, sur le tremblement de terre de la Martinique, le 11 janvier 1839, *Comptes-Rendus de l'Academie des Sciences de Paris* **8**, 364 (in French).

Paiewonsky, I. (1979) "History Corner," *Virgin Islands Daily News,* St. Thomas, July 30; August 6, 13, 20 & 27; November 5, 12, 19 & 26; December 3, 10, 17, & 24.

Paiewonsky, I. (1981) "History Corner," *Virgin Islands Daily News,* St. Thomas, November 2 & 9.

Paige, S. (1930) The earthquake at Cumaná, Venezuela, January 17, 1929, *Bulletin of the Seismological Society of America* **20**(1), 1–10.

Palgrave, W.G. (1874) Earthquake in St. Thomas, *Nature* **9**, 483–484.

Panamá Star and Herald (1873) Earthquake and tsunami on October 14 at Colón Harbor, Panamá, October 16 (in Spanish).

Panamá Star and Herald (1882) Effects of September 7 tsunami at Archipiélago de San Blas, Panamá, September 8–14; October 11 (in Spanish).

Panamá Star and Herald (1916) Knee-deep tsunami on April 25 at Archipiélago de Bocas del Toro, Panamá, April 26 (in Spanish).

Parker, W.E. (1922) Unusual tidal registration of earthquakes, *Bulletin of the Seismological Society of America* **12**(1), 28–30.

Parsons, J. (1868) *Report on the Effects of the Earthquake Wave, which Occurred in the West Indies, in November 1867,* by J. Parsons, Staff Commander, R.N., of the West India Survey (QE–536), 268.

Peraldo, G. and Montero, W. (1994) *Los Temblores del Período Colonial de Costa Rica, 1538–1821,* Editorial Tecnológica de Costa Rica, Cartago (in Spanish).

Pereira, E.J. (1888) The Great Earthquake of Lisbon, *Transactions of the Seismic Society of Japan* **12**, 5–19.

Pereira de Sousa, Sir F.L. (1911) La raz-de-marée du grand tremblement de terre de 1755 en Portugal, *Comptes-Rendus de l'Academie des Sciences de Paris* **152**, 1129–1131 (in French).

Pereira de Sousa, Sir F.L. (1913) Sur les effets au Maroc du grand tremblement de terre en Portugal (1755), *Comptes-Rendus de l'Academie des Sciences de Paris* **157**, 805–808 (in French)

Pereira de Sousa, Sir F.L. (1914) Sur les effects en Portugal du mégaséisme du 1er novembre 1755, *Comptes-Rendus de l'Academie des Sciences de Paris* **158**, 2033–2035.

Pereira de Sousa, Sir F.L. (1919) O Terremoto do 1° de Novembro de 1755 em Portugal e um Estudo Demográfico, 2 Volumes, 473 pp. (in Portuguese).

Pérez, O. (1971) Notes on the tropical cyclones of Puerto Rico, 1508–1970.

Perrey, A. (1843) Note historique sur les tremblements de terre aux Antilles, *Comptes-Rendus (Hebdomadairies) de l'Academie des Sciences de Paris* **14**, 1283–1312 (in French).

Perrey, A. (1847a) *Annales Société d'Emulation des Vosges* **6**, 402, 536, 557 (French).

Perrey, A. (1847b) Nouveau catalogue des tremblements de terre aux Antilles 1530–1846, et suppléments à la liste des secousses ressentis en 1844, *Mémoires de l'Academie des Sciences, Artes, et Belle-Lettres de Dijon, Années 1845–1846,* 325–479 (in French).

Perrey, A. (1853) Tremblements de terre ressentis en 1852, *Bulletin de l'Academie Royale Mémoires Couronées de Bruxelles* **20**, 1e Partie, 39–69 (in French).

Perrey, A. (1854b) Note sur les tremblements de terre en 1853, *Bulletin de l'Academie Royale Mémoires Couronées de Bruxelles* **21**, 1e Partie, 457–489 (in French).

Perrey, A. (1855c) Note sur les tremblements de terre en 1854, avec les suppléments pour les années antérieures, *Bulletin de l'Academie Royale Mémoires Couronées de Bruxelles* **22**(6), 1e Série, 526–572 (in French).

Perrey, A. (1856b) Note sur les tremblements de terre en 1855, avec les suppléments pour les années antérieures, *Bulletin de l'Academie Royale Mémoires Couronées de Bruxelles* **23**(7), 1e Partie, 2e Série, 23–68 (in French).

Perrey, A. (1858) Documents sur les tremblements de terre au Pérou, dans la Colombie et dans le bassin de l'Amazone, les suppléments années antérieures, *Mémoires Couronées et autres Mémoires publiées de l'Academie Royale des Sciences, Lettres et Beaux Arts de Belgique* **7**, 8°, Bruxelles, 1–134 (in French).

Perrey, A. (1859) Note sur les tremblements de terre en 1856, avec les suppléments années antérieures, *Mémoires Couronées et autres Mémoires publiées de l'Academie Royale des Sciences, Lettres et Beaux Arts de Belgique* **8**, Bruxelles, 1–79 (in French).

Perrey, A. (1862c) Note sur les tremblements de terre en 1860, avec les suppléments pour les années antérieures, *Mémoires Couronées et autres Mémoires publiées par l'Academie Royale de Sciences, de Lettres et de Beaux Arts de Belgique* **14**, Bruxelles, 1–74 (in French).

Perrey, A. (1864a) Note sur les tremblements de terre en 1861, avec les suppléments pour les années antérieures, *Mémoires Couronées et autres Mémoires publiées par l'Academie Royale de Sciences, de Lettres et de Beaux Arts de Belgique* **16**, 1e Partie, Bruxelles, 1–112 (in French).

Perrey, A. (1864b) Note sur les tremblements de terre en 1862, avec les suppléments pour les années antérieures, *Mémoires Couronées et autres Mémoires publiées par l'Academie Royale de Sciences, de Lettres et de Beaux Artes de Belgique* **16**, 1e Partie, Bruxelles, 1–179 (in French).

Perrey, A. (1867) Note sur les tremblements de terre en 1865, avec les suppléments années antérieures de 1843 à 1864, *Mémoires Couronées et autres Mémoires publiées par l'Academie Royale de Sciences, de Lettres et de Beaux Arts Belgique* **19**, Bruxelles, 1–125 (in French).

Perrey, A. (1870) Note sur les tremblements de terre en 1866 et 1867, avec les suppléments pour les années antérieures de 1843 à 1865, *Mémoires Couronées et autres Mémoires publiées par l'Academie Royale de Sciences, de Lettres et de Beaux Arts de Belgique* **21**, Bruxelles, 1–223 (in French).

Perrey, A. (1872) Note sur les tremblements de terre en 1869, avec les suppléments pour les années antérieures de 1843 à 1868, *Mémoires Couronées et autres Mémoires publiées par l'Academie Royale de Sciences, de Lettres et de Beaux Arts de Belgique* **22**, Bruxelles, 1–116 (in French).

Perrey, A. (1873) Suppléments aux notes sur les tremblements de ressentis de 1843 à 1868, *Mémoires Couronées et autres Mémoires publiées par l'Academie Royale de Sciences, de Lettres et de Beaux Arts de Belgique* **23**, Bruxelles, 1–70 (in French).

Perrey, A. (1875a) Note sur les tremblements de terre en 1870, avec les suppléments pour 1869, *Mémoires Couronées et autres Mémoires publiées par l'Academie Royale de Sciences, de Lettres et de Beaux Arts de Belgique* **24**, 1e Partie, Bruxelles, 1–146 (in French).

Perrey, A. (1875b) Note sur les tremblements de terre en 1871, avec les suppléments pour les années antérieures de 1843 à 1868, *Mémoires Couronées et autres Mémoires publiées par l'Academie Royale de Sciences, de Lettres et de Beaux Arts de Belgique* **24**, 2e Partie, Bruxelles, 1–143 (in French).

Pezuela, J. de la (1866) *Diccionario Geográfico, Estadístico e Histórico de la Isla de Cuba,* La Habana (in Spanish).

Pichardo, E. (1854) *Geografía de la Isla de Cuba,* Parte Segunda, Habana (in Spanish).

Piddington, H. (1852) *The Sailor's Horn-Book for the Law of Storms,* Smith, Elder and Company, London, 360 pp.

Pinkerton, J. (1812) *General Collection of Voyages and Travels,* London.

Plafker, G. (1976) Tectonic aspects of the Guatemala earthquake of 4 February 1976, *Science* **193**, 1201–1208.

Plafker, G. and Ward, S. (1992) Backarc thrust faulting and tectonic uplift along the Caribbean Sea Coast during April 22, 1991 Costa Rica earthquake, *Tectonics* **11**, 709–718.

Poëy, A.M. (1855a) Catalogue chronologique avec supplément au tableau chronologique des tremblements de terre ressentis à l'Île Cuba de 1530 à 1855, *Nouvelles Annales des Voyages* **2**, Rédigées par M. Malte-Brun, 6e Série, Paris, 301 (in French).

Poëy, A.M. (1855b) Supplément au tableau chronologique des tremblements de terre ressentis à l'île de Cuba de 1551–1855, accompagnée d'une note additionnelle, *Nouvelle Annales des Voyages* **4**, 6e Série, Paris, 286–292 (in French).

Poëy, A.M. (1857) Une revue bibliographique avec tous les travaux relatifs aux tremblements de terre des Antilles, avec Resume du Catalogue chronologique des tremblements de terre ressentis dans les Indes-Occientales, de 1530 à 1857, *Annuaire de la Société Météorologique de France, Bulletin des Séances* **5**, 1e Partie, Paris, 75–127, 227–252, 286–292 (in French).

Porter, G.H.A. (1867) Porter's Diary, Administrator of the British Virgin Islands; also cited by Sir A. Rumbold, President of the British Virgin Islands, in Dispatch No. 96 of 23rd December.

Port-of-Spain Gazette (1825) Severe earthquake damage on September 20 at Port-of-Spain, Trinidad, September.

Port-of-Spain Gazette (1831) Report of the sea being in a state of violent agitation on December 3 at Trinidad, December 7.

Port-of-Spain Gazette (1867) Tsunami on November 18, 1.2–1.5 meters high invaded the harbor at Grenada, damaging boats and buildings, December 11.

Port-of-Spain Gazette (1874) Shock on August 17 at Trinidad, August 22.

Port-of-Spain Gazette (1900) Very severe shocks on October 29 at Trinidad, October 30.

Port-of-Spain Gazette (1918) Damage and deaths on October 11 at Puerto Rico, October 13.

Preliminary Determination of Epicenters (1969) PDE Monthly Listing for December 25, U.S. Department of Interior, Geological Survey, National Earthquake Information Service, Golden.

Preliminary Determination of Epicenters (1985) PDE Monthly Listing for March 16, U.S. Department of Interior, Geological Survey, National Earthquake Information Service, Golden.

Preliminary Determination of Epicenters (1989) PDE Monthly Listing for November 11, U.S. Department of Interior, Geological Survey, National Earthquake Information Service, Golden.

Preliminary Determination of Epicenters (1991) PDE Monthly Listing for April 22, U.S. Department of Interior, Geological Survey, National Earthquake Information Service, Golden.

Ramírez, Br. J.E. (1957) *Mapa Sísmico y Tectónico de Colombia, Catálogo y Escala de Intensidades,* Bogotá (in Spanish).

Ramírez, Br. J.E. (1969) *Historia de los Terremotos en Colombia,* publicado por el Instituto Geográfico Agustín Codazzi, Bogotá (in Spanish).

Rappaport, E.N. and Fernández-Partagas, J. (1995) *The deadliest Atlantic tropical cyclones, 1492–present,* NOAA Technical Memorandum NWS–NHC–47, National Hurricane Center.

Raupach, O.F. (1868) Letter dated 28th December 1867 to U.S. Secretary of State in Washington, *American Journal of Science* **45**, 134–135; also in *Comptes-Rendus de l'Academie des Sciences de Paris* **66**, 280–281 (in French).

Reboul, A. (1843) Extrait du rapport de M.A. Reboul, envoyé par la ville de Port-Royal (Martinique) auprès de la Pointe-à-Pitre, au sujet du tremblement de terre de 8 février 1843, *Le Politique de Bruxelles* (27 mars) (in French).

Reid, H.F. (1914) The Lisbon earthquake of November 1, 1775, *Bulletin of the Seismological Society of America* **4**(2), 53–80.

Reid, H.F. (1917) Note on the earthquakes at Almirante, Panamá, in April 1916, *Bulletin of the Seismological Society of America* **7**(1), 27–30.

Reid, H.F. and Taber, S. (1919a) *The Porto Rico Earthquake of 1918 with Descriptions of Earlier Earthquakes,* Report of the Earthquake Investigation Commission, United States House of Representatives Document 269, 66th Congress, 1st Session, Washington.

Reid, H.F. and Taber, S. (1919b) The Porto Rico earthquakes of October–November 1918, *Bulletin of the Seismological Society of America* **9**(4), 94–127.

Reid, H.F. and Taber, S. (1920) The Virgin Islands earthquakes of 1867–1868, *Bulletin of the Seismological Society of America* 10(1), 9–30.

Reid, W.: (1841) *Law of Storms,* John Weale, London, 566 pp.

Richter, C.F. (1958) *Elementary Seismology,* W.H. Freeman, San Francisco, 768 pp.

Riise-Baumann, J.A. and Baumann, V. (1910) A.H. Riise family memoirs regarding the extraordinary events of November, 1867 at St. Thomas.

Roberts, W.O. (1829) *Narrative of Voyages and Excursions on the Coast and the Interior of Central America* (Collection Montemont y Maussion de Conde) (in French).

Robson, G.R. (1964) An earthquake catalogue for the eastern Caribbean, 1530–1960, *Bulletin of the Seismological Society of America* **54**(2), 785–832.

Robson, G.R. (1966) The pyroclastic deposits of the 1902 and 1903 eruptions of the Soufrière on St. Vincent, *Transactions of the 4th Caribbean Geological Conference*, Trinidad (1965).

Robson G.R. and Tomblin, J.F. (1966) *Catalogue of the Active Volcanoes of the World Including Solfatara Fields,* Part XX, West Indies, International Association of Volcanology, Istituto di Geologia Applicata, Rome, 56 pp.

Rockstroch, E. (1902) Earthquake in Guatemala, *Nature* **66**(98), London, 150.

Rockwood, C.G., Jr. (1883) Letter about the earthquake of 7 September 1882, *American Journal of Science* **26**, 155.

Rockwood, C.G., Jr. (1886) Notices of recent American Earthquakes, *American Journal of Science* **32**(15), 7.

Rodríguez, J.A. (1998) Information from FUNVIS personnel about a tsunami report for 1997 at Cariaco 1997, personal communication, May 1.

Rojas, W., Bungun, H., and Lindholm, C. (1993) *A Catalog of Historical and Recent Earthquakes in Central America,* NORSTAR Technical Report.

Romer, T. (1932) Tsunami and storm waves at Martinique from the 17th century to the present, *Annales de la Commission de l'Étude de Raz-de-Marée* (2), Union Géodésique et Géophysique International, Paris, 28 (in French).

Rothman, R.L. (1968) A note on the New England earthquake of November 18, 1755, *Bulletin of the Seismological Society of America* **50**(5), 1501–1502.

Rubio, M. (1982) Ocurrencia de tsunamis en el Caribe, *Investigaciones Sismológicas en Cuba* (2), Academia de Ciencias de Cuba, Instituto de Geofísica y Astronomía, 170–180 (in Spanish).

Rudolph, E. (1887) Über submarine Erdbeben und Eruptionen, *Beiträge zur Geophysik* **1**, 131–365 (in German).

Rudolph, E. (1895) Über submarine Erdbeben und Eruptionen, *Beiträge zur Geophysik* **2**, 537–666 (in German).

Rudolph, E. and Szirtes, S. (1906) Das Kolumbianische Erdbeben am 31 Januar 1906, *Gerlands Beiträge zur Geophysik* **11**, Teilen i–iii (in German).

Rumbold, Sir A. (1868) Accounts of the earthquake which visited the Virgin and adjacent islands on November 18, 1867, from Tortola, by President Rumbold of the British Virgin Islands, *Proceedings of the Meteorological Society* **4**; also in Rumbold's Dispatch No. 83 of 25th November & No. 96 of 23rd December.

Saint Croix Avis (1867) Tribulations from the November 18 tsunami and earthquake described, November 29; December 3, 10, 13, 17 & 20.

Sainte-Claire Deville, C. (1843b) *Observations sur le tremblement de terre éprouvé à la Guadeloupe, le 8 février 1843,* 4°, Basse-Terre, 52 pp. (in French).

Sainte-Claire Deville, C. (1848b) Note sur un mémoire de M. Duchassaing, relatif à des observations sur le tremblement de terre de la Guadeloupe, du 8 février 1843, *Comptes-Rendus de l'Academie des Sciences de Paris* **27**, 294–295 (in French).

Sainte-Claire Deville, C. (1867) Sur le tremblement de terre du 18 novembre 1867 aux Antilles, *Extrait des Comptes-Rendus des Séances de l'Academie des Sciences* **65**, Séance du 30 décembre, Institut Imperial de France, Paris, 1039, 1110–1114 (in French).

Saint-Guily, B. (1960) "Danger! 500 mph tidal wave coming," *The UNESCO Courier* (7–8), Paris.

Saint Vincent Observer (1843) Sugar mills and all stone buildings severely damaged or destroyed on February 8, at Basseterre, St. Kitts, and at Nevis and Montserrat, February 14.

Salas, C.J. (1971) *Tierra Firma (Venezuela y Colombia), Estudios sobre Ethnología e Historia,* Segunda Edición, Universidad de Los Andes, Mérida (in Spanish).

Salcedo, R.J. (1640) *Historical and Geographical Report of the Province of Panamá* (in Spanish); English translation by A.E. Wetman, Panamá Canal Collection (1947).

Salivia, L.A. (1972) *Historia de los Temporales de Puerto Rico y las Antillas,* Editorial Edil, San Juan, 385 pp. (in Spanish).

Salmons (no date) *Modern History of Jamaica,* 3°, 579.

Salterrain y Viñes, P. (1883) Ligera reseña de los temblores de tierra ocurridos en la Isla de Cuba, *Boletín Comisión del Mapa Geológico de España* **10**, Madrid, 371–385 (in Spanish).

Sanct Thomæ Tidende (1867a) Calamities from the November 18, 1867 earthquake and tsunami, November 23 & 30; December 4, 7, 14 & 28.

Sanct Thomæ Tidende (1868) Waves and a strong shock on March 17 at St. Thomas, considered the strongest of many aftershocks from the November 1867 earthquake, March 18.

Sarasola, S. (1928) *Los Huracanes de las Antillas, Gutierrez-Lanza Appendix: Ciclones que Han Pasado por la Isla de Cuba, o Tan Cerca que Hayan Hecho Sentir en ella sus Efectos con Alguna Fuerza, desde 1865 a 1926,* Clásica Española, Madrid, Spain, 254 pp. (in Spanish).

Scherer, Rev. J. (1911) Earthquakes in Hayti, *Bulletin of the Seismological Society of America* **1**(4), 171.

Scherer, Rev. J. (1912) Great earthquakes in the island of Haiti, *Bulletin of the Seismological Society of America* **2**(3), 161–180.

Scherer, Rev. J. (1913) Catalogue chronologique des tremblements de terre ressentis dans l'Île d'Haïti de 1551 à 1900, *Bulletin Semestriel de l'Observatoire Météorologique du Séminaire-Collège St.-Martial,* Port-au-Prince, Haiti (July–December), 136–137, 147–151 (in French).

Schmidt, J.F.J. (1881) *Studien über Vulcane and Erdbeben,* Part ii, Alwin Georgi, Leipzig, 612 pp. (in German).

Schubert, C. (1994) Tsunamis in Venezuela: Some observations on their occurrence, *Journal of Coastal Research,* Special Issue No. 12, Coastal Hazards, Chapter 13, 189–195.

Seismological Notes (1918) *Bulletin of the Seismological Society of America* **8**(4), 137–138.

Seismological Notes (1929) *Bulletin of the Seismological Society of America* **19**(1), 55.

Seismological Notes (1939) *Bulletin of the Seismological Society of America* **29**(4).

Seismological Notes (1955) *Bulletin of the Seismological Society of America* **45**(2), 160.

Seon, K. (no date) Preliminary disaster catalog—Jamaica (unpublished report), 31 pp.

Shepherd, J.B. and Lynch, L.L. (1992) An earthquake catalogue for the Caribbean, Part I, The pre-instrumental period, 1502–1900, Paper Presented to the Steering Committee, Latin American and Caribbean Seismic Hazard Programme (April 1992), Florida, 60 pp.

Sieberg, A. (1932) Erdbeben geographie: Mittelamerika samt Westindien, in B. Gutenberg (ed.), *Handbuch der Geophysik* **4**, Berlin, 948–960 (in German).

Sigurdsson, H.R. (1996) Volcanic Tsunamis, *Summary Report of the UNESCO IOC IOCARIBE Tsunami Warning Workshop*, UNESCO International Conference on Tsunamis, St. John, U.S. Virgin Islands (May 1996), Paris.

Sigurdsson, H.R. (1997) Concern about activity at the Soufrière Hills on Montserrat and Kick-'em-Jenny submarine volcano mount, personal communication, September.

Sigurdsson, H.R., Sparks, S.J., Carey, S.N., and Huang, T.C. (1980) Volcanogenic sedimentation in the Lesser Antilles arc, *Journal of Geology* **88**, 523–540.

Silgado, E.F. (1985) Terremotos destructivos en América del Sur, 1530–1894, centro región de sismología para América del Sur, *Programa para Mitigación de Efectos de los Terremotos en la Región Andina* **10**, CERESIS, Lima, Peru (in Spanish).

Simkin, T., Siebert, L., McClelland, L., Bridges, D., Newhall, C., and Latter, J.H. (1981) *Volcanoes of the World,* Smithsonian Institution, Hutchinson Ross, Stroudsburg, Pennsylvania.

Singer, A., Rojas, C., and Lugo, M. (1983) *Inventario Nacional de Riesgos Geológicos, Estado Preliminar, Mapa, Glosario y Comentarios,* Fundación Venezolana Investigaciones Sismológicas (FUNVISIS), Serie Técnica 03–83, Caracas, 128 pp. (in Spanish).

Sloane, Sir H. (1694) A letter from Hans Sloane M.D. and S.R.S. with several accounts of the Earthquakes in Peru, October the 20th 1687; And at Jamaica, February the 19th 1688; and June the 7th 1692, *Philosophical Transactions* **18**(209), Letter I, Royal Philosophical Society of London, 77–100; also in *Philosophical Transactions 1665–1800* **3**, Abridged, London (1809), 624–632; also in *Gibilin* **1**, 43–45; also in *Ludrini,* 46–58.

Small, W.A. (1948) A short description of the general geology of the Dominican Republic, with notes on the earthquake of August 4, 1946, *Bulletin of the Seismological Society of America* **38**(1), 19–32.

Smallwood, J.R. (1980) Earthquakes, *Encyclopedia of Newfoundland and Labrador* **1**, 663.

Smith, Rev. W. (1745) *A Natural History of Nevis and the Rest of the English Leeward Charibee Islands in America,* Cambridge (Seismic Research Unit, University of the West Indies, St. Augustine, Trinidad).

Snow, E.R. (1952) *Great Gales and Dire Disasters,* Dodd, Mead and Company, New York.

Solov'ev, S.L. (1990) Tsunamigenic zones of the Mediterranean Sea, *Natural Hazards* **3**, 183–202.

Solov'ev, S.L. and Go, C.N. (1984a) *A Catalogue of Tsunamis on the Western Shore of the Pacific Ocean (1513–1968),* English translation by Canadian Institute of Scientific and Technical Information, National Research Council (5077), Ottawa, Ontario, 439 pp.

Soloviev, S.L., Solovieva, O.N., Go, C.N., Kim, K.S., and Shchetnikov, N.A. (2000) *Tsunamis in the Mediterranean Sea 2000 B.C.–2000 A.D.,* Kluwer Academic Publishers, Dordrecht, 110, 191, 205.

Soulan, I. (1994) *Cyclones tropicaux les plus meurtriers de l'Atlantique,* 2 pp. (in French)

Southey, Capt. T.V. (1827) *Chronological History of the West Indies* **2**, 3 Volumes, 8 °, printed for Longman, Rees, Orme, Brown and Green, London, 407, 422, 472–473 (British Museum Library).

Surinaamsche Courant (1939) Earthquake on January 11 at Martinique; also felt in the Lesser Antilles, Surinam, and Guyana where bells set ringing in Georgetown, January (in Dutch).

Surinaamsche Courant (1943) Violent earthquake at Paramaribo, Surinam, February 10 (in Dutch).

Sutch Osieki, P.L. (1981) Estimated intensities and probable tectonic sources for the historic (pre-1898) earthquakes, *Bulletin of the Seismological Society of America* **71**, 865–881.

Tabor, L.L. (1967) *Compilation on seismic data: Panamá and adjoining sea area, NVO–99–22,* John A. Blume Earthquake Engineering Service Research Division, San Francisco, 40 pp.

Taber, S. (1920) Jamaica earthquakes and the Bartlett Trough, *Bulletin of the Seismological Society of America* **10**(2), 55–89.

Taber, S. (1922a) The great fault troughs of the Antilles, *Journal of Geology* **30**, 89–114.
Taber, S. (1922b) The seismic belt in the Greater Antilles, *Bulletin of the Seismological Society of America* **12**(4), 199–219.
Tannehill, I.R. (1950) *Hurricanes,* Princeton University Press, Princeton.
Taylor, C.E. (1888) *Leaflet from the Danish West Indies,* William Dawson and Sons, London; Reprinted by Negro University Press, Greenwood Press, New York (1970).
Tazieff, H. (1964) *When the Earth Trembles,* Harcourt, Brace & World, New York, 96.
Tegg, W. (1854) Dictionary of Chronology: Or Historical and Statistical Register from the Birth of Christ to the Present Time, 5th Edition enlarged, London, 766 pp.
Texas A&M Nautical Archives Website (2000) Background of Port Royal and the history of the famous earthquake and tsunami of 1692, Texas A&M University, on-line at website: *http://nautarch.tamu.edu/projects/PR-project/PRhist.htm.*
The New York Times (1967) "Waves at San Juan," Puerto Rico, December 5, 64:7.
The Sun (1999) Article warns of the potential for eruption at Kick-'em-Jenny Submarine Volcano Mount and its consequences for Barbados, February 13.
Thibault de Chanvalon, J.-B. (1763) Sur les Tremblements de Terre à la Martinique, *Voyage à la Martinique, Contenant Diverses Observations sur la Physique, l'Histoire Naturelle, l'Agriculture, les Mœrs et les Usages de cette Île, Faites en 1751 et dans les Années Suivantes* **8**, Paris, 192 pp. (in French)
Thomson, C.W. and Murray, J. (1885) *Report on the Scientific Results of the Voyage of the H.M.S. Challenger during the Years 1873–1876, Narrative* **1**, Longmans & Company, London, 126.
Tippenhauer, Sir L. (1893) Liste der Erdbeben auf Haïti, in *Die Insel Haïti,* Leipzig, 170–175 (in German).
Tomblin, J.M. and Robson, G.R. (1977) *A Catalogue of Felt Earthquakes for Jamaica with Reference to Other Islands in the Greater Antilles, 1564–1971,* Special Publication No. 2, Ministry of Mining and Natural Resources, Mines and Geology Division, Kingston, 243 pp.
van Housel, L. (1878) An earthquake experience, *Scribner's Monthly* **XV**(5), New York, 662–672.
Verbeek, R.D.M. (1886) *Krakatau,* Publie par Ordre de Son Exellence *[sic]* le Gouverneur-Général des Indies Néerlandaises, Imprimerie de l'État, Batavia, 396–461 (in French).
Víquez, V. and Toral, J. (1987) Sismicidad histórica sentida en el Istmo de Panamá, *Geofísica* **27**, 135–165 (in Spanish).
Vivenzio (1788) *Istoria de Terremoti,* 22 (in Italian).
Von Hake, C.A. and Cloud, W.K. (1971) Tsunamis, *United States Earthquakes, 1969,* National Oceanic and Atmospheric Administration, National Ocean Survey, Rockville, Maryland, 53.
von Hoff, K.E.A. (1840) Chronik der Erdbeben und Vulkan-Ausbrüch von 3460 A.C. bis 1832 P.C., mit vorausgehender Abhandlung über die Natur dieser Erscheinungen, in *Geschicte der durch Überlieferung nach Gewiesenen Natürlichen Veränderungen der Erdoberfläche* **4**, Gotha (in German).
Warner, Br. S. (1876) *Periodical Accounts Relating to the [Moravian] Missions of the Church of the United Brethren, Established Among the Heathen (St. Croix, November 26th, 1867)* **26**, Moravian Brethren's Society for the Furtherance of the Gospel Among the Heathen, London.
Watlington, R.A. and Lincoln, S.H. (1997) *Disaster and Disruption in 1867: Hurricane, Earthquake and Tsunami in the Danish West Indies: A Collection of Accounts and Reports,* Eastern Caribbean Center, University of the Virgin Islands, 134 pp.
Watts, P., Lander, J.F., and Sokolowski, T. (1995) Update: Skagway, Alaska, landslide and tsunami of November 3, 1994, *Tsunami Newsletter* **27**(2), 7–8.
Weissert, T.P. (1990) Tsunami travel time charts for the Caribbean, *Science of Tsunami Hazards* **8**(2), 67–78.
West Indies Pilot (1942) Submarine eruption five miles north of Grenada **2**, 275.
Wolfall (1809) Account of the earthquake at Lisbon, November 1, 1755, *Philosophical Transactions* **10**, Royal Philosophical Society of London, Abridged, 656–669.
Wood, H.O. (1917) *On Cyclical Variations in Eruptions at Kilauea,* Second Report of the Massachusetts Institute of Technology and the Hawai'ian Volcano Research Association, Cambridge, Massachusetts.

Woodring, W.P., Brown, J.S., and Burbank, W.S. (1924) *Geology of the Republic of Haiti*, Department of Public Works, Port-au-Prince.
Woodworth, H.G. (2001a), Articles cited from the *Elizabeth* [New Jersey] *Daily Journal* (May 7–17) about tsunamis resulting from the 1902 eruptions on Mont Pelée, Martinique, and on The Soufrière, St. Vincent, personal communication, April 26.
Woodworth, H.G. (2001b) Further reports on 1902 volcanic tsunamis at Mont Pelée, Martinique, and The Soufrière, St. Vincent, from articles in the *Elizabeth Daily Journal* (May 19–22, 1902), personal communication, April 29.
Woodworth, H.G. (2001c) Hurricane-produced undersea landslide generated by Camille (1969) probably produced a tsunami, personal communication, March 19.
Young, C., Lay, T., and Lynnes, C. (1989) Rupture of the 4 February 1976 Guatemalan earthquake, *Bulletin of the Seismological Society of America* **79**, 670–689.

Related Source Materials and Other Relevant Works

Abbad y Lasierra, I. (1866) *Histórica Geográfica, Civil y Natural de la Isla de San Juan Bautista de Puerto Rico,* Nueva Edición, Anotada por José Julián Acosta y Calbo, Puerto Rico (in Spanish).
Abe, K. (1979) Size of great earthquakes 1837–1974 inferred from tsunami data, *Journal of Geophysical Research* **84**, 1561–1568.
Abe, K. (1989) Quantification of tsunamigenic earthquakes by the M_t scale, *Tectonophysics* **166**.
Abe, T. (1973) Tsunami mechanisms of great earthquakes, *Physics of the Earth and Planets International* **7**(2), 143–153.
Adams, W.M. and Nakashizuka, N. (1985) A working vocabulary for tsunami study, *Science of Tsunami Hazards* **3**, 45–51.
A full Account of the late dreadful Earthquake at Port-Royal in Jamaica, Written in two Letters from the Minister of that Place (1692) From aboard the *Grenada* in Port-Royal Harbour, June 22, 1692, R. Baldwin, 1692 Folio, London, Shelf mark 719–m–17(16) (British Museum Manuscript).
Agamemnone, G. (1899) *Bollettino della Sociéta Sismologica Italiana, Anno 1899* (in Italian).
A genuine Account of the late Earthquakes, especially at Oxford, in the Year 1683, and of an other Terrible One at Port-Royal in Jamaica, in the Year 1692 (1750) **8**, 8°, London, 24 pp.
Aki, K. and Richards, P.G. (1980) *Quantitative Seismology* **1**, W.H. Freeman, San Francisco.
Alarcon, J.C. (1898) *Compendio Historia Departamento de Magdalena desde 1525 hasta 1895,* Bogotá, 462 pp. (in Spanish).
Alcedo, A. de (1815) *The Geographical and Historical Dictionary of America and the West Indies*; English translation with additions and compilations by G.A. Thompson, 5 Volumes, London.
Alexander, W.H. (1902) Hurricanes: Especially those of Porto Rico and St. Kitts, *Weather Bureau Bulletin 32,* U.S. Department of Agriculture, 79.
Alvarenga, E., Barquero, R., Boschini, I., Escobar, J., Fernández, M., Mayol, P., Havskov, J., Galvez, N., Hernandez, Z, Ottomoller, L., Pacheco, J., Redondo, C., Rojas, W., Vega, F., Talavera, E., Taylor, W., Tapia, A., Tenorio, C., and Toral, J. (1998) Central American Seismic Center (CASC), *Seismological Research Letters* **59**(5), 394–399.
Alvarez, L., Vaccari, F., and Panza, G. (1999) Deterministic seismic zoning of eastern Cuba, *Pure and Applied Geophysics* **156**(3), 469–486.
Anderson, J. (1785) An account of Morne-Garou, a mountain in the island of Saint Vincent in the Caribees, with description of the volcano on its summit, *Philosophical Transactions* **75**, Royal Philosophical Society of London, 16–31.
Anderson, T. (1908) Report on the eruptions of the Soufrière in St. Vincent, in 1902, and on a visit to Montagne Pelée in Martinique, Part II, The changes in the districts and the subsequent history of the volcanoes, *Philosophical Transactions* **208**, Series A, Royal Philosophical Society of London, 275–303.

Arago, Fr. (1857) On the Phænomena of the Sea, *Complete Works of Arago* **9**, J.A. Barral, Paris (in French).

Arboleda, G. (1918) *Historia Contemporánea de Colombia,* 2 Volúmenes, Bogotá (in Spanish).

Ardouin, B. (1853) Études sur l'Histoire Haïti, *Histoire d'Haïti* **9**, 8°, Paris (in French).

Ardouin, B. (1860) Sur le tremblement de terre senti à Saint-Domingue le 7 mai 1842, Études sur l'Histoire Haïti, *Histoire d'Haïti* **2**, Paris, 221–225 (in French).

A Sad and Terrible Relation of the dreadful Earth-quake that happened at Jamaco in the West-Indies, on the 7th of June 1692, &c. (1692) 12°, 16.

Asenjo y Arteaga, F. (1876) *Efemérides de la Isla de Puerto Rico (1867–1876)* (in Spanish).

A True and Perfect Relation of the Earthquake at Port-Royal in Jamaica, 7th June 1692 (1692) Folio with lithograph, London, Shelf mark 719–m–17(15) (British Museum Manuscript).

Atwood, T. (1791) History of the Island of Dominica, London.

Baker, B.B., Jr., Deebel, W.R. and Geisenderfer (1966) Glossary of Oceanic Terms, 2nd Edition, SP–35, U.S. Naval Oceanographic Office, Washington, 204 pp.

Baker, P.E. (1963) The geological history of Mt. Misery volcano, St. Kitts, *Bulletin of the Overseas Geology and Mineralogy Research.*

Ballet, J. (1899) *La Guadeloupe: Renseignements sur l'histoire, la flore, la fauna, la géologie, la minéralogie, etc.,* 5 Tommes, Imprimerie du gouvernement, Basse-Terre (in French).

Banbuck, C.A. (1935) *Histoire de la Martinique,* Paris (in French).

Barham, H. (1718) On the weather, earthquakes, &c. of Jamaica, *Philosophical Transactions* **6**, Royal Philosophical Society of London, Abridged.

Barrio, D.M. (1946) Consideraciones sobre la historia sísmica de la República Dominicana, *Publicaciones de Universidad de Santo Domingo* **44**, 38 (in Spanish).

Bayley, F.W.N. (1833) *Four Years Residence in the West Indies,* 3rd Edition, London, 742 pp.

Bedford, W. (1790) *Descriptive Account of the Island of Jamaica,* London.

Bernard, E.N. (1991) Assessment of Project THRUST (Tsunami Hazards Reduction Utilizing Systems Technology): Past, present, future, *Natural Hazards* **4**, 285–292.

Bernard, E.N. (1993) Tsunamis, Stop Disasters (12) (March–April), 1920.

Bernard, E.N., Behn, R.R., Hebenstreit, G.T., González, F.I., Krumpe, P., Lander, J.F., Lorca, E., McManamon, P.M., and Milburn, H.B. (1988) On mitigating rapid onset natural disasters: Project THRUST, *EOS Transactions* **69**(24), American Geophysical Union, 649–661.

Bernard, J. and Lambert, J. (1988) Subduction and Seismic Hazard in the Northern Lesser Antilles: Revision of the historical seismicity, *Bulletin of the Seismological Society of America* **78**, 1965–1983.

Bernard, J. and Lambert, J. (1992) Reply to Shepherd, *Bulletin of the Seismological Society of America* (July).

Bernstein, J. (1954) Tsunamis, *Scientific American* **191**(2), New York, 60–64.

Bertrand, E. (1757) *Mémoires Historiques et Physiques sur les Tremblements de Terre,* The Hague (in French).

Bird, M.B. (1869) Notes Historiques sur l'Independence Haïtienne, *The Republic of Hayti and Its Struggles,* From historical notes, Issued under the auspices of the Haytian Government, 8°, London (in French).

Blackford, M.E. and Kanamori, H. (1995) Executive Summary, in *Tsunami Warning Workshop Report,* NOAA Technical Memorandum ERL–PMEL–105, National Oceanic and Atmospheric Administration, Seattle.

Blome, R. (1672) *Description of the Island of Jamaica,* London.

Blond, P.H. le and Mysak, L.A. (1977) *Ocean Waves,* Elsevier, Amsterdam.

Booy, T. de and Faris, J.T. (1918) *The Virgin Islands, Our New Possessions, and the British Islands,* J.B. Lippincott, Philadelphia and London.

Bose, E. (1910) *Earthquakes,* Zickfeldt (ed.), 130–134 (in German).

Bouasse, H. (1924) *Swells, Ripples, Seiches, and Tides,* Delagrave, Paris (in French).

Boussingault, J.-B. and Roulin, F. (1849) *Viajes científicos a los Andes Ecuatoriales; Colección de Memorias sobre Física, Química e Historia Natural de la Nueva Granada, Ecuador y Venezuela* (Paris), Traducción de Joseph de Acosta de la Edición de Paris, 322 pp. (in Spanish).

Boyer-de-Peyreleau, E.-E. (1815) *Les Antilles Françaises, Particulièrement la Guadeloupe,* Paris (in French).

Bracey, D.R. and Vogt, P.R. (1970) Plate tectonics in the Hispaniola area, *Bulletin of the Geological Society of America* **81**, 2855–2860.

Breen, H.H. (1844) *Saint Lucia: Historical, Statistical, and Descriptive,* London.

Bridges, Rev. G.W. (1828) An account of the Earthquake of June 7th 1692, at Jamaica, *Annals of Jamaica* **1**, London, 3 Volumes, 303–442.

Brown, C.W. (1907) The Jamaica Earthquake, *Scottish Geographical Magazine* **23**, Edinburgh, 535–543; also in *Popular Science Monthly* **70**, 385–403 (1907).

Brown, P. (1789) *The Civil and Natural Histori of Jamaica,* London.

Browne, R.-B. (1694) *The general History of Earthquakes, &c. from the Creation to This Time and Particularly Those Lately in Naples, Smyrna, Jamaica, and Sicily,* 12°, London, 176 pp.

Burdon, K.J. (1920) *A Handbook of St. Kitts and Nevis,* West India Committee, London, 248 pp.

Burney, J. (1949) *History of the Buccaneers of America,* London.

Burns, A.C. (1954) *History of the British West Indies,* Barnes & Noble, New York and London

Burrage, H.S. (1932) *Early English and French Voyages, 1534–1608,* New York.

Byrne, D.B. and McCann, W. (1985) Muertos Trough subduction—microplate tectonics in the northern Caribbean, *Nature* **317**, 420–421.

Caiger, S.L. (1951) *British Honduras, Past and Present,* London.

Callejas, J.M. (1911) *Historia de Santiago de Cuba, Compuesta y Redactada en Vista de los Manuscritos Originales e Inéditos de 1823 y Precedida de un Prólogo per Fernándo Ortiz,* La Habana (in Spanish).

Campbell, P. (1828) *Nouvelles Annales des Voyages,* 254–256 (in French).

Cardona-Bonet, W.A. (1989) *Shipwrecks in Puerto Rico's History: I, 1502–1650,* Model Offset Printing, 371 pp.

Carmichael, G. (1961) *The History of the West Indian Islands of Trinidad and Tobago 1498–1900,* Alvin Redman, London.

Carnegie, Hon. Capt. (1843) An account of the late Earthquake at the Islands of Antigua and Guadeloupe on the 8th of February 1843, *Report of the British Association for the Advancement of Science,* Part 2, 59.

Case, J.E. and Holcombe, T.L. (1980) Geologic-tectonic map of the Caribbean region, *Miscellaneous Investigations Map 1–110,* U.S. Geological Survey, Reston, Virginia.

Charlevoix, Fr. P.-F.-X. de (1730b) *Histoire l'isle Espagnole de Saint-Domingue; Écrite Particulièrement des Mémoires Manuscrits du Jean-Baptiste Labat le Pers, Jésuite, Missionaire à Saint-Domingue, et sur les Pièces Originales, qui se Conservent au Dépôt de la Marine* **2**, 6 Tommes, 12°, Paris, 244 pp. (British Museum Library) (in French).

Chaunu, H. and Chaunu, P. (1955) *Seville et l'Atlantique (1504–1560),* Multiple Vols., Paris (Librairie Armand Colin) (in French).

Chevignard de la Palle (1788) Description à peu près incomplète du tremblement de 7 juin 1692 à la Jamaïque, *Idée du Monde ou Idées Générales des Choses,* 3e Édition, 12°, Paris (in French).

Chuy, T. and Rodríguez, M. (1980) La actividad sísmica de Cuba basada en datos históricos, *Investigaciones Sismológicas en Cuba* (1), La Habana (in Spanish).

Cleve, P.T. (1871) On the geology of north-eastern West India Islands, *Kgl. Svenska Vetenskaps Akademie Handelingar* **9**(12) (1870), 1–48 (in Dutch).

Cline, I.M. (1926) *Tropical Cyclones,* The Macmillan Company, New York.

Coll y Toste, C. (1926) *Boletín Histórico de Puerto Rico,* San Juan (in Spanish).

Condit, D.D. and Ross, C.P. (1921) Geology of the provinces of Barahona and Azua, in T.W. Vaughan, W. Cooke, D.D. Condit, C.P. Ross, W.P. Woodring, and F.C. Calkins, *A Geological Reconnaissance of the Dominican Republic,* Washington, 206.

Cortés, J. (1810) Mémoires sur la géologie des Antilles, lettre à M. de Humboldt sur *Journal de Physique* **70**, 129–134 (in French).

Costello, J.A. (1985) Tsunamis: Hazard Definition and Effects on Facilities, Technical Report Draft of Subcommittee III, "Evaluation of Site Hazards," Interagency Committee on Seismic Safety in Construction, U.S. Geological Survey, Reston, Virginia.

Cox, D.C. (1968) *Performance of the Seismic Sea Wave Warning System, 1948–1967,* HIG–68–2, Hawai'i Institute of Geophysics, University of Hawai'i, Honolulu, 69 pp.

Cox, D.C. (1987) Tsunami Casualties and Mortality in Hawai'i, Joint Institute for Marine and Atmospheric Research, University of Hawai'i, Honolulu.

Cotte, le P. (1809) Tableau chronologique des principaux phénomènes météorologiques observés en différents pays depuis 33 ans de 1774 a 1806—et comparés avec les températures correspondantes de Paris, *Journal de Physique* **68**, Paris, 331–335 (in French).

Cotte, le P. (1810) Tableau chronologique des principaux phénomènes météorologiques observés en différents pays depuis 33 ans de 1774 a 1806—et comparés avec les températures correspondantes de Paris, *Journal de Physique* **70**, Paris, 199 (in French).

Crouse, N.M. (1940) *French Pioneers in the West Indies, 1624–1664,* New York.

Cullum, T.M. (1805) *Voyage to Trinidad,* Liverpool.

Curtis, G.D. (1990) A methodology for developing tsunami inundation and evacuation zones, *Proceedings of the Pacific Congress on Marine Technology* (July 1990) Tokyo.

Curtis, G.D. (1992) Tsunamis—seismic sea waves, in S.K. Majumdar, G.S. Forbes, E.W. Miller, and R.F. Schmalz (eds.) *Natural and Technological Disasters: Causes, Effects, and Preventative Measures,* The Pennsylvania Academy of Science, Easton.

Curtis, G.D. and Mader, C.L. (1987) Real-time monitoring and modeling of tsunami threat evaluation, *Science of Tsunami Hazards* **5**(1), 49–56.

Cuvier, G. (1809) *Voyages dans l'Amérique Méridionale depuis 1781 jusqu'à en 1801, avec atlas,* 4 Tommes, 8°, Paris (in French).

d'Anghiera, P.M. [Peter Martyr] (1555) *Decades,* Translation by R. Eden; also translated by Lock (1612).

d'Anghiera, P.M. (1912) *De Orbe Novo,* Translation by F.A. MacNutt, New York and London.

Davison, C. (1907) The Kingston earthquake, *Nature* **75**, 296.

Davison, C. (1921) *A Manual of Seismology,* Chapter 6, Cambridge Geological Series, Cambridge University Press, London.

Dean, R.G. and Dalrymple, R.A. (1984) *Water Wave Mechanics for Engineers and Scientists,* World Scientific Publishing Company, Singapore, 353 pp.

Dengler, L. and Moley, K. (1995) *Living on Shaky Ground,* California Office of Emergency Services, Federal Emergency Management Agency, and Humboldt State University, 23 pp.

DeWeese, W. (1953) Letter to Isidor Paiewonsky about the 21 December 1867 Issue of the Army and Navy Journal, The Museum, U.S. Naval Academy, Annapolis, March 6.

Dillon, W.P., Austin, J.A., Jr., Scanlon, K.M., Edgar, N.T., and Parson, L.M. (1992) Hispaniola: Morphology, structure and development of part of the northern Caribbean plate boundary, *Marine Petrol. Geology* **9**(1), 70–88.

Dookhan, I. (1974) *A History of the Virgin Islands,* for the College of the Virgin Islands, Cambridge University Press, Epping, U.K.

Dorn, W.G. van (1960) Tsunamis, *EOS Transactions* **41**(2), American Geophysical Union, Richmond, 265–266.

Dorvo-Soulastre (1809) *Voyage par terre de Santo-Domingo, au Cap-Français, &c,* 8°, Paris, 407 pp. (in French).

Douglas, B.C. (1991) Global sea level rise, *Journal of Geophysical Research* **96**(C4), 6981–6992.

Doxsee, W.W. (1948) The Grand Banks earthquake of November 18, 1929, *Publications of the Dominion Observatory* **7**(7), Ottawa, 323–335.

d'Ulloa, Don A. (1748) *Relación Histórica del Viaje Hecho por Orden de Su Majestad a la América Meridional,* 4 Volúmenes, Madrid (in Spanish).

d'Ulloa, Don J. and d'Ulloa, Don A. (1807) *A Voyage to South America* **2**, 5th Edition, London, Adams, 82.

Dunbar, P.K., Lockridge, P.A., and Whiteside, L.S. (1992) *Catalog of Significant Earthquakes (2150 B.C.–1991 A.D.),* NOAA Report SE–49, National Oceanic and Atmospheric Administration, National Geophysical Data Center, Boulder.

Du Terte, J.-B. (1667) *Histoire générale des Antilles habitée par les Français,* Paris (in French).

Earle, K.W. (1923) *Geological Survey of Grenada and the (Grenada) Grenadines,* St. George's.

Edwards, B. (1801a) *An Historical Survey of the Island of Santo Domingo, and also a Tour Through The Several Islands of Barbadoes, Saint Vincent, Antigua, Tobago, and Grenada* **19**, London, 443 pp.

Edwards, B. (1801b) *The History Civil and Commercial of the British Colonies in the West Indies*, 3 Volumes, 8°, Legger, London.

Ellms, C. (1860) *Shipwrecks and Disasters at Sea*, I.J. Rouse, New York, 428 pp.

El-Sabh, M.I. (1995) The role of public education and an awareness in tsunami hazard management, in Y. Tsuchiya and N. Shuto (eds.), *Tsunami: Progress in Prediction, Disaster Prevention and Warning*, Kluwer Academic Publishers, Dordrecht, 336 pp.

Esguerra, J. (1879) *Diccionario Geográfico de los Estados Unidos de Colombia*, Bogotá, 284 pp. (in Spanish).

Fels, G. (1903) Ein Anorthitauswürfling von der Insel S. Christopher, *Zentbl. Kristallographische und Mineralogische* **37**, 450–460 (in German).

Fernández, M. (1998) Proposal to establish a tsunami warning system in Central America, *Technical Report II 1–10*, Reduction of Natural Disasters in Central America, Earthquake Preparedness and Hazard Mitigation, Institute of Solid Earth Physics, University of Bergen, Norway.

Feuillard, M. (1990) *Séismicité des Caraïbes—Décennie 1980–1989* (in French).

Feuillard, M. and Dorel, J. (1963) La Soufrière dans le complexe volcanique de la Guadeloupe (unpublished typescript), 8 pp. (in French).

Fiedler, G. (1972) La liberación de energía sísmica en Venezuela, Volúmenes sísmicos y mapa de isosistas, *Boletín Geología*, Memorias del IV Congreso Geológico Venezolano, Ministeria de Minas e Hidrocarburos, Caracas (in Spanish).

Fiedler, G. (1980) *Mapa Sísmico, 1530–1980*, República de Venezuela, Dirección Nacional de Cartografía, Caracas (in Spanish).

Fillias, A. (1935) *Proofs of the Oscillations of the Earth, the True Cause of Tsunamis, Earthquakes, and Volcanoes*, Édition Jouve, Paris (in French).

Finch, R.H. (1924) On the Prediction of Tidal Waves, *Monthly Weather Review* **52**(3), U.S. Weather Bureau.

Flores, J. (1993) The Guam 8.1 earthquake—A hazard perception study of human experience and adjustment, Term Paper for Geography of Micronesia, University of Guam, Fall Semester.

Fowle, W. (1781) General Account of the calamities occasioned by the tremendous Hurricanes and Earthquakes in the West India Islands, &c., 8°, London.

Francis, P. and Self, S. (1983) The eruption of Krakatau, *Scientific American* **249**, 172–187.

Fryer, G. (1995) The most dangerous wave, *The Sciences* (July/August), 38–43

Fujiwara, S. (1935) On the meaning of words: Long wave, tsunami, and surge, *Science* **5**, Japan, 308 (in Japanese).

Fukao, Y. (1979) Tsunami earthquakes and subduction processes near deep-sea trenches, *Journal of Geophysical Research* **84**, 2303–2314.

Fukuoka, J., Véliz, C., and Velázquez, R. (1992) La propagación de las ondas del maremoto (tsunami) en el mar, *Lagena* **30**, 29–32 (in Spanish).

Funck-Brentano (1930) *L'Île de la Tortue, Haiti*, Paris.

Furumoto, A.S. (1991) Source Parameters of Destructive Tsunamis, *Science of Tsunami Hazards* **9**(2), 95–113.

Furumoto, A.S. (1996) Using M_W or M_t to forecast tsunami heights, *Science of Tsunami Hazards* **14**, 107–118.

Furumoto, A.S., Tatehata, H., and Morioka, C. (1999) Japanese tsunami warning system, *Science of Tsunami Hazards* **17**(2), 85–105.

Galbis Rodríguez, J. (no date) *Catálogo Sísmico de la Zona Compendida entre los Meridianos 5° E y 20° W*, Spain (in Spanish).

García, J.G. (1879) *Compendio de la Historia de Santo Domingo*, Segunda Edición, Augmentada, 2 Volúmenes, 8°, Santo Domingo (in Spanish).

Garriott, E.B. (1900) West Indian Hurricanes, with Notes by Alexander in 1902, *Weather Bureau Bulletin H*, U.S. Department of Agriculture, 66 pp.

Gazaryan, Y.L (1955) Surface waves in the ocean excited by submarine earthquakes, *Akusticheskiy Zhurnal Akademiya Nauk* (3), Moscow, 203–217 (in Russian).

Gibbs, A.R. (1883) *British Honduras: An Historical and Descriptive account of the Colony from Its Settlement, 1670,* London.

Gilii, F.S. (1784) *Saggio di Storia Americana,* 4 Volumes (1780–1817), Roma (in Italian).

Glover III, L. (1971) *Geology of the Coamo Area, Puerto Rico, and its Relations to the Volcanic Arc,* Professional Paper 636, Trench Association, U.S. Geological Survey, Washington.

González, F.I., and Kulikov, Y.A. (1991) On frequency modulation observed in two PACTOP deep ocean tsunami records, in A.M. Brennan and J.F. Lander (eds.), *Proceedings of the Second UJNR Tsunami Workshop,* National Oceanic and Atmospheric Administration, National Geophysical Data Center, Boulder.

González, F.I., and Kulikov, Y.A. (1993) Tsunami dispersion observed in the deep ocean, in S. Tinti (ed.), *Tsunamis of the World,* Kluwer Academic Publishers, Dordrecht, 7–16.

Gornitz, V. (1991) Global coastal hazards from future sea level rise, *Palaeogeography, Palaeoclimatology, Palaeoecology (Global and Planetary Change Section)* **89**, 379–398.

Grases, J. (1979a) *Investigatión sobre los Sismos Destructores Que Han Afectado el Orient de Venezuela, Delta del Orinoco y Regiones Adyacentes,* Informe e Anexos, INTEVEP, Los Teques (in Spanish).

Grases, J. (1979b) *Riesgos Catastróficos Naturales en el Caribe y Evaluación de Amenaza Sísmica a Nivel Regional,* Anteprojecto de investigatión, Universidad Central de Venezuela, Caracas (in Spanish).

Groot, J.M. (1889) *Historia Eclesiástica y Civil de Nueva Granada,* 5 Volúmenes, Bogotá (in Spanish).

Günther, S. (1897) *Handbuch der Geophysik* **1**, Stuttgart (in German).

Gusiakov, V, Marchuk, A., and Osipova, A. (1997) Expert Tsunami Database for the Pacific: Motivation, Design, and Proof-of-Concept Demonstration, in G.T. Hebenstreit (ed.), *Perspectives on Tsunami Hazard Reduction, Observations, Theory and Planning,* Kluwer Academic Publishers, Dordrecht 218 pp.

Gutenberg, B. (1932) *Handbuch der Geophysik* **4**, Berlin (in German).

Gutenberg, B. (1939) Tsunamis and Earthquakes, *Bulletin of the Seismological Society of America* **29**(4), 517–526.

Gutenberg, B. and Richter, C.F. (1936) On Seismic Waves, *Gerlands Beiträge zur Geophysik* **47**, Leipzig (in German).

Gutenberg, B. and Richter, C.F. (1954) *Seismicity of the Earth and Associated Phenomena,* 2nd Edition, Princeton University Press, Princeton, 310 pp.; also Hafner Publishing Company, New York (facsimile edition) (1965).

Hakluyt, R. (1907) *The Principal Navigations, Voyages, Traffiques and Discoveries of the English Nation, &c.,* New York.

Hall, D.A. (1971) *Caribbean History Monographs–Five of The Leewards, 1834–1870,* Caribbean University Press, 209 pp.

Halliday (1837) Natural and Physical History of the West Indies, with a Map of the supposed Course of the Volcanic Current, 8°, London.

Harrison, J.B. (1896) *The rocks and soils of Grenada and Carriacou,* Waterlow & Sons, London.

Hartog, J. (1961) *Aruba, Past and Present,* Translated by J.A. Verleum, Oranjestad, Aruba.

Hartwig, G. (1893) *Volcanoes and Earthquakes.*

Hatori, T. (1995) Magnitude scale for Central American tsunamis, *Pure and Applied Geophysics* **144**(3–4), 471–479.

Hay, L.G. (1884) Colony of Tobago Handbook (unpublished excerpts), Scarborough, Tobago.

Hebenstreit, G.T. (1997) *Perspectives on Tsunami Hazard Reduction: Observations, Theory and Planning,* Kluwer Academic Publishers, Dordrecht.

Heck, N.H. (1943) Seismic sea waves, *Bulletin of the Seismological Society of America, Earthquake Notes* **15**(1–2)

Heck, N.H. and Bodle, R.R. (1931) The Grand Banks earthquake of November 18, *United States Earthquakes, 1929,* Serial No. 511, Coast and Geodetic Survey, Washington, 28–29.

Hedervari, P. (no date) Catalog of Submarine Volcanoes and Hydrological Phenomena Associated with Volcanic Events, Georgiana Observatory, Budapest (unpublished manuscript).

Heezen, B.C., Nesteroff, W.D., Rawson, M., and Freeman-Lynde, R.P. (1985) Visual evidence for subduction in the western Puerto Rico Trench, *Symposium de la Géodynamique de Caraïbes* (février 1985), Éditions Technip, Paris, 287–304 (in French).

Herrera, D. (1847) Temporales, huracanes, tormentas y terremotos que había en esta Isla de Cuba, *Memoria Sobre los Huracanes en la Isla de Cuba,* Imprenta Barcina, La Habana, 72 pp. (in Spanish).

Herrera y Tordesillas, A. de (1730) *Descripción de las Indias Occidentales,* Oficina Real, Madrid (British Museum Library) (in Spanish).

Hewitt, R. (1957) *From Earthquake, Fire and Flood,* Charles Scribner's Sons, New York, 215 pp.

Ho, F.P. (1989) *Extreme Hurricanes in the Nineteenth Century,* NOAA Technical Memorandum HYDRO–43A, National Oceanic and Atmospheric Administration, National Weather Service, Silver Spring, Maryland.

Holcombe, T.L. and Sharman, G.F. (1983) Post-Miocene Cayman Trough evolution: A speculative model, *Geology* **11**, 714–717.

Holcombe, T.L., Vogt, P.R., Mathews, J.E., and Murchison, R.R. (1973) Evidence for sea-floor spreading in the Cayman Trough, *Earth Planetary Science Letters* **20**(3), 357–371.

Hovey, E.O. (1902) Martinique and St. Vincent; A preliminary report upon the eruptions of 1902, *Bulletin of the American Museum of Natural History* **16**, 333–372.

Hovey, E.O. (1905) Volcanoes of St. Vincent, St. Kitts and Statia, *Report of the 8th International Geographic Congress,* Washington, 452–454.

Howell, B.F., Jr. (1959) *Introduction to Geophysics,* McGraw-Hill, New York, Chapter 7, 98–99.

Hubert, H. (1935) The rapid inundation of coastal areas by the sea in the French colonies, *Annales de Physique de la Globe de France d'Outre-Mer* (10), 97 (in French).

Hughes, G. (1750) *The Natural History of the Island of Barbadoes,* London; also in *Journal Oeconom,* 174–182 (1755).

Hughes, P. (1987) Hurricanes haunt our history, *Weatherwise* **40**(3), 134–140.

Humboldt, Baron A. von (1956) *Viaje a las Regiones Equinocciales del Nuevo Continente,* 5 Volúmenes, Buenos Aires (in Spanish).

Huntress, K.G. (1979) *A checklist of Narratives of Shipwrecks and Disasters at Sea to 1860,* Iowa State University Press.

Hwang, L.-S., Butler, H.L., and Divoky, D.J. (1972) Tsunami model: Generation and open sea characteristics, *Bulletin of the Seismological Society of America* **62**(6), 1579–1596.

Iida, K. (1940) A study on the relation between the direction of the initial motion of the earthquake and that of the tsunami, *Zisin* **14**(1) (in Japanese).

Iida, K. (1958) Magnitude and energy of earthquakes accompanied by tsunamis and tsunami energy, *Journal of Earth Science* **6**(2), Nagoya University, Japan, 101–112.

Iida, K. (1984) *Catalog of Tsunamis in Japan and Its Neighboring Countries,* Aichi Institute of Technology, Yachigusa, Yak-Cho, Toyota-Shi 470–03, Japan.

Ikonnika, L.N. (1963) *Atlas of Tsunamis,* Moscow DVNIGMI, 53 pp. (in Russian).

Imamura, A. (1905) Causes of tsunami, *Journal of Geology* **17** Tokyo (in Japanese).

Imamura, A. (1935) How to escape from tsunamis, *Disasters from Water and Snow, Science for Disaster Prevention* **3**, Japan.

Imamura, A. (1944) Earthquake phenomena in ancient times in Japan, *Zisin* **16**(4), 4–9 (in Japanese).

Imamura, F., Shuto, N., Ide, S., Yoshida, Y., and Abe, K (1993) Estimate of the tsunami source of the 1992 Nicaragua earthquake from tsunami data, *Geophysical Research Letters* **20**, 1515–1518.

Iwasaki, S.I. (1997) The wave forms and directivity of a tsunami generated by an earthquake and a landslide, *Science of Tsunami Hazards* **15**(1), 23–40.

Jaggar, T.A. (1925b) Tidal waves, *Volcano Letters* **50**, U.S. Geological Survey, Hawai'i.

Jaggar, T.A. (1930) Ocean waves from submarine earthquakes, *Volcano Letters* **274**, U.S. Geological Survey, Hawai'i.

Jorgensen, K.M. (1966) *Seismological History of the Canal Zone and Panamá,* IOC Memorandum PCC–3, Intergovernmental Oceanographic Commission, (February), 12 pp.

Kaistrenko, V. (1989), Probability model for tsunami run-up, *Proceedings of the 1989 International Tsunami Symposium,* Novosibirsk (July–August), 249–253.

Kaistrenko, V., Klyacho, M., Nudner, I. and Pelinovsky, E. (2001), A new paradigm of tsunami safety solution, *Proceedings of the 2001 International Tsunami Symposium,* **1**(1–7), 303–313.

Kaye, C.A. (1959) *Geology of Isla Mona, Puerto Rico, and Notes on the Age of Mona Passage,* Professional Paper 317–C, U.S. Geological Survey, Washington.

Kirkwood, J.G. and Seeger, R.J. (1948) Surface Waves from an underwater explosion, *Journal of Applied Physics* **19**(4), American Institute of Physics, New York, 346–360.

Korgen, B.J. (1995) Seiches, *American Scientist* **83**, 330–341.

Kowalik, Z. (1997) Landslide-generated tsunami in Skagway, Alaska, *Science of Tsunami Hazards* **15**(2), 89–106.

Koyama, J. (1988) Focal mechanisms of tsunami and tsunamigenic earthquakes, in E.N. Bernard and R.L. Whitney (eds.), *Proceedings of the 1987 International Tsunami Symposium,* National Oceanic and Atmospheric Administration, Pacific Marine Laboratory, Seattle, 125–133.

Krümmel, O. (1911) *Handbuch der Ozeanographie* **2**, Stuttgart, 155 pp. (in German).

Lacroix, A. (1904) *La Montagne Pelée et ses Éruptions,* Masson et Cie, Paris, 662 pp. (in French).

Lacroix, A. (1926) Les caractéristiques lithographiques des Petites Antilles, *Livre Jubilaire de la Société de Géologie de Belge,* 387–405 (in French).

Ladd, J.W., Holcombe, T.L., Westbrook, G.K., and Edgar, N.T. (1990) Caribbean Marine Geology: Active margins of the plate boundary, in G. Dengo and J.E. Case (eds.), *The Geology of North America, Caribbean Region* **H**, Geological Society of America, Boulder, 261–290.

La Guadalupa, la Bassa-Terra, la Solfatara e la Punta-a-Pitre Istoria, Descrizione e Tremuoto dell'8 Febrajo 1842 (1843) Napoli, 12°, 67 pp. (in Italian).

Lander, J.F. (1996) *Tsunamis Affecting Alaska 1737–1996, Key to Geophysical Research Documentation,* National Geophysical Data Center, Boulder, 155 pp.

Lander, J.F. (1997) Caribbean tsunamis: An initial history, *Proceedings of the 22nd Caribbean Conference on Natural Hazards Disaster,* R. Ahmad (ed.), Publication 3, Unit for Disaster Studies, Department of Geography and Geology, University of the West Indies, Kingston, Jamaica (October 1997), 1–18.

Lander, J.F. and O'Loughlin, K.F. (1998b) Caribbean tsunamis: Newly recognized hazard, UNESCO International Conference on Tsunamis (May 1998), Paris.

Lander, J.F. and O'Loughlin, K.F. (1999a) Caribbean Natural Hazard, Poster presentation at the National Natural Hazard Center Meeting (July 1999), Boulder.

Lander, J.F. and O'Loughlin, K.F. (1999d) Proposed Caribbean Tsunami Warning System, Invited presentation to the International Meeting of the Tsunami Society (May 1999), Honolulu.

Lander, J.F., Lockridge, P.A., and Kozuch, M.J. (1993) *Tsunamis Affecting the West Coast of the United States 1806–1992,* NOAA Publication KGRD–29, National Oceanic and Atmospheric Administration, National Geophysical Data Center, Boulder, 242 pp.

Lander, J.F., Lockridge, P.A., and Meyer, H. (1991) Subaerial and submarine landslide-generated tsunamis, in N.J. Raufaste (ed.), *Wind and Seismic Effects,* NIST Special Publication 820, National Institute of Standards and Technology, Washington.

Lapparent, A. de (1912) *Volcans et Tremblements de Terre,* Paris (in French).

La Roncière, C. de (1932) *Histoire de la Marine Française,* 6 Tommes, Paris (in (French).

Lasserre, G. (1961) *La Guadeloupe,* 2 Tommes, Union Française d'Impression, Bordeaux (in French).

Latter, J.H. (1981) Tsunamis of volcanic origin: Summary of causes, with particular reference to Krakatoa, 1883, *Bulletin of Vulcanology* **44**(3), 467–490.

Leet, L.D. (1948) *Causes of Catastrophe,* McGraw-Hill, New York–London.

Leewards Islands Almanack (1987) First Issue, Roseau, Dominica.

Lefort de Latour (1787) *Description générale et particulière de l'Île de Sainte Lucie* (in French).

Lewisohn, F. (1970) *St. Croix under Seven Flags,* Dukane Press, Hollywood, Florida.

L'herminier, F. (1848) Rapport des tremblements de terre à la Point-a-Pitre, *Voyage Géologique aux Antilles et aux Îles de Ténériffe et de Fogo,* 1e Série, Paris (in French).

Lloyd's List (1969) Extant issues from 1741–1784 and 1790–1797, Gregg International Publishers Limited, Westmead, Farnborough, Hampshire, England.

Lockridge, P.A. and Smith, R.H. (1983) *Tsunamis in the Pacific Basin (1990–1983),* National Oceanic and Atmospheric Administration, National Geophysical Data Center, Boulder.

Lonsdale, P. and Klitgord, K. (1978) Structure and tectonics of the eastern Panamá Basin, *Bulletin of the Geological Society of America* **89**, 981–999.

Loomis, H.G. (1983a) *A Primer on Tsunamis Written for Boaters in Hawai'i,* NOAA Technical Memorandum ERL–PMEL–16, National Oceanic and Atmospheric Administration, Seattle

Loomis, H.G. (1983b) The nonlinear response of a tide gauge to a tsunami, in E.N. Bernard (ed.), *Proceedings of the 1983 Tsunami Symposium,* Hamburg, 177.

Lynch, D.K. (1978) Tidal bores, *Scientific American* **238**, 146–156.

Macdonald, G.A. (1972) *Volcanoes,* Prentice-Hall, Englewood Cliffs, New Jersey, 510 pp.

MacElwane, J.B. (1947) *When the Earth Quakes,* Chapter 7, Bruce Publishing Company, Milwaukee, Wisconsin, 132–135.

MacGregor, A.G. (1938) The volcanic history and petrology of Montserrat, with observations on Mont Pelée in Martinique, *Philosophical Transactions* **229**, Series B, Royal Philosophical Society of London, 1–90.

MacGregor, A.G. (1949) Prediction in relation to seismo-volcanic phenomena in the Caribbean volcanic arc, *Bulletin Volcanologique* **8**, Série II, 69.

Mader, C.L. (1988) *Numerical Modeling of Water Waves,* Los Alamos Series in Basic and Applied Sciences, University of California Press, Los Angeles.

Maggio, L. (1571) *Del Terræmoto Dialogo del Signor Lucio Maggio, Gentilhuomo Bolognese,* 4°, Bologna (in Italian).

Mahuma, S. (1994) Hurricanes and Tropical Storms of the Netherlands Antilles and Aruba, *Meteorological Series, Netherlands Antilles and Aruba,* Curaçao, 40 pp.

Martin-Kaye, P.H.A. (1959) *Report on the Geology of the Leeward and British Virgin Islands,* Voice Publishing Company, St. Lucia, 117 pp.

Martin-Kaye, P.H.A. (1961) Geological survey of the Windward Islands: Progress reports of the government geologist 1956–1961 (1–2) (unpublished).

Mason, D.G. and Scanlon, K.M. (1991) The neotectonic setting of Puerto Rico, *Bulletin of the Geological Society of America* **103**, 144–145.

Master Plan for the Tsunami Warning System in the Pacific (1987) IOC–INF–730, UNESCO Intergovernmental Oceanic Commission, Paris.

McCann, W.R. and Sykes, L.R. (1984) Subduction of aseismic ridges beneath the Caribbean plate: Implications for the tectonics and seismic potential of the northeastern Caribbean, *Journal of Geophysical Research* **89**, 4493–4519.

McCredie, S. (1994) When nightmare waves appear out of nowhere to smash the land, *Smithsonian* **24**, 28–39.

McNally, K.C. and Minster, J.B. (1981) Nonuniform seismic slip rates along the Middle America Trench, *Journal of Geophysical Research* **86**, 4949–4959.

Memorial Monthly (1938) Hurricane and Tidal Wave [November 1867], First Year (2) (April).

Miller, D.J. (1960) *Giant Waves in Lituya Bay, Alaska,* Professional Paper 354–C, U.S. Geological Survey, Washington, 51–86.

Milne, J. (1886) *Earthquakes and Other Earth Movements,* 2nd Edition, Chapter 9, Kegan-Paul, London, 165 pp.

Milne, J. (1897a) Sub-ocean changes, *Geographical Journal* **10**(2) 129–146; **10**(3) 259–289, London.

Milne, J. (1897b) Sub-ocean changes, *British Association for the Advancement of Science,* 2nd Report of the Seismic Investigation Committee, London, 181–206.

Milne, J. (1898a) *Seismology,* Kegan-Paul, London.

Milne, J. (1898b) Sub-ocean changes in relation to earthquakes, *British Association for the Advancement of Science,* 3rd Report of the Seismic Investigation Committee, London, 251–254.

Miloh, T. and Striem, H.L. (1978) Tsunami effects at coastal sites due to offshore faulting,

Tectonophysics **46**, 347–356.
Mitchell, C.L. (1924) West Indian hurricanes and other cyclones of the North Atlantic Ocean, *Monthly Weather Review*, Supplement 24, U.S. Weather Bureau, 47 pp.
Mitchell, G.J. (1922) Geology of the Ponce District, *Scientific Survey of Porto Rico and the Virgin Islands* **1**, New York Academy of Science, Part III, 266.
Mitchell J. (1818) Conjectures on the cause and observations upon the phænomena of earthquakes: Particularly of that great earthquake of the first of November 1755, *Royal Society Philosophical Magazine* **52**, 186–195, 254–270, 323–340.
Miyoshi, H. (1960) Tsunamis without earthquakes, *Science* **30**(6), Japan (in Japanese).
Miyoshi, H. (1984) Reconsiderations on huge tsunamis—utility of the seawall, in E.N. Bernard (ed.), *Proceedings of the 1983 Tsunami Symposium*, Hamburg, 107–115.
Miyamura, S. (1976) Provisional magnitudes of Middle America earthquakes not listed in the magnitude catalogue of Gutenberg and Richter, *Bulletin of the International Institute of Seismic Earthquake Engineering*, **14**, 41–46b.
Miyamura, S. (1980) *Sismicidad de Costa Rica*, Editorial Universidad de Costa Rica (in Spanish).
Molengraaff, G.A.F. (1886) *De Geologie van het eiland Sint Eustatius, Eene bijdrage tot de kennis der Nederlandsche koloniën*, Akademie Thesis, Leiden, 62 pp. (in Dutch).
Molnar, P. (1977) Gravity anomalies and the origin of the Puerto Rico Trench, *Journal de la Royal Astronomique Société de Géophysique* **51**, 701–708.
Molnar, P. and Sykes, L.R. (1969) Tectonics of the Caribbean and middle America regions from focal mechanisms and seismicity, *Bulletin of the Geological Society of America* **80**, 1639–1684.
Montero, W. (1986b) Períodos de recurrencia y tipos de secuencia sísmica de los temblores intraplaca en la región de Costa Rica, *Revista Geológica de América Central* (5), 35–72 (in Spanish).
Montero, W. (1990) Tsunamis en Costa Rica, *Boletín Informativo del Colegio de Geólogos de Costa Rica* (in Spanish).
Montessus de Ballore, F. de (1895) Relation entre le relief et la séismicité, *Comptes-Rendus de l'Academie des Sciences de Paris* **120**,1183–1187 (in French).
Montessus de Ballore, F. de (1903) On the unstable oceanic regions and the coasts struck by seismic sea waves, *Archive des Scientifique, Physique et Naturelles*, Genève (in French).
Montessus de Ballore, F. de (1906) *La Géographie Séismologique, Les Tremblements de Terre*, Paris (Librairie Armand Colin), 273, 278, 377–378, 385, 387 (in French).
Montessus de Ballore, F. de (1907) *La Science Séismologique: Les Tremblements de Terre*, Chapitre VII, Paris (Librairie Armand Colin) (in French).
Montessus de Ballore, F. de (1923) *Ethnographie Séismologique et Volcanique*, Honoré Champion, Paris, 206 pp. (in French).
Montessus de Ballore, F. de (1924) *La Géologie Séismologique: Les Tremblements de Terre*, Paris, 488 pp. (Librairie Armand Colin) (in French).
Montoulieu, E. (no date) Catálogo de datos macrosísmicos históricos (inédito) (in Spanish).
Montoulieu, E. (1968) Notas acerca de terremotos occuridos en Cuba (inédito) (Seismic Research Unit, University of the West Indies, Trinidad) (in Spanish).
Morales, L.D. (1984) Los temblores sentidos en Costa Rica durante 1973–1983 y su relación con la sismicidad del país, *Revista Geológica de América Central* (1) (Octubre) (in Spanish).
Moreau de Jonnès, A. (1822) Tableau chronologique des tremblements de terre des Antilles, *Histoire Physique des Antilles Françaises, Savoir: La Martinique et les îles de Guadeloupe* **1**, 2 Tommes, 8°, Paris (British Museum Library) (in French).
Morland (1810) Observations on the partial inundations which have devastated the world, *Mémoires de l'Academie de Science de Dijon, Séance Publication* (22), 65 (in French).
Murphy, A., Sykes, L. and Donnelly, T. (1970) Preliminary survey of the microseismicity of the northeastern Caribbean, *Bulletin of the Geological Society of America* **81**, 2459.
Nagano, O., Imamura, F., and Shuto, N. (1991) A numerical model for far-field tsunamis and its application to predict damages done to aquaculture, *Natural Hazards* **4**, 235–255.
Nakamori, H. (1994) Tsunami warning and broadcasting, *Collected Papers on Sociology* (120),

Nihon University, Tokyo.
Napier, A.K. (1997) Landslide!, *Honolulu Magazine* (February), 28–34.
Nicholls, H.A.A. (1880) The volcanic eruption in Dominica, *Nature* **21**, 372–373.
Nicholson, R.P. le (1776) *Essai sur l'Histoire Naturelle de l'Isle de Saint-Domingue, avec des Figures en Taille-Douce,* Paris, 376 pp. (in French).
Nugent, Lady N. (1811) An account of "The Sulphur" or "Soufrière" of the island of Montserrat, *Transactions of the Geological Society of London* **1**, 185–190.
Nuñez, S. (1969) Catalogue of Earthquakes felt at Cuba, DIC thesis, Imperial College, London.
Obee, B. (1989) Tsunami!, *Canadian Geographic* **109**, 46–53.
Odonel Gómez, L. (1993) *Seismicity and Seismic Hazard in the Dominican Republic.*
Okada, M. (1991) Ocean bottom pressure gauge for Tsunami Warning System in Japan, in A.M. Brennan and J.F. Lander (eds.), *Proceedings of the Second UJNR Tsunami Workshop*, National Oceanic and Atmospheric Administration, National Geophysical Data Center, Boulder, 219–227.
Okal, E.A. (1988) Seismic parameters controlling far-field tsunami amplitudes: A review, *Natural Hazards* **1**, 67–96.
Okal, E.A. (1994) Tsunami warning: Beating the waves to death and destruction, *Endeavor* **18**, 38.
Oliver, V.L. (1894) *The History of the Island of Antigua, From the First Settlement in 1635,* 3 Volumes, London.
Orlov, A.P. (1887) *Earthquakes and Their Connection to with Other Natural Phenomena,* Kazan State University, Kazan, 170 pp. (in Russian).
Oviedo y Valdes, G.F. de (1847) Sumario de la natural historia de las Indias, in B.C. Aribau (ed.) *Biblioteca de Autores Españoles desde la Formación del Lenguaja Hasta Nuestros Días* **22**, Madrid, 471–515 (in Spanish).
Oviedo y Valdes, G.F. de (1851) *Historia General y Natural de las Indias, Islas y Tierra Firma de la Mar Océano,* Madrid (in Spanish).
Papadopoulos, G.A. and Chalkis, B.J. (1984) Tsunamis observed in Greece and the surrounding area from antiquity up to the present times, *Marine Geology,* 56309–56317.
Papadopoulos, G.A., Murty, T., Venkatesh, S., and Blong, R. (2000) *Natural Hazards, State-of-the-art at the end of the Second Millennium,* Kluwer Academic Publishers, Dordrecht/Boston/ London.
Pararas-Carayannis, G. (1969) *Catalog of Tsunamis in the Hawai'ian Islands,* Coast and Geodetic Survey, Word Data Center–A, 94 pp.
Pararas-Carayannis, G. (1986) *The Pacific Tsunami Warning System, Earthquakes and Volcanoes* **18**(3), U.S. Geological Survey, 125–126.
Patton, R.S. and Marmer, H.A. (1932) The Waves of the Sea, *Physics of the Earth V, Oceanography Bulletin of the Natural Research Council* (85), National Academy of Science, Washington, 222–225.
Pelayo, A. and Wiens, D. (1992) Tsunami earthquakes: Slow thrust-faulting events in the accretionary wedge, *Journal of Geophysical Research* **97**, 15321–15337.
Pelinovsky, E.N. (1996) Models of tsunami waves, *Proceedings of the International Conference on Computational Techniques and Applications,* CTAC–95, World Scientific, Singapore (1996).
Pennington, W.D. (1981) Subduction of the eastern Panamá Basin and seismotectonics of northern South America, *Journal of Geophysical Research* **86**(10), 753–770.
Perrey, A. (1845) Mémoires sur les tremblements de terre ressentis en France, Belgique et Hollande, *Bulletin de l'Academie Royale Mémoires Couronées de Bruxelles* **18**, 110 pp. (in French).
Perrey, A. (1847c) Sur les tremblements de terre de la Péninsule Iberique, *Annales de la Société d'Agriculture de Lyon* **10**, 461–510 (in French).
Perrey, A. (1847d) Tremblements de terre en Algérie et dans l'Afrique septentrionale, *Mémoires de l'Academie des Sciences, Artes, et Belle-Lettres de Dijon, Années 1845–1846,* 299–479 (in French).
Perrey, A. (1848a) Documents sur les tremblements de terre et les éruptions volcaniques dans le bassin de l'Océan Atlantique, *Mémoires de l'Academie des Sciences, Artes, et Belle-Lettres de Dijon,* 1–67 (in French).

Perrey, A. (1848b) Mémoires sur les tremblements de terre de la Péninsule Italique, *Bulletin de l'Academie Royale Mémoires Couronées de Bruxelles* **22**, 143 pp. (in French).

Perrey, A. (1848c) Tremblements de terre ressentis en 1847, *Mémoires de l'Academie des Sciences, Artes, et Belle-Lettres de Dijon, Années 1847–1848,* 69–115 (in French).

Perrey, A. (1849) Sur les tremblements de terre dans les Îles Britanniques, *Annales de la Société d'Agriculture de Lyon* **1**, 115–177 (in French).

Perrey, A. (1850) Rapport des tremblements de terre au Mexique et dans l'Amérique Centrale, *Annuaire de la Société d'Emulation des Vosges* **7**, 2e Cahier, 341, 402 (in French).

Perrey, A. (1852a) Note sur les tremblements de terre en 1850, avec les suppléments pour les années antérieures, *Mémoires de l'Academie des Sciences, Artes, et Belle-Lettres de Dijon, Année 1851,* 1–36 (in French).

Perrey, A. (1852b) Supplément à la Note sur les tremblements de terre en 1851, *Bulletin de l'Academie Royale Mémoires Couronées de Bruxelles* **19**, 1e Partie, 21–28 (in French).

Perrey, A. (1852c) Tremblements de terre ressentis en 1851, *Bulletin de l'Academie Royale Mémoires Couronées de Bruxelles* **19**, 1e Partie, 353–396 (in French).

Perrey, A. (1854a) *Annales de la Société d'Agriculture de Lyon* **6**, 232 (in French).

Perrey, A. (1854c) Note sur les tremblements de terre en 1851, avec les suppléments pour les années antérieures, *Mémoires de l'Academie des Sciences, Artes, et Belle-Lettres de Dijon, Années 1852–1853* **2**, 1–65 (in French).

Perrey, A. (1854d) Note sur les tremblements de terre en 1852, avec les suppléments pour les années antérieures, *Mémoires de l'Academie des Sciences, Artes, et Belle-Lettres de Dijon, Années 1852–1853* **2**, 79–128 (in French).

Perrey, A. (1855a) Bibliographie séismique: Catalogue de livres, mémoires et notes sur les tremblements de terre et les phénomènes volcaniques, *Mémoires de l'Academie des Sciences, Artes, et Belle-Lettres de Dijon* **4**, 2e Série, 1–112 (in French).

Perrey, A. (1855b) Note sur les tremblements de terre en 1853, avec les suppléments pour les années antérieures, *Mémoires de l'Academie des Sciences, Artes, et Belle-Lettres de Dij*on, *Année 1854,* 55 (in French).

Perrey, A. (1856a) Bibliographie séismique: Catalogue de livres, mémoires et notes sur les tremblements de terre et les phénomènes volcaniques, *Mémoires de l'Academie des Sciences, Artes, et Belle-Lettres de Dijon* **5**, 2e Série, 113–253 (in French).

Perrey, A. (1857) Note sur les tremblements de terre en 1856, avec les suppléments pour les années antérieures, *Bulletin de l'Academie Royale Mémoires Couronées de Bruxelles* **24**(1), 2e Partie, 2e Série, 64–128 (in French).

Perrey, A. (1860) Note sur les tremblements de terre en 1857, avec les suppléments années antérieures, *Mémoires Couronées et autres Mémoires publiées de l'Academie Royale des Sciences, Lettres et Beaux Arts de Belgique* **10**, Bruxelles, 1–114 (in French).

Perrey, A. (1861) Bibliographie séismique: Catalogue de livres, mémoires et notes sur les tremblements de terre et les phénomènes volcaniques, *Mémoires de l'Academie des Sciences, Artes, et Belle-Lettres de Dijon* **9**, 2e Série, 3–18 (in French).

Perrey, A. (1862a) Bibliographie séismique: Catalogue de livres, mémoires et notes sur les tremblements de terre et les phénomènes volcaniques, *Mémoires de l'Academie des Sciences, Artes, et Belle-Lettres de Dijon* **10**, 2e Série, 109–161 (in French).

Perrey, A. (1862b) Note sur les tremblements de terre en 1859, avec les suppléments pour les années antérieures, *Mémoires Couronées et autres Mémoires publiées par l'Academie Royale de Sciences, de Lettres et de Beaux Arts de Belgique* **13**, Bruxelles, 1–77 (in French).

Perrey, A. (1865a) Bibliographie séismique: Catalogue de livres, mémoires et notes sur les tremblements de terre et les phénomènes volcaniques, *Mémoires de l'Academie des Sciences, Artes, et Belle-Lettres de Dijon* **13**, 2e Série, 1–70 (in French).

Perrey, A. (1865b) Note sur les tremblements de terre en 1863, avec les suppléments pour les années antérieures de 1843 à 1862, *Mémoires Couronées et autres Mémoires publiées par l'Academie Royale de Sciences, de Lettres et de Beaux Artes de Belgique* **17**, Bruxelles, 1–213 (in French).

Person, W.J. (1986) Significant Earthquakes of the World, 1986, *Preliminary Determination of Epicenters* (PDE), U.S. Geological Survey, National Earthquake Information Center, Golden, Colorado, 3 pp.

Pezuela, J. de la (1842) *Ensayo Histórico de la Isla de Cuba,* New York (in Spanish).

Picard, X. and Goddard, D. (1975) Geomorfología y sedementación de la costa entre Cabo Codera y Puerto Cabello, *Boletín Informativo de la Asociación Venezolana de Geología, Minería y Petroquímica* **18**, 39–106 (in Spanish).

Planning for Risk: Comprehensive Planning for Tsunami Hazard Areas (1988) Urban Regional Research prepared for the National Science Foundation, Seattle.

Poëy, A.M. (1862) *Table Chronologique de Quatre Cents Cyclones, depuis 1493 jusqu'en 1855,* Paris, 49 pp. (in French).

Pons, Fr. R.J. de (1806) *Voyage à la Partie Orientale de Terre Ferme, dans l'Amérique Méridionales, Fait Pendant les Années 1801, 1802, 1803 et 1804* **1**, Paris, 125, 245 (in French).

Ponyavin, I.D. (1965) *Tsunami Waves,* Gidrometeozdat, Leningrad, 110 pp. (in Russian).

Powell, C.F. (1938) The Royal Society expedition to Montserrat, British West Indies, Final Report, *Philosophical Transactions* **237**, Series A, Royal Philosophical Society of London, 1–34.

Press, F. (1956) Volcanoes, Ice, and Destructive Waves, *Engineering and Science* **20**(2), California Institute of Technology, 26–30.

Ramming, H.G. and Kowalik, Z. (1980) *Numerical Modeling of Marine Hydrodynamics,* Elsevier, New York, 368 pp.

Ramsey, G.D. (1960) A compilation of extracts from newspapers dated 1812–1953, filed in Bridgetown Public Library, Barbados, describing earthquakes and volcanic eruptions in the Caribbean, Paynes Bay, Barbados (unpublished, bound typescript, unnumbered) (Library of Seismic Research, Trinidad), 650 pp.

Raper, S.C.B., Wigley, T.M.L., and Warrick, R.A. (1996) Global sea-level rise: past and future, in J.D. Milliman and B.U. Haq (eds.), *Sea-Level Rise and Coastal Subsidence,* Kluwer Academic Publishers, Dordrecht.

Raubenheimer, B. and Guza, R.T. (1996) Observations and predictions of runup, *Journal of Geophysical Research* **101**, 25575–25587.

Ray, J. (1721) *Three Physico-Theological Discourses Concerning Primitive Chaos, the Deluge, and the Dissolution of the World; Also an Historical Account of those Two Late Remarkable Ones in Jamaica and England,* 4th Edition, 8°, London.

Redfield, W.C. (1846) On the several hurricanes of the American seas and their relations to the Northers, so called, of the Gulf of Mexico and the Bay of Honduras, with charts, *American Journal of Science* **52**, 162–187, 311–334.

Reid, H.F. (1912) *Strong Shocks in the United States and its Dependencies,* Report of the 81st Meeting of the British Association for the Advancement of Science, Portsmouth (1911), London, 94 pp.

Reid, S.B. (1997) Tsunamis, in *Introduction to Hazards,* 3rd Edition, United Nations Disaster Program–Department of Humanitarian Affairs, InterWorks.

Reyes, J.-R. de los (no date) De los temblores de tierra en la Isla de Cuba, *Revista de la Habana* **1**, 26 (in Spanish).

Roberts, E.B. (1960) The day the earth ran over, *Proceedings of the U.S. Naval Institute* (May), Washington, 62–66.

Roberts, E.B. (1961) History of a tsunami, *Smithsonian Institution Annual Report, 1960,* Publication No. 4442, Washington, 327–340.

Robinson, C. (1848) *An Account of Discoveries in the West until 1519,* Virginia Historical Philosophical Society, 491 pp.

Robson, G.R. and Willmore, P.L. (1955) Some heat measurements in the West Indian soufrières, *Bulletin Volcanologique* **17**, Série II, 13–39.

Robson, G.R., Barr, K.G., and Smith, G.W. (1962) Earthquake series in St. Kitts–Nevis, 1961–1962, *Nature* **195**, 972–974.

Rochefort, C. de (1658) *Histoire Naturelle et Morale des Îles Antilles de l'Amérique,* Rotterdam (in French).

Rockwood, C.G., Jr. (1880) Notices of recent American Earthquakes, *American Journal of Science* **19**, Series III, 299.

Rodríguez Férrer, M. (1876) *Naturaleza y Civilización de la Grandiosa Isla de Cuba,* J. Noguera, Madrid (in Spanish).

Roig de Lauschenring, E. (1658) Historia de La Habana, Desde sus primeros días hasta 1565 **1**, La Habana (in Spanish).

Rojas, A. (1907) Orígenes Venezolanos, in *Obras Escogidas,* Paris, 787 pp. (in Spanish).

Romer, M. (1931) La Dernière éruption de la Montagne Pelée, *Bulletin Volcanologique,* Série I (in French).

Rothé, J.P. (1958) Chronique Séismologique, *Révue pour l'Étude des Calamités, Bulletin de la Union Internationale Sécours* (36), Genève (in French).

Rothé, J.P. (1969) *The Seismicity of the Earth, 1953–1969,* UNESCO, Paris, 336 pp.

Rudolph, E. (1898) Über submarine Erdbeben und Eruptionen, *Beiträge zur Geophysik* **3**, 273–366, Stuttgart (in German).

Rue, E. de la (1937) Reports of the Commission for the Study of Tsunamis, *Bulletin Volcanologique* **1**, Série II, Assemblée d'Edinbourg, Naples (in French).

Ruff, L.J. and Kanamori, H. (1980) Seismicity and subduction process, *Phys. Earth Planet. Inter.* **23**, 240–252.

Rutten, L.M.R. (1938) *Bibliography of West Indian Geology,* Verhandelingen der Nederlandsche Akademie, Utrecht, University of Utrecht, Geographische en Geologische Mededeelingen, Physiographische, Geologische Reeks (16), Oosthoek (in Dutch).

Rutten, L.M.R. and Raadshooven, B. van (1940) On earthquake epicenters and earthquake shocks between 1913 and 1938 in the region between 0° and 30° N Latitude and 56° and 120° W Longitude, *Verhandelingen der Nederlandsche Akademie van Wetenschappen, Afdeeling Natuurkundige* **39**(4), Tweede Sectie, Amsterdam (in Dutch).

Sagra, R. de la (1842) *Histoire Physique, Politique et Naturelle de l'Île d Cuba,* Paris (in French)

Sainte-Claire Deville, C. (1843a) Extrait à des observations sur le tremblement de terre de la Guadeloupe, du 8 février 1843, *Comptes-Rendus de l'Academie des Sciences de Paris* **17**, 1283–1288 (in French).

Sainte-Claire Deville, C. (1848a) Les tremblements de terre à la Guadeloupe, *Voyage Géologique aux Antilles et aux Îles de Ténériffe et de Fogo,* 1e Série, Paris, 283–336 (in French).

Sainte-Claire Deville, C. (1856) *Bulletin de la Société de Géologie de France* **14**, 254 (in French).

Sánchez Devora, A.J. and Farreras Sanz, S.F. (1993) *Catalog of Tsunamis on the Western Coast of Mexico,* NOAA Report SE–50, National Oceanic and Atmospheric Administration, National Geophysical Data Center, Boulder, 79 pp.

Sapper, K. (1902) *In den Vulkangebieten Mittelamerikas und Westindien, Reiseschildernen und Studien über die Vulkanausbrüchen, der Jahre 1902 bis 1903, ihre Geologischen, Wirtschaftlichen und Sozialen Folgen,* E. Schweizerbartsche (ed.), Verlagsbuchhandlung, Stuttgart, 334 pp. (in German).

Sapper, K. (1903a) Ein Busuch der Insel Grenada, *Zentbl. Mineralogische, Geologische und Paläontologische,* 182–186 (in German).

Sapper, K. (1903b) Ein Busuch der Insel Montserrat (Westindien), *Zentbl. Mineralogische, Geologische und Paläontologische,* 279–283 (in German).

Sapper, K. (1903c) Ein Busuch der Inseln Nevis und Sanct Kitts (Sanct Christopher), *Zentbl. Mineralogische, Geologische und Paläontologische,* 284–287 (in German).

Sapper, K. (1903d) Ein Busuch von Dominica, *Zentbl. Mineralogische, Geologische und Paläontologische,* 305–314 (in German).

Sapper, K. (1903e) Ein Busuch von Sanct Eustatius und Saba, *Zentbl. Mineralogische, Geologische und Paläontologische,* 313–318 (in German).

Sapper, K. (1903f) Zur kenntnis der Insel Sanct Lucia in Westindien, *Zentbl. Mineralogische, Geologische und Paläontologische,* 273–278 (in German).

Sapper, K. (1904) Die vulkanischen Kleineren Antillen und die Ausbrüche der Jahre 1902 und 1903, *Neues Jahresbericht Mineralogische, Geologische und Paläontologische* **2**, 1–70 (in German).

Satake, K. (1988) Effects of bathymetry on tsunami propagation: Application of ray-tracing to tsunamis, *Pure and Applied Geophysics* **126**, 27–36.

Satake, K. (1992) Tsunamis, in *Encyclopedia of Earth System Science* **4**, Academic Press.

Satake, K. and Imamura, F. (1995) Introduction to "Tsunamis: 1992–1994, Their generation, dynamics, and hazard, *Pure and Applied Geophysics* **145**, 373–379.

Satake, K. and Okal, E.A. (1996) Fresh data on tsunami earthquakes roll in, *EOS, Transactions,* American Geophysical Union (April 23), 167.

Schmidt, J.F.J. (1879) *Studien über Vulcane and Erdbeben,* Alwin Georgi, Leipzig (in German).

Schmidt, O. von (1938) Propagation of waves due to explosion in liquids and solids, *Zeitschrift für Technische Physik* **19**(12), 554 (in German).

Scholte, J.G. (1943) The relations between sea waves and microseisms, *Natuurkundige Verlags* **52**(10), Akademie Wetenschappelijk Amsterdam, 669–683 (in Dutch)

Schomburgk, Sir R.H. (1848) *The History of Barbados,* Longman, Brown, Green, & Longman, London, 695 pp.

Schuchert, C. (1935) *Historical Geology, Antillean–Caribbean Region,* John Wiley & Sons, New York.

Schuchert, C. (1988) Are the Venezuelan fault systems part of the Caribbean plate boundary? *Geologische Rundschau* **70**, 542–551.

Schwab, W.C., Danforth, W.W., Scanlon, K.M. and Masson, D.G. (1991) A giant slope failure on the northern insular slope of Puerto Rico, *Marine Geology* **96**, 237–246.

Self, S. (1992) Krakatau revisited: The course of events and interpretation of the 1883 eruption, *Geojournal* **28**(2), 109–121.

Shapiro, N., Singh, K., and Pacheco, J. (1998) A fast and simple diagnostic method for identifying tsunamigenic earthquakes, *Geophysical Research Letters* **25**(20), 3911–3914.

Shepard, C. (1831) *An Historical Account of the Island of Saint Vincent,* London.

Shepard, F.P. (1948) *Submarine Geology,* Harper and Brothers, New York, 47–53, 59–60, 240–241.

Shepard, F.P., Macdonald, G.A., and Cox, D.C. (1950) The Tsunami of April 1, 1946, *Bulletin of the Scripps Institution of Oceanography* **5**(6), University of California, La Jolla, 391–470.

Shepherd, E.S. and Merwin, H.E. (1927) Gases of the Mont Pelée lavas of 1902, *Journal of Geology* **35**, 97–116.

Shepherd, J.B. (1971) *A Study of Earthquake Risk in Jamaica and its Influence on Physical Development Planning,* Ministry of Planning and Natural Development, Kingston, 49 pp.

Shepherd, J.B. and Rogers, I. (1985) *Catalogue of Earthquakes for South America: Trinidad and Tobago and Adjacent Regions* **8**, CERESIS, Lima.

Shepherd, J.B., Tomblin, J.F., and Woo, D.A. (1971) Volcano-seismic crisis in Montserrat, 1966–1967, *Bulletin Volcanologique* **35**, Série I, 143–163.

Shower, J. (1693) *Practical Reflections on the Late Earthquakes in Jamaica, England, Sicily, Malta, &c. A.D. 1692,* 8°, London.

Shurygin, A.M. (1980) *On Long-Term Forecast of Major Tsunamis, Theory and Operation of Tsunami Forecast,* Nauka, Moscow (in Russian).

Shuto, N. (1991) Numerical simulations of tsunamis—Its present and near future, *Natural Hazards* **4**, 171–191.

Shuto, N. and Imamura, F. (1997) *Numerical Method of Tsunami Simulation with Leap-frog Theme,* Intergovernmental Oceanic Commission, Manuals and Guides (35).

Shuto, N., Suzuki, T., Hasegawa, K., and Inagaki, K. (1985) A study of numerical techniques on tsunami propagation and run-up, in T.S. Murty and W.J. Rapatz (eds.), *Proceedings of the International Symposium of the Institute of Ocean Science,* British Columbia, Canada, 88–92.

Sieberg, A. (1923) Die Erdbeben, *Geologische, Physikalische und Angewandte Erdbeben Kunde,* Jena, 572 pp. (in German).

Simon, Fr. P. (1882) *Noticias Históricas de las Conquistas de Tierra Firme en las Indias Occidentales,* Edición hecha sobre de la Cuenca de 1626 en Bogotá, 5 Volúmenes (in Spanish).

Simon, Fr. P. (1962) *Noticias Históricas de Venezuela,* Biblioteca de la Academia Nacional de la Historia, Caracas, Reedición; Primera Edición, Cuenca (1627) (in Spanish).

Singer, A. and Rojas, C. (1981) *Inventario Nacional de Riesgos Geológicos Escala 1:1.000.000, Resultados Preliminares,* FUNVISIS, Caracas (in Spanish).

Sloane, Sir H. (1770) *A Voyage to the Islands, Madera, Barbadoes, Nieves, Saint Christophers, and Jamaica,* 2 Volumes, London.

Smith, R. (1976) Giant waves, *Journal of Fluid Mechanics* **77**, 417–431.

Snodgrass, F.R., Munk, W.H., and Tucker, M.J. (1958) Off-shore recording of low-frequency waves, *EOS Transactions* **39**(1), American Geophysical Union, 114–120.

Solov'ev, S.L. and Go, C.N. (1974) *A Catalogue of Tsunamis on the Western Shore of the Pacific Ocean (1513–1968),* Academy of Sciences of the USSR, Nauka Publishing House, Moscow, 308 pp. (in Russian)

Solov'ev, S.L. and Go, C.N. (1975) *A Catalogue of Tsunamis on the Eastern Shore of the Pacific Ocean (1513–1968),* Academy of Sciences of the USSR, Nauka Publishing House, Moscow, 203 pp. (in Russian)

Solov'ev, S.L. and Go, C.N. (1984b) *A Catalogue of Tsunamis on the Western Shore of the Pacific Ocean (1513–1968),* English translation from Russian by Canadian Institute of Scientific and Technical Information, National Research Council (5077), Ottawa, Ontario, 439 pp.

Solov'ev, S.L., Go, C.N., and Kim, K.S. (1986) *Catalogue of Tsunamis in the Pacific (1969–1982),* Academy of Sciences of the USSR Publishing House, Moscow, 164 pp. (in Russian)

Solov'ev, S.L., Go, C.N., and Kim, K.S. (1992) *Catalogue of Tsunamis in the Pacific (1969–1982),* English Translation from Russian by Amerind Publishing Company, Limited, New Delhi (1988); Academy of Sciences of the USSR Publishing House, Moscow, 208 pp.

Soloviev, S.L., Go, C.N., Kim, K.S., Solovieva, O.N., and Shchetnikov, N.A. (1997) *Tsunamis in the Mediterranean Sea 2000 B.C.–1991 A.D.,* Moscow, 139 pp. (in Russian)

Steinbrugge, K.V. (1982) *Earthquakes, Volcanoes, and Tsunamis: An Anatomy of Hazards,* Skandia America Group, New York.

Sternbeck, A. (1930) *Filibusters and Buccaneers,* Translated by E. Hill and D. Mudie, New York.

Stewart, A.T. (no date) Time of the first buying of the Islands by the United States [meaning the Danish West Indies, which included the islands of St. Thomas, St. John, and St. Croix in the Virgin Islands] (unpublished).

Stone, O.M. (1889) *Tenerife and its satellites or the Canary Islands Past and Present: Tenerife, Gomera, Hierro, Palma* **1**, London.

Stothers, R.B. (1989) Volcanic eruptions and solar activity, *Journal of Geophysical Research* **94**(B12), 17371–17382.

Stromeyer, C.E. (1895) Abnormal Atlantic waves: Submarine volcanic activity, *Nature* **51**, London, 437–438.

Sturm, H. (1974) Giant Waves, *Ocean* **2**(3).

Sykes, L.R. and Ewing, M. (1965) The seismicity of the Caribbean Region, *Journal of Geophysical Research* **70**(20), 5065–5074.

Sykes, L.R., McCann, W. and Kafka, A. (1982) Motion of Caribbean Plate during the last 7 million years and implications for earlier cenozoic movements, *Journal of Geophysical Research* **87**, 246–265.

Synolakis, C. (1995) Tsunami prediction, *Science* **270**(15).

Talwani, M., Sutton, G.H., and Warsdell, J.L. (1959) A crustal section across the Puerto Rico Trench, *Journal of Geophysical Research* **64**(10), 1545–1555.

Tanioka, Y. and Satake, K. (1996) Tsunami generation by horizontal displacement of ocean bottom, *Geophysical Research Letters* **23**(8), 861–864.

Tanioka, Y., Satake, K., and Ruff, L. (1995) Analysis of seismological and tsunami data from the 1993 Guam earthquake, *Pure and Applied Physics* **144**(3/4), 823–837.

The truest and largest Account of the late Earthquake in Jamaica, June 7th 1692, (1693) London, 8°, Shelf mark T–1762(18) (British Museum Manuscript).

Tilloch (1812) Earthquakes and volcanic phænomena in the West Indies, *Philosophical Magazine* **39**(170), 474–477.

Tomblin, J.M. (1981) Earthquakes, volcanoes, and hurricanes: A review of national hazards and vulnerability in the West Indies, *Ambio* **10**, 340–345.

Torre, J.M. de la (1854) *Geografía de la Isla de Cuba,* La Habana (in Spanish).

Travagini, F. (1679) *Super Observationibus a se Partis, Tempore Ultimorum Terræmotuum, ac Potissimum Ragusani: Physica Disquisitio, seu Terræ Diurni Indicium,* 4°, Lug. Bat. (in Latin); Venetian Edition (1683) (British Museum).

Tremblement de Terre du 8 février 1843, qui Renversa la Ville de la Pointe-à-Pitre, à la Guadeloupe (1855) En vers, 8°, Imprimerie de Tinterlin à Paris (in French).

Tsuchiya, Y. and Shuto, N. (1995) *Tsunami: Progress in Prediction, Disaster Prevention and Warning,* Kluwer Academic Publishers, Dordrecht.

United States National Archives and Records Administration (no date) *Naval Records Collection of the Office of Naval Records and Library* Record Group 45, Washington.

United States National Archives and Records Administration (no date) *Records of the Bureau of Naval Personnel,* Record Group 24, Washington.

Vaissière, P. de (1909) *Saint-Domingue 1629–1789,* Paris (in French)

Valiente, A. (1853) *Tabla Cronológica de los Sucesos Ocurridos en Santiago de Cuba,* New York (in Spanish).

van Dijk, G. (1919) Sea Quakes, *De Zee, De Bilt* (4) (in Dutch).

Vaughan, T.W., Cooke, W., Condit, D.D., Ross, C.P., Woodring, W.P., and Calkins, F.C. (1921) *A Geological Reconnaissance of the Dominican Republic,* Washington.

Vedia, E. de (1853) *Historios primitivos de Indias,* Madrid (in Spanish).

Vega, F. and Havskov, J. (1998) Magnitude determination for Central America, *Technical Report II 1–11,* Reduction of Natural Disasters in Central America, Earthquake Preparedness and Hazard Mitigation, Institute of Solid Earth Physics, University of Bergen, Norway.

Villiers, A. (1957) *Wild Ocean,* McGraw-Hill, New York.

von Hillebrandt-Andrade, C.G. (1998) Tsunami possibility during hurricane season, personal communication, June 1.

von Hoff, K.E.A. (1841) Chronik der Erdbeben und Vulkan-Ausbrüch von 3460 A.C. bis 1832 P.C., mit vorausgehender Abhandlung über die Natur dieser Erscheinungen, in *Geschicte der durch Überlieferung nach Gewiesenen Natürlichen Veränderungen der Erdoberfläche* **5**, Gotha (in German).

von Hoff, K.E.A. (1835) *Chronik die Vulkanischen Ereignisse des Jahres 1821,* **19**, Leipzig (in German).

Wager, L.R. (1962) Igneous cumulates from the 1902 eruption of Soufrière, St. Vincent, *Bulletin Volcanologique* **24**, Série II, 93–99.

Walker, D.A. (1993) Pacific-wide tsunamis reported in Hawai'i from 1819 through 1990: Runups, magnitudes, and implications for warning systems, School of Ocean and Earth Science and Technology Report 93–2, University of Hawai'i, Honolulu, 117 pp.

Walker, D.A. (1994) *Tsunami Facts,* School of Ocean and Earth Science and Technology Report 93–2, University of Hawai'i, Honolulu.

Warneford, F.H.S. (1956) *An Introduction to the History of Volcanic and Seismic Activity in the West Indies, with Special Reference to the Lesser Antilles,* Antigua Printery Limited, Antigua.

Westergaard, W.C. (1917) *The Danish West Indies Under Company Rule (1671–1754),* New York.

Westermann, J.H. and Kiel, H. (1961) The geology of Saba and Sint Eustatius, *Uitg. Natuurw. Werkgrp. Nederlandsche Antillen* **24**, 1–175 (in Dutch).

Whiteside, L.S., Dater, D., and Dunbar, P. (1996) Seismicity Catalog CD-ROM, National Oceanic and Atmospheric Administration, National Geophysical Data Center, Boulder, and U.S. Geological Survey, Golden, Colorado; also at *http://www.ngdc.noaa.gov/seg/hazard/tsu.shtml.*

Whiteside, L.S., Lander, J.F., and Hattori, P. (1999) The tsunami history of Guam, 1849–1993 (unpublished manuscript).

Whitmore, P.M. (1993) Expected tsunami amplitudes and currents along the North American Coast for Cascadia Zone earthquakes, *Natural Hazards* **8**.

Wichmann, A. (1877a) *Die Erdbeben des Indishen Archipels* **1**, to 1857 (in German).

Wichmann, A. (1877b) *Die Erdbeben des Indishen Archipels* **2**, from 1858–1877 (in German).

Wiegel, R.L. (1970) Tsunamis, *Earthquake Engineering,* Prentice-Hall, Englewood Cliffs, New Jersey.

Williams, N. (1992) *A history of the Cayman Islands,* 94 pp.
Willmore, P.L. (1952) The earthquake series in St. Kitts–Nevis, 1950–1951, *Nature* **169**, 770–772.
Wilson, B.W. and Mehaute, B. le (1964) *Propagation and Runup of Tsunami Waves,* U.S. Coast and Geodetic Survey, Technical Report SN–166, National Engineering Science Company, Pasadena, California.
Winds and Weather of the West Indian Region (1940) U.S. Weather Bureau, 190 pp.
Wright, I.A. (1929) *Spanish Documents Concerning English Voyages to the Caribbean, 1527–1568,* The Hakluyt Society, London.
Yeh, H. (1991) Tsunami bore runup, *Natural Hazards* **4**(2–3), 209–220.
Yeh, H., Liu, P., Briggs, M., and Synolakis, C. (1994) Propagation and amplification of tsunamis at coastal boundaries, *Nature* **372**, 353–355.
Yoshida, I. (1944) Ainu folk-tale about tsunamis, *Anthropological Journal* **50**(11) (in Japanese).
Young, T. (1807) A catalogue of works relating to earthquakes, and agitations to subterraneous fires and volcanoes, and to geology, in *Course of Lectures on Natural Philosophy,* London.
Zabriskie, L.K. (1918) *The Virgin Islands of the United States of America,* G.P. Putnam & Sons, New York and London.

津波

INDEX OF GEOGRAPHIC NAMES

The Index of Geographic Names includes locations affected by tsunamis, earthquakes, volcanoes, tectonic settings, and other locations of interest in the Caribbean Tsunami History. For listings with "Saint" see also listings for "St." Country affiliations for islands are in Appendix A.

A

B

K

L

M

T

U

V

W

Y

ACKNOWLEDGEMENT OF COPYRIGHT EXCLUSION

MIX
Papier aus verantwortungsvollen Quellen
Paper from responsible sources
FSC® C105338

If you have any concerns about our products,
you can contact us on
ProductSafety@springernature.com

In case Publisher is established outside the EU,
the EU authorized representative is:
Springer Nature Customer Service Center GmbH
Europaplatz 3, 69115 Heidelberg, Germany

Printed by Libri Plureos GmbH
in Hamburg, Germany